Benchmark Papers in Geology Series

Editor: Rhodes W. Fairbridge—Columbia University

PALYNOLOGY, PART I: Spores and Pollen / *Marjorie D. Muir and William A. S. Sarjeant*
PALYNOLOGY, PART II: Dinoflagellates, Acritarchs, and Other Microfossils / *Marjorie D. Muir and William A. S. Sarjeant*
NORTH AMERICAN GEOLOGY: Early Writings / *Robert Hazen*
TERRESTRIAL TRACE-FOSSILS / *William A. S. Sarjeant*
NANNOFOSSIL BIOSTRATIGRAPHY / *Bilal U. Haq*
CALCAREOUS NANNOPLANKTON / *Bilal U. Haq*
STRATIGRAPHY: Foundations and Concepts / *Barbara M. Conkin and James E. Conkin*

Related Title

THE ENCYCLOPEDIA OF PALEONTOLOGY / *Rhodes W. Fairbridge and David Jablonski*

VERTEBRATE PALEONTOLOGY

Edited by
ROBERT M. SCHOCH
Yale University and Boston University

A Hutchinson Ross Benchmark® Book

VNR VAN NOSTRAND REINHOLD COMPANY

Benchmark Papers in Geology, Volume 83
Library of Congress Catalog Card Number: 84-13133
ISBN: 0-442-28052-1

Manufactured in the United States of America.

Published by Van Nostrand Reinhold Company Inc.
135 West 50th Street
New York, New York 10020

Van Nostrand Reinhold Company Limited
Molly Millars Lane
Wokingham, Berkshire RG11 2PY, England

Van Nostrand Reinhold
480 Latrobe Street
Melbourne, Victoria 3000, Australia

Macmillan of Canada
Division of Gage Publishing Limited
164 Commander Boulevard
Agincourt, Ontario MIS 3C7, Canada

15 14 13 12 11 10 9 8 7 6 5 4 3 2 1

Library of Congress Cataloging in Publication Data
Main entry under title:
Vertebrate paleontology.
"A Hutchinson Ross Benchmark book."
Bibliography: p.
Includes index.
1. Vertebrates, Fossil—Addresses, essays, lectures.
I. Schoch, Robert M.
QE841.V39 1984 566 84-13133
ISBN 0-442-28052-1

CONTENTS

SERIES EDITOR'S FOREWORD

The philosophy behind the Benchmark Papers in Geology is one of collection, sifting, and rediffusion. Scientific literature today is so vast, so dispersed, and, in the case of old papers, so inaccessible for readers not in the immediate neighborhood of major libraries that much valuable information has been ignored by default. It has become just so difficult, or so time consuming, to search out the key papers in any basic area of research that one can hardly blame a busy person for skimping on some of his or her "homework."

This series of volumes has been devised, therefore, as a practical solution to this critical problem. The geologist, perhaps even more than any other scientist, often suffers from twin difficulties—isolation from central library resources and immensely diffused sources of material. New colleges and industrial libraries simply cannot afford to purchase complete runs of all the world's earth science literature. Specialists simply cannot locate reprints or copies of all their principal reference materials. So it is that we are now making a concerted effort to gather into single volumes the critical materials needed to reconstruct the background of any and every major topic of our discipline.

We are interpreting "geology" in its broadest sense: the fundamental science of the planet Earth, its materials, its history, and its dynamics. Because of training in "earthy" materials, we also take in astrogeology, the corresponding aspect of the planetary sciences. Besides the classical core disciplines such as mineralogy, petrology, structure, geomorphology, paleontology, and stratigraphy, we embrace the newer fields of geophysics and geochemistry, applied also to oceanography, geochronology, and paleoecology. We recognize the work of the mining geologists, the petroleum geologists, the hydrologists, and the engineering and environmental geologists. Each specialist needs a working library. We are endeavoring to make the task of compiling such a library a little easier.

Each volume in the series contains an introduction prepared by a specialist (the volume editor)—a "state of the art" opening or a summary of the object and content of the volume. The articles, usually some twenty to fifty reproduced either in their entirety or in significant extracts, are selected in an attempt to cover the field, from the key papers of the last century to fairly recent work. Where the original works are in foreign languages, we have endeavored to locate or commission translations. Geologists, because of their global subject, are often acutely aware of the oneness of our world. The selections cannot therefore be restricted to any one country, and whenever possible an attempt is made to scan the world literature.

To each article, or group of kindred articles, some sort of "highlight commentary" is usually supplied by the volume editor. This commentary should serve to bring that article into historical perspective and to emphasize its particular role in the growth of the field. References, or citations, wherever possible, will be reproduced in their entirety—for by this means the observant reader can assess the background material available to that particular author, or, if desired, he or she too can double check the earlier sources.

A "benchmark," in surveyor's terminology, is an established point on the ground that is recorded on our maps. It is usually anything that is a vantage point, from a modest hill to a mountain peak. From the historical viewpoint, these benchmarks are the bricks of our scientific edifice.

RHODES W. FAIRBRIDGE

PREFACE

Vertebrate paleontology is a popular subject. An exciting dinosaur find or the discovery of another "missing link" in the evolutionary tree of man can make front-page headlines in newspapers around the globe. Every preschooler seems to know and love dinosaurs, which with cavemen, star on a regular basis in comics, movies, and cartoons. The data of vertebrate paleontology continue to fuel the ongoing public debate between evolutionists and creationists (Futuyma, 1983). In addition, the practitioners of the discipline have often been extremely colorful personalities, lending interest to the field. Nearly a century has passed since the infamous Cope-Marsh feuds over fossil vertebrate collecting sites in the American West. This debate continues to be a popular, often repeated story (Casanova, 1982; Wheeler, 1960).

The discoveries of vertebrate paleontologists have helped shape the way modern man thinks about himself and the world around him (e.g., Chardin, 1959; Greene, 1959; Huxley, 1904; Rudwick, 1976; Simpson, 1949, 1953, 1960). Perhaps more than any other scientific discipline, vertebrate paleontology has helped place man—the species *Homo sapiens*—in perspective within the larger context of life on earth through the ages. It has seriously been argued that if dinosaurs had not become extinct 65 million years ago that they might have evolved into large-brained, bipedal, humanoid (or "dinosauroid") forms remarkably similar to our own species (Russell and Séguin, 1982).

Even though vertebrate paleontology is a popular and influential subject, there have been relatively few professional vertebrate paleontologists. The older paleontological literature is widely scattered, obscure, and increasingly more difficult to obtain. It is also often extremely abstruse. In this volume I have brought together some of the classic papers in the field, covering a period of over 150 years. In this collection the main emphasis is on the central core of vertebrate paleontological literature—that is, the morphology, distribution,.and relationships of vertebrates—rather than on methodology, general historical geology, or biology as related to fossil vertebrates (these subjects are covered comprehensively in Simpson, 1980). Some of the papers give general overviews, while others are relatively technical and detailed. All are classics. In selecting these papers I have tried to give all major groups of vertebrates reasonable coverage. Thus, papers are included on fishes, amphibians, reptiles, dinosaurs, birds, and mammals. It is hoped that this volume may serve both as an introduction to vertebrate paleontology for the novice and as a source of reference and continuing interest to the most advanced senior researchers. Of course, many additions have been made to our knowledge of ancient vertebrate life since the papers reprinted in this volume were written. Space does not permit all of these papers to be corrected and updated (indeed, this is unnecessary for present purposes). However, in a few instances I noted important corrections or additions in the editorial comments. In the Introduction I present a new, revised classification of the vertebrates that is intended to give the reader an idea of some current hypotheses as to the phylogenetic relationships of major vertebrate groups.

I thank the series editor, Professor Rhodes W. Fairbridge, for his interest and helpful suggestions. His first paid job, he tells me, was working on Canadian dinosaurs with Charles Sternberg. Professor Mott T. Greene very kindly provided me with biographical notes and references on some vertebrate paleontologists. My wife, Cynthia, provided unfailing encouragement for which I am most grateful.

ROBERT M. SCHOCH

REFERENCES

Casanova, R. L., 1982, The Battle of Bridger Basin—The Cope-Marsh Wars, *Fossils Quart.* **1:**4-21.

Chardin, P. Teilhard de, 1959, *The Phenomenon of Man, with an Introduction by Sir Julian Huxley,* Harper & Row, New York, 320p.

Futuyma, D. J., 1983, *Science on Trial: The Case for Evolution,* Pantheon Books, New York, 264p.

Greene, J. C., 1959, *The Death of Adam,* Iowa State University Press, Ames, 398p.

Huxley, T. H., 1904, *Man's Place in Nature,* T. A. Hill and Company, New York, 200p.

Rudwick, M. J. S., 1976, *The Meaning of Fossils,* Neale Watson Academic Publications, New York, 296p.

Russell, D. A., and R. Séguin, 1982, Reconstructions of the Small Cretaceous Theropod *Stenonychosaurus inequalis* and a Hypothetical Dinosauroid, *Natl. Mus. Nat. Sci., Natl. Mus. Canada, Syllogeus* **37:**1-43.

Simpson, G. G., 1949, *The Meaning of Evolution: A Study of the History of Life and Its Significance for Man,* Yale University Press, New Haven, Conn., 380p.

Simpson, G. G., 1953, *The Major Features of Evolution,* Columbia University Press, New York, 444p.

Simpson, G. G., 1960, On the Remarkable Testament of the Jesuit Paleontologist Pierre Teilhard de Chardin, *Sci. American* **202:**201-202, 204, 206-207.

Simpson, G. G., 1980, *Why and How: Some Problems and Methods in Historical Biology,* Pergamon Press, Oxford, England, 272p.

Wheeler, W. H., 1960, The Uintatheres and the Cope-Marsh War, *Science* **131:**1171-1176.

CONTENTS BY AUTHOR

VERTEBRATE PALEONTOLOGY

INTRODUCTION

Paleontology is the study of ancient life. Vertebrate paleontology is the study of ancient vertebrates and their nearest nonvertebrate chordate relatives. The central theme of the discipline is the history of vertebrate life on our planet. The scope of vertebrate paleontology is worldwide and extends through half a billion years or more of earth history. The foundation of vertebrate paleontology is comparative anatomy, and its basic language is a standardized nomenclature for the organisms and their morphologies. The discipline relies upon fossils and their geological context for the bulk of its data.

It has been estimated that there are approximately 400 practicing vertebrate paleontologists in the world today (Hopson and Radinsky, 1980), most of whom find themselves associated with geology, biology, anatomy, or anthropology departments in universities or museums. Historically, whereas invertebrate paleontologists have generally been more geologically oriented (the main emphasis is on the use of invertebrate fossils in the practical, often commercial, application of correlating strata and reconstructing depositional environments), vertebrate paleontologists on the whole have been more biologically oriented (Simpson, 1961). Indeed, the great anatomist Sir Richard Owen considered vertebrate paleontology a branch of comparative anatomy (Owen, 1866–1868). (He was actually referring to paleochordatotomy, which is the study of the anatomy of ancient chordates.) This is not to imply that vertebrate paleontology has taken from geology and given nothing in return. Vertebrates have played, for example, an important role in establishing refined biostratigraphic correlations, most notably in continental Cenozoic deposits, and have contributed to paleobiogeographic analysis (see Part II). On the biological side, there is a continuum from extant vertebrates to those long extinct. No vertebrate zoologist, be he or she a paleontologist or a neontologist, can afford to ignore either the extant or the extinct groups. Vertebrate paleontology is dependent upon vertebrate neontology for extant analogues of fossil forms, but elucidates for the neontologist an otherwise unsuspected diversity of once-living forms.

Vertebrate paleontology is highly dependent on material objects—

fossils—for its basis. This dependence in turn necessitates field expeditions to collect the fossils, which must be carefully prepared, accurately identified, and thoroughly described. Then, this information must be published in order to disseminate the raw data to other workers worldwide. Repositories are also required where the specimens can be stored safely to serve as references for later generations of workers.

The beginnings of vertebrate paleontology as a distinct discipline can be traced back to the work of Georges Cuvier (see Paper 1) in France during the late eighteenth and early nineteenth centuries. Throughout the nineteenth century, the infant science progressed by leaps and bounds as new discoveries were made, such as that of the first-known dinosaurs in the 1820s (Man, 1982). Vertebrate paleontology moved to the forefront of science and was also large in the public eye. After 1859, with the publication of Charles Darwin's *The Origin of Species,* vertebrate paleontology added fuel to the debate over evolution. In North America, vertebrate paleontology was firmly established as a scientific discipline by Joseph Leidy in the 1840s and reached its heyday later in the nineteenth century when the arch rivals Edward Drinker Cope (Papers 3 and 8) and Othniel Charles Marsh (Papers 2 and 12) sent out numerous expeditions to the wild and virgin American West to collect ever more fossils (Gregory, 1979; Howard, 1975; Wheeler, 1960).

Over the last century, vertebrate paleontology has steadily progressed and expanded. Through good years and bad, wars and revolutions, a few intrepid souls have continued to work. Today the field is more active than ever and shows no signs of abating. Field expeditions continue to be sent out and more fossils are collected, prepared, and analyzed. Subjects of particular concern to vertebrate paleontologists, both past and present, include biostratigraphy and biogeography (Part II), comparative anatomy, classification and phylogeny (Part III), and functional morphology, paleobiology, and paleoecology (Part IV).

To the uninitiated, much of the literature on vertebrate paleontology, including the majority of papers reprinted in this volume, can seem like nothing more than a senseless mass of zoological names. Therefore, I have included a classification at the end of this introduction (Table 1) showing the approximate phylogenetic relationships of the major groups of vertebrates. I believe that the immediate future of vertebrate paleontology lies within the area of phylogenetic analysis. In recent years, the enormous flaws in past classifications and phylogenies of vertebrates have become patently apparent, and great advances have been made in phylogenetic analysis using cladistic methodology and principles. (For an introduction to cladistics, and also for various views on the pros and cons of this methodology, see articles in Cracraft and Eldredge, 1979; see also McKenna, 1981, and Szalay, 1981.) If we do not know the basic phylogenetic relationships of the animals being considered, studies depending on such information cannot be carried out. Examples of such studies are the paleobiogeographic analysis of the distribution of higher taxa and examinations of macroevolutionary patterns, trends, and processes (see Cracraft, 1983, and references cited therein; Croizat, 1964; see also articles reprinted in Smith, 1982).

SOURCES AND BIBLIOGRAPHIES OF FOSSIL VERTEBRATES

The field of vertebrate paleontology is fortunate to have included students who were meticulous bibliographers. It is through their efforts that the *Bibliographies of Fossil Vertebrates* have been, and continue to be, compiled (Gregory et al., 1983; all bibliographies published to date, covering the literature from 1509 to 1980, are cited in this volume). The *Zoological Record,* a massive bibliography published annually by BioSciences Information Service (Philadelphia) and The Zoological Society of London, also contains references to vertebrate fossils. Although now outdated in many details, Romer's (1966) textbook, *Vertebrate Paleontology,* remains a basic standard reference and is perhaps the best introduction to the field (although focusing primarily on anatomy). This book also contains a good bibliography of the older literature (see also Romer, 1968). Likewise, although written over half a century ago, Goodrich's *Studies on the Structure and Development of Vertebrates* (1958, a reprint of the 1930 edition) is still a good, basic reference on comparative vertebrate anatomy. The book is profusely illustrated and discusses both extant and extinct forms. Other useful, basic references, sources, and compilations that span the entire field of vertebrate paleontology are Müller (1966–1970), Orlov (1964), Piveteau (1952–1969), and Zittel (1900–1925). Moy-Thomas and Miles (1971) provide a thorough introduction to fossil fishes. The multivolume, multi-authored irregular series *Handbuch der Paläoherpetologie—Encylopedia of Paleoherpetology* (O. Kuhn and P. Wellnhofer, eds., 1969–present, Gustav Fischer Verlag, Stuttgart) contains reviews of various fossil groups of lower tetrapods. Romer (1956) deals exhaustively with fossil reptiles. Feduccia (1980) deals with fossil birds. Lillegraven, Kielan-Jaworowska, and Clemens (1979), Savage and Russell (1983), and Kurtén and Anderson (1980) discuss respectively Mesozoic, Cenozoic, and Pleistocene mammals and mammalian faunas. Although now somewhat outdated, another extremely useful work on Tertiary faunas is

Thenius (1959). Szalay and Delson (1979) give a detailed summary of all fossil primates known through 1977. Beneš and Burian (1980) give valuable life restorations of a number of extinct vertebrates. A practical guide to the collecting, preparation, and conservation of vertebrate fossils is supplied by Rixon (1976). Sarjeant's (1983) recent volume, *Terrestrial Trace Fossils,* reprints a number of papers that deal with vertebrate trace fossils (e.g., coprolites and footprints).

The history of vertebrate paleontology is recounted in Bowler (1977), Camp (1946, 1969), Clemens (1976, 1977, 1978), Desmond (1976, 1982), Dutro (1973), Emry and Schultze (1984), Fairbridge (in prep.), Gilmore (1941), Gregory (1974, 1975, 1979), Hopson and Radinsky (1980), Howard (1975), Lanham (1973), Lull (1918), Man (1982), Marsh (1879), Mayr (1982), McKenna, Hopson, and Schultze (1981), McKenna and Schultze (1982), Merriam (1921), Osborn (1931), Romer (1941, 1959), Romer et al. (1968), Rudwick (1976), Sarjeant (1980), Schneer (1969), Scott (1939), Shor (1971), Simpson (1942, 1961, 1980, 1983), Steel and Harvey (1979), Stock (1946), Webb (1979, 1980), Wheeler (1960) and Zittel (1901). The otherwise unreferenced biographical notes concerning vertebrate paleontologists that appear on the following pages are based on these sources.

CLASSIFICATION OF THE VERTEBRATES

The classification of the vertebrates is a topic of continued controversy among vertebrate zoologists. First, there is the question of whether classification should be natural (i.e., be composed of only monophyletic taxa as defined below), thereby reflecting the evolutionary history of the organisms classified rather than arbitrarily grouping organisms together on the basis of overall similarity and grade of evolution. Second, there is the problem of whether extinct organisms should, or even can, be classified together with extant organisms (Patterson and Rosen, 1977). Finally, many groups remain too poorly known to assess their phylogenetic relationships either because the fossils themselves are incomplete or poorly preserved or because they have not been adequately described and analyzed.

Since Romer's (1966) classification of the vertebrates, no comprehensive classification of all vertebrates has been attempted. Romer's classification included many traditional grade groups rather than monophyletic taxa, and thus did not reflect the phylogenetic relationships among the taxa. I hold the view, which is being adopted by more and more workers, that classification should reflect phylogeny as closely as possible. The following classification is my synthesis of recent work (since Romer, 1966) on various vertebrate groups. I have attempted to include only strictly monophyletic (holophyletic) taxa, in which all members of a given taxonomic group are descended from a single ancestor—stem species—that is itself included in the taxon. The same taxon also includes all descendants of any member of the taxon (Hennig, 1966, p. 73). But, due to inadequacies of analyses of some groups, some of the taxa used here may be unnatural (i.e., nonmonophyletic). Within any higher taxon, the two taxa of next lower rank indicate the initial dichotomy within the group. For example, within the Craniata it is hypothesized that the Hyperotreti represent a separate offshoot from all other craniates, here termed the Vertebrata; and, within the Vertebrata, the Pteraspidomorphi represent a branch on the "phylogenetic tree" distinct from all other vertebrates, here termed the Myopterygii. When, within a taxon of given rank, more than two taxa of the next lower rank are listed, a presently unresolved multicotomy is indicated. In the classification presented here, I have not used the concept of sequencing plesiomorphous (i.e., "primitive" and potentially "ancestral") fossil taxa as plesions within the classification (as proposed and discussed by Patterson and Rosen, 1977). However, I do believe that the concept of plesions and sequencing in classification may be useful to reflect detailed phylogenetic hypotheses of low-level, monophyletic taxa.

As far as possible, I have used previously coined names. (Original references for these names can be found by consulting the *Bibliographies of Fossil Vertebrates* cited above and monographic studies of various groups cited therein.) In a few cases, I have found it necessary to coin new names. These names are indicated as "new" and are coined in an unambiguous fashion, in accordance with the recommendations of Brothers (1983) for nomenclature at the ordinal and higher levels, by listing all subordinate taxa at the next lower level. Rather than add to the proliferation of categorical ranks (e.g., kingdom, phylum, class, series, division, legion, cohort, and the prefixes and suffixes that can be used with these ranks), I have used a numerical code system (lower numbers pertain to higher, more inclusive, ranks) as suggested, but not implemented, by McKenna (1975). Within the classification that follows, I have also included familiar names or brief descriptions of the animals involved, and I have included selected lower-level or alternative taxonomic names that are commonly known or that specifically appear in the papers reprinted in this volume.

Remarks

Traditionally, the Chordata, the Echinodermata, the Hemichordata, and sometimes various other invertebrate groups, depending on the author, have been associated together in the superphylum Deuterostomia (Anderson, 1982; Gans and Northcutt, 1983). Hemichordata has also been considered a subphylum of the phylum Chordata (Young, 1981). Løvtrup (1977) has suggested that the concept of Deuterostomia may be an unnatural group and that chordates (Craniata) may find their closest relatives among arthropods or molluscs. Løvtrup (1977) also suggested that the Urochordata and Cephalochordata (which together are sometimes referred to as the "Protochordata") be removed from the Chordata. Jefferies (e.g., Jefferies, 1968; Jefferies and Lewis, 1978) has argued that the "calcichordates" are chordate (*sensu lato*) "ancestors," but others (e.g., Philips, 1979; Ubaghs, 1975) believe they are true echinoderms that superficially resemble chordates.

The classification used here of the Hyperotreti, Pteraspidomorphi, and Hyperoartii (which together form the "Agnatha") is based on the work of Janvier (1981). The classification of the Gnathostomata, in general, has been influenced especially by the recent work of Løvtrup (1977), Patterson (1980), Rosen et al. (1981), Gardiner (1982), Gaffney (1979*a*, 1979*b*), Schultze and Trueb (1981), and Jarvik (1980, 1981) in conjunction with the traditional synthetic classifications of Romer (1966) and Moy-Thomas and Miles (1971). I have not attempted a synthesis of recent work on actinopterygian phylogeny and classification (e.g., Patterson, 1973; Patterson and Rosen, 1977), but instead use the three grades of Chondrostei, Holostei, and Teleostei. The placement of the living amphibians (Salientia, Caudata, and Gymnophiona, often all placed in the Lissamphibia) is extremely uncertain. I follow approximately Løvtrup's (1977) phylogeny of these forms, rather than associating them closely with the Temnospondyli as is often done (e.g., Gaffney, 1979 *a*). The concept and internal classification of the "Reptilia" used here is a mixed "wastebasket" of derived (advanced) and primitive forms, and is essentially the Reptilia of Romer (1966) with the removal of Archosauria, Dinosauria, Theropoda, and Synapsida. The Reptilia of the present classification are in dire need of a careful, modern, phylogenetic analysis. It should be noted that in his analysis of tetrapod phylogeny and classification, Gardiner (1982) dropped the use of the term "Reptilia" altogether. Here I take the conservative view and continue to associate the Pterosauria with the Archosauria, rather than with the Aves (Gardiner, 1982; Seeley, 1866).

The concepts of the Dinosauriiformes and Therornithes are based on the work of many students and are summarized principally in Bakker and Galton (1974), Ostrom (1976), and Padian (1982). The classification of birds adopted here is based on the work of Walker (1981). The concept of the Mammaliiformes is based on the almost universally accepted close relationship between the Synapsida (the mammallike reptiles) and mammals (for recent work, see Kemp, 1982, 1983, and for a contrary opinion see Gardiner, 1982). The classification of Mammalia used here is based primarily on the recent work of McKenna (1975; see also Domning, Morgan, and Ray, 1982, pp. 59–60), Szalay (1977, 1982), Novacek (1982), and Kemp (1983), none of which agree.

Table 1. A Classification of the Vertebrates

1. Animalia
 2. Deuterostomia
  3. Echinodermata
  3. Hemichordata—acorn worms
  3. Chordata
   4. Urochordata (=Tunicata)
    5. Ascidiacea—ascidians, sea squirts
    5. Thaliacea—salps and doliolids
    5. Larvacea—appendicularians
   4. Cephalochordata—Acrania, amphioxus, lancelots
   4. Craniata
    5. Hyperotreti—Agnatha, in part
     6. Myxinoidea—extant hagfishes and slime hags; Cyclostomata, in part
    5. Vertebrata
     6. Pteraspidomorphi (=Diplorhina)—primitive jawless fishes; Agnatha, in part
      7. Heterostraci (=Pteraspida)
      7. Thelodonti (=Coelolepida)
      7. Galeaspidida
     6. Myopterygii
      7. Hyperoartii—Agnatha, in part
       8. Osteostraci (=Cephalaspida)—armored jawless fishes
       8. Anaspida—armored jawless fishes
       8. Petromyzonida—lampreys; Cyclostomata, in part
      7. Gnathostomata—jawed vertebrates
       8. Elasmobranchiomorphi
        9. Placodermi—bizarre, archaic jawed fishes with head and trunk bony armor; Antiarchi, Arthrodira, *Bothryolepis (=Bothriolepis), Pterichthys,* Ptyctodontida
        9. Chondrichthyes—cartilaginous, jawed fishes
         10. Elasmobranchii—sharks, skates, rays, sawfishes; Cladoselachii, Pleuracanthodii, Selachii, *Xenacanthus*
         10. Holocephali—chimaeras; ?Bradyodonti, *Myledaphus*
       8. Teleostomi
        9. Acanthodii—"spiny sharks" of Paleozoic; *Acanthodes*
        9. Osteichthyes—higher bony fishes and tetrapods
         10. Actinopterygii (=Osteichthyes of some classifications)
          11. Chondrostei—sturgeons, paddlefishes; *Acipenser, Elonichthys (=?Pariostegus), Polypterus, Saurichthys*
          11. Holostei—bowfin, gars; Amiidae, "Dapediidae," Lepisosteidae (=Lepidosteidae), *Lepisosteus (=Clastes),* Semionotidae, *Protosphyraena,* Pycnodontidae
          11. Teleostei—higher bony fishes; *Benthesikyme, Beryx, Empo, Enchodus, Ichthyodectes, Leptolepis, Leptotrachelus, Pachyrhizodus, Portheus, Saurocephalus, Saurodon, Spaniodon, Stratodus, Thrissops*
         10. Sarcopterygii—unclassified rhipidistians (Osteolepiformes and Porolepiformes); Crossopterygii, in part; *Ectosteorhachis, Eusthenopteron, Holoptychus, Megalichthys*
          11. Actinistia (=Coelacanthiformes)—Crossopterygii, in part

(continued)

Table 1. (continued)

11. Choanata
 12. Dipnoi—lungfishes; *Ceratodus, Dipterus, Gnathorhiza, Heliodus,* Lepidosirenidae
 12. Tetrapoda—unclassified ichthyostegalians
  13. Lepospondyli—Batrachians=amphibians, in part
   14. Nectridea—*Diplocaulus, Keraterpeton (=Ceraterpeton), Ptyonius, Sauropleura (=Oestocephalus), Urocordylus*
   14. Aistopoda
   14. Microsauria—*Cocytinus, Euryodus, Hylerpeton, Lysorophus, Molgophis, Pariotichus*
   14. Lissamphibia
    15. Salientia (anurans)—frogs and toads
    15. Caudata (=Urodela)—salamanders
  13. Anthracoamniota, new
   14. Labyrinthodontia—Batrachians=amphibians, in part
    15. Temnospondyli (questionably assigned here)—Stegocephalia, in part; *Amphibamus (=Raniceps), Baphetes,* Branchiosauria, *Cacops, Dendrerpeton, Dictyocephalus, Eryops,* Rhachitomi, *Saurerpeton (=Pelion), Trematops, Trimerorhachis*
    15. Anthracosauria—Stegocephalia, in part; *Anthracosaurus (=Leptophractus), Diadectes, (=Empedocles), Diplovertebron,* Embolomeri, *Seymouria*
    15. Gymnophiona (Apoda)—caecilians
   14. Amniota
    15. Sauropsida
     16. Reptilia (=Eureptilia, new)
      17. Anapsida
       18. Cotylosauria—*Bolosaurus, Captorhinus (="Pariotichus"), Hylonomus (=Hylerpeton), Labidosaurus, Pareiasaurus,* Procolophonia
       18. Mesosauria
       18. Chelonia—turtles and relatives; *Adocus,* Cheloniidae, Chelydridae, *Compsemys, Polythorax,* Propleuridae
      17. Lepidosauria
       18. Eosuchia—*Champsosaurus (=Simoedosaurus)*
       18. Squamata
        19. Lacertilia—lizards, Mosasauridae (=Pythonomorpha)
        19. Ophidia (=Serpentes)—snakes
       18. Rhynchocephalia—tuatara *(Sphenodon)*
      17. Euryapsida (=Synaptosauria)
       18. Araeoscelidia (=Protorosauria)
       18. Sauropterygia—reptiles specialized for aquatic life
        19. Nothosauria
        19. Plesiosauria—*Elasmosaurus, Pliosaurus, Polycotylus*
       18. Placodontia
      17. Ichthyopterygia—"fish reptiles"
       18. Ichthyosauria
     16. Archosauria
      17. Thecodontia—*Belodon, Euparkeria, Phytosaurus (=Clepsysaurus)*

Table 1. (continued)

17. Crocodilia—*Goniopholis, Steneosaurus (=Streptospondylus), Teleosaurus*
17. Pterosauria—flying reptiles; *Dimorphodon, Pteranodon, Pterodactylus, Rhamphorhynchus*
17. Dinosauriiformes, new
 18. Dinosauria—dinosaurs, in part
  19. Ornithischia—Ankylosauria, *Camptosaurus,* Ceratopsia, *Claosaurus, Dryosaurus, Dysganus, Hadrosaurus, Hylaeosaurus, Hypsilophodon, Iguanodon, Laosaurus, Monoclonius,* Ornithopoda, *Palaeoscinus, Priodontosaurus (=Omosaurus), Scelidosaurus,* Stegosauria, *Stegosaurus, Thespesius (=Cionodon, =Diclonius), Triceratops*
  19. Saurischia—*Amphicoelias, Brontosaurus, Camarasaurus, Cetiosaurus, Diplodocus, Epantherias (=?Epanterias), Pelorosaurus (=Caulodon, =Chondrosteosaurus, =Ornithopsis),* Prosauropoda, Sauropoda, *Thecodontosaurus, Yaleosaurus (=Amphisaurus, =?Megadactylus)*
 18. Therornithes, new—dinosaurs, in part, and birds
  19. Theropoda—*?Palaeosauriscus (=?Palaeosaurus)*
   20. Coelurosauria—*Compsognathus*
   20. Carnosauria—*Antrodemus, Ceratosaurus, Dryptosaurus (=Laelaps), Megalosaurus, Ornithosuchus, Tyrannosaurus*
  19. Aves—birds
   20. Archaeornithes (Sauriurae)
    21. Archaeopterygiformes—*Archaeopteryx*
   20. Veraves, new
    21. Enantiornithes
    21. Neornithes
     22. Odontognathae (=Odontoholcae, =Odontornithes, in part)
      23. Hesperornithiformes—*Hesperornis*
     22. Euornithes, new
      23. Palaeognathae—Tinamidae, many "ratites"
      23. Neognathae—Odontornithes, in part; *Apatornis, Enaliornis, Ichthyornis*

15. Mammaliiformes, new (=Theropsida)
 16. Synapsida—mammallike "reptiles"
  17. Pelycosauria (Theromorpha)—*Bathygnathus, Casea, Clepsydrops, Dimetrodon, Edaphosaurus*
  17. Therapsida—Anomodontia, Cynodontia, *Deuterosaurus, Dicynodon,* Dinocephalia, Dromasauria, *Endothiodon,* Ictidosauria, *Lycosaurus,* Therocephalia
 16. Mammalia—mammals
  17. Eotheria
   18. Triconodonta
   18. Docodonta

(continued)

Table 1. (continued)

17. Apotheria, new
 18. Kuehneotheria
 18. Ornithotheria, new—unclassified pantotheres and ?symmetrodonts
  19. Ornithodelphia (=Prototheria)
   20. Monotremata—platypus
   20. Multituberculata (=Allotheria)
  19. Theria (=Tribosphenida)
   20. Marsupialia (=Metatheria)—marsupials
    21. Ameridelphia—*Borhyaena,* Didelphidae (=Didelphyidae), *Microbiotherium, Palaeothentes (=Epanorthus), Prothylacynus, Thylacodictis (=Amphiproviverra)*
    21. Australidelphia—Dasyuridae, Diprotodonta, Peramelina
   20. Eutheria—placental mammals
    21. Edentata—Cingulata, ?Palaeanodonta, Pilosa *(Pelecyodon),* ?Pholidota
    21. Epitheria—unclassified: Rodentia *(Acaremys,* Caviidae, Caviomorpha, Dasyproctidae, *Echimys* [=*Echinomys*], Eocardiinae, Octodontidae, *Sciamys)*
     22. Ernotheria—anagalids, *Kennalestes,* Lagomorpha (rabbits and hares), leptictids, Macroscelidea
     22. Preptotheria
      23. Deltatheridia—*Deltatheridium, Deltatheroides*
      23. Tokotheria
       24. Ferae—Apatotheria, Carnivora, Creodonta, Didelphodonta, Pantodonta (*Coryphodon*; Amblypoda, in part), Pantolesta, Taeniodonta
       24. Insectivora—Erinaceomorpha, Soricomorpha
       24. Archonta—Chiroptera (bats), Dermoptera, Primates *(Australopithecus, Homo, Homunculus,* Lemuroidea), Scandentia (tree shrews)
       24. Ungulata
        25. Eparctocyona—pigs and relatives; Arctocyonia, Artiodactyla, "Condylarthra," in part, Dinocerata (*Uintatherium;* Amblypoda, in part), Embrithopoda, Tillodontia, Tubulidentata
        25. Cete—whales, dolphins and relatives; Acreodi (Mesonychidoidea), Cetacea
        25. Meridiungulata—mostly extinct South American ungulates; *Astrapotherium,* Litopterna (Macraucheniidae, Proterotheriidae, *Theosodon, Thoatherium*), Notoungulata (Homalodotheriidae, *Notohippus,* Toxodontia), Pyrotheria, Trigonostylopoidea, Xenungulata
        25. Phenacodonta—horses, tapirs, rhinos and relatives; Condylarthra (Meniscotheriidae, *Periptychus*), Hyracoidea, Perissodactyla (Ceratomorpha, Chalicotherioidea, Hippomorpha)
        25. Tethytheria—elephants and relatives; Desmostylia, Proboscidea, Sirenia (*Ribodon*)

The classification given in Table 1 is an attempt to summarize the basic phylogenetic relationships of the major groups of vertebrates as currently understood. Rather than use named categorical ranks (e.g., kingdom, phylum, class, order), I have used a numerical code system of rankings in which lower numbers pertain to higher, more inclusive, groupings. Within any higher taxon, the two taxa of next lower rank indicate the initial evolutionary dichotomy of organisms within the group in question. More than two taxa of the next lower rank within a given higher taxon indicates an unresolved multicotomy (i.e., the phylogenetic relationships of the taxa concerned are unclear).

This classification groups extant reptiles with extant birds relative to extant mammals. An example of an alternative classification that groups extant birds with extant mammals relative to extant reptiles might include (with the lower taxa renumbered appropriately): 14. Amniota, 15. Reptilia, 15. Haemothermia, 16. Dinosauriiformes, 16. Mammaliiformes. Whether birds are more closely related to living mammals or to certain living reptiles (especially crocodiles) is an ongoing controversy in vertebrate paleontology. It can also be suggested that the Archosauria (Archosauria minus Dinosauriiformes) should be associated with the Dinosauriiformes within the Haemothermia. Some workers argue that the Crocodilia or Pterosauria are the sister group of the Aves or that the Thecodontia are ancestral to the Dinosauriiformes. The Cotylosauria are often considered ancestral to all other reptiles, Archosauria, Dinosauriiformes, and Mammaliiformes. I do not strongly favor any one of these views over the others.

REFERENCES

Anderson, D. T., 1982, Origins and Relationships among the Animal Phyla, *Linnean Soc. New South Wales Proc.* **106:**151-166.

Bakker, R. T., and P. M. Galton, 1974, Dinosaur Monophyly and a New Class of Vertebrates, *Nature* **248:**168-172.

Beněs, J., and Z. Burian, 1980, *Tiere der Urzeit,* Werner Dausien, Hanau, Germany, 360p.

Bowler, P. J., 1977, Edward Drinker Cope and the Changing Structure of Evolutionary Theory, *Isis* **68:**249-265.

Brothers, D. J., 1983, Nomenclature at the Ordinal and Higher Levels, *Systematic Zoology* **32:**34-42.

Camp, C. L., 1946, Trends and Developments in Vertebrate Paleontology, *Geol. Soc. America Bull.* **57:**541-546.

Camp, C. L., 1969, Paleontology in the West, *Jour. of the West* **8:**165-290.

Clemens, W. A., 1976, Vertebrate Paleontology, *Geotimes* **21:**32-33.

Clemens, W. A., 1977, Vertebrate Paleontology, *Geotimes* **22:**35-36.

Clemens, W. A., 1978, Vertebrate Paleontology: A Look Back on 1977, *Geotimes* **23:**19-20.

Cracraft, J., 1983, Cladistic Analysis and Vicariance Biogeography, *Am. Scientist* **71:**273-281.

Cracraft, J., and N. Eldredge, eds., 1979, *Phylogenetic Analysis and Paleontology,* Columbia University Press, New York, 233p.
Croizat, L., 1964, *Space, Time, Form: The Biological Synthesis,* published by the author, Caracas, Venezuela, 900p.
Darwin, C. R., 1859, *On the Origin of Species by Means of Natural Selection, or the Preservation of Favored Races in the Struggle for Life,* Murray, London, 500p.
Desmond, A. J., 1976, *The Hot-Blooded Dinosaurs: A Revolution in Paleontology,* The Dial Press, New York, 238p.
Desmond, A. J., 1982, *Archetypes and Ancestors: Paleontology in Victorian London 1850–1875,* Blond and Briggs, London, 287p.
Domning, D. P., G. S. Morgan, and C. E. Ray, 1982, North American Eocene Sea Cows (Mammalia: Sirenia), *Smithsonian Contr. Paleobiology* **52:**1–69.
Dutro, J. T., Jr., 1973, Paleontology, *Geotimes* **27:**28.
Emry, R. J., and H. -P. Schultze, 1984, Vertebrate Paleontology Reviewed, *Geotimes* **29:**14–16.
Fairbridge, R. W., ed., in preparation, *Geological Biography,* Encyclopedia of Earth Sciences Series, Van Nostrand Reinhold Company Inc., Stroudsburg, Pa.
Feduccia, A., 1980, *The Age of Birds,* Harvard University Press, Cambridge, Mass. and London, 208p.
Gaffney, E., 1979*a*, An Introduction to the Logic of Phylogeny Reconstruction, in *Phylogenetic Analysis and Paleontology,* J. Cracraft and N. Eldredge, eds., Columbia University Press, New York, pp. 79–111.
Gaffney, E., 1979*b*, Tetrapod Monophyly: A Phylogenetic Analysis, *Carnegie Mus. Nat. History Bull.* **13:**92–105.
Gans, C., and R. G. Northcutt, 1983, Neural Crest and the Origin of Vertebrates: A New Head, *Science* **220:**268–274.
Gardiner, B. G., 1982, Tetrapod Classification, *Linnean Soc. London Jour. Zoology* **74:**207–232.
Gilmore, C. W., 1941, History of Vertebrate Paleontology in the U.S. National Museum, *U.S. Natl. Mus. Proc.* **90:**305–377.
Goodrich, E. S., 1958, *Studies on the Structure and Development of Vertebrates,* 2 vols., Dover Publications, New York, 906p.
Gregory, J. T., 1974, Vertebrate Paleontology, *Geotimes* **19:**29–30.
Gregory, J. T., 1975, Vertebrate Paleontology, *Geotimes* **20:**36.
Gregory, J. T., 1979, North American Vertebrate Paleontology, 1776–1976, in *Two Hundred Years of Geology in America,* C. J. Schneer, ed., University Press of New England, Hanover, N.H., pp. 305–335.
Gregory, J. T., J. A. Bacskai, G. V. Shkurkin, and L. J. Bryant, 1983, *Bibliography of Fossil Vertebrates: 1980,* American Geological Institute, Falls Church, Va., 480p.
Hennig, W., 1966, *Phylogenetic Systematics,* University of Illinois Press, Chicago, 263p.
Hopson, J. A., and L. B. Radinsky, 1980, Vertebrate Paleontology: New Approaches and New Insights, *Paleobiology* **6:**250–270.
Howard, R. W., 1975, *The Dawnseekers: The First History of American Paleontology,* Harcourt Brace Jovanovich, New York and London, 330p.
Janvier, P., 1981, The Phylogeny of the Craniata, with Particular Reference to the Significance of Fossil "Agnathans," *Jour. Vert. Paleontology* **1:**121–159.

Jarvik, E., 1980, 1981, *Basic Structure and Evolution of Vertebrates,* Academic Press, London and New York, vol. 1 (1980), 592p.; vol. 2 (1981), 350p.

Jefferies, R. P. S., 1968, Some Fossil Chordates with Echinoderm Affinities, *Zool. Soc. London Symp.* **20:**163-208.

Jefferies, R. P. S., and D. N. Lewis, 1978, The English Silurian Fossil *Placocystites forbesianus* and the Ancestry of Vertebrates, *Royal Soc. London Philos. Trans.,* ser. B, **282:**205-323.

Kemp, T. S., 1982, *Mammal-like Reptiles and the Origin of Mammals,* Academic Press, London, 378p.

Kemp, T. S., 1983, The Relationships of Mammals, *Linnean Soc. London Jour. Zoology* **77:**353-384.

Kurtén, B., and E. Anderson, 1980, *Pleistocene Mammals of North America,* Columbia University Press, New York and London, 460p.

Lanham, U., 1973, *The Bone Hunters,* Columbia University Press, New York and London, 296p.

Lillegraven, J. A., Z. Kielan-Jaworowska, and W. A. Clemens, eds., 1979, *Mesozoic Mammals: The First Two-thirds of Mammalian History,* University of California Press, Berkeley, 322p.

Løvtrup, S., 1977, *The Phylogeny of Vertebrata,* John Wiley & Sons, London, 342p.

Lull, R. S., 1918, On the Development of Vertebrate Paleontology, *Am. Jour. Sci.,* ser. 4, **46:**193-221.

McKenna, M. C., 1975, Toward a Phylogenetic Classification of the Mammalia, in *Phylogeny of the Primates,* W. P. Luckett and F. S. Szalay, eds., Plenum Press, New York and London, pp. 21-46.

McKenna, M. C., 1981, Review of "Comparative Biology and Evolutionary Relationships of Tree Shrews," edited by W. P. Luckett, *Internat. Jour. Primatol.* **2:**97-101.

McKenna, M. C., and H. -P. Schultze, 1982, Vertebrate Paleontology, *Geotimes* **27:**56-57.

McKenna, M. C., J. A. Hopson, and H. -P. Schultze, 1981, Vertebrate Paleontology, *Geotimes* **26:**56-57.

Man, J., 1982, *The Day of the Dinosaur,* Bison Books Ltd, London, 192p.

Marsh, O. C., 1879, History and Methods of Palaeontological Discovery, *Nature* **20:**494-499, 515-521.

Mayr, E., 1982, *The Growth of Biological Thought,* Harvard University Press, Cambridge, Mass., and London, 986p.

Merriam, J. C., 1921, An Outline of Progress in Paleontological Research on the Pacific Coast, *Univ. California Dept. Geol. Sci. Bull.* **12:**237-266.

Moy-Thomas, J. A., and R. S. Miles, 1971, *Palaeozoic Fishes,* W. B. Saunders Co., Philadelphia and Toronto, 270p.

Müller, A. H., 1966-1970, *Lehrbuch der Paläozoologie, Vertebraten,* Gustav Fischer, Jena, Germany, vol. 1 (1966), 654p.; vol 2 (1968), 674p.; vol. 3 (1970), 870p.

Novacek, M. J., 1982, Information for Molecular Studies from Anatomical and Fossil Evidence on Higher Eutherian Phylogeny, in *Macromolecular Sequences in Systematic and Evolutionary Biology,* M. Goodman, ed., Plenum Press, New York and London, pp. 3-41.

Orlov, Yu. A., general ed., 1964, *Osnovy Paleontologii* [*Fundamentals of Paleontology*], Nauka, Moscow, 15 vols. (Available as: Orlov, Yu. A., general ed., 1967, 1968, *Fundamentals of Paleontology,* Israel Program for Scientific Translations, Jerusalem, and Smithsonian Institution and National Science Foundation, Washington, D.C., vol. 11: *Agnatha, Pisces* [1967], 836p.; vol. 13: *Mammals* [1968], 592p.)

Osborn, H. F., 1931, *Cope, Master Naturalist,* Princeton University Press, Princeton, N.J., 756p.

Ostrom, J. H., 1976, *Archaeopteryx* and the Origin of Birds, *Linnean Soc. Jour. Biology* **8:**91-182.

Owen, R., 1866-1868, *On the Anatomy of the Vertebrates,* Longmans, Green, and Co., London, vol. 1 (1866), 694p.; vol. 2 (1866), 600p.; vol. 3 (1868), 926p.

Padian, K., 1982, Macroevolution and the Origin of Major Adaptations: Vertebrate Flight as a Paradigm for the Analysis of Patterns, *North American Paleontol. Conv., Third, Proc.* **2:**387-392.

Patterson, C., 1973, Interrelationships of Holosteans, in *Interrelationships of Fishes,* P. H. Greenwood, R. S. Miles, and C. Patterson, eds., Academic Press, London and New York, pp. 233-305.

Patterson, C., 1980, Origin of Tetrapods: Historical Introduction to the Problem, in *The Terrestrial Environment and the Origin of Land Vertebrates,* A. L. Panchen, ed., Academic Press, London and New York, pp. 159-175.

Patterson, C., 1981, Significance of Fossils in Determining Evolutionary Relationships, *Ecol. Syst. Ann. Rev.* **12:**195-223.

Patterson, C., and D. E. Rosen, 1977, Review of Ichthyodectiform and Other Mesozoic Teleost Fishes and the Theory and Practice of Classifying Fossils, *Am. Mus. Nat. History Bull.* **158:**81-172.

Philips, G. M., 1979, Carpoids-Echinoderms or Chordates? *Biol. Rev.* **54:**439-471.

Piveteau, J., ed., 1952-1969, *Traité de Paléontologie,* Masson et Cie, Paris, vol. 1 (1952), 782p.; vol. 2 (1952), 790p.; vol. 3 (1953), 1063p.; vol. 4, pt. 1 (1964), 387p.; vol. 4, pt. 2 (1969), 790p.; vol. 4, pt. 3 (1966), 442p.; vol. 5 (1955), 1113p.; vol. 6, pt. 1 (1961), 1138p.; vol. 6, pt. 2 (1958), 962p.; vol. 7 (1957), 675p.

Rixon, A. E., 1976, *Fossil Animal Remains: Their Preparation and Conservation,* Athlone Press (University of London), London, 312p.

Romer, A. S., 1941, Vertebrate paleontology, in *Geological Society of America Fiftieth Anniversary Volume,* C. P. Berkey, ed., Geological Society of America, Boulder, Colo., pp. 107-135.

Romer, A. S., 1956, *Osteology of the Reptiles,* The University of Chicago Press, Chicago and London, 794p.

Romer, A. S., 1959, Vertebrate Paleontology, 1908-1958, *Jour. Paleontology* **33:**915-925.

Romer, A. S., 1966, *Vertebrate Paleontology,* The University of Chicago Press, Chicago and London, 478p.

Romer, A. S., 1968, *Notes and Comments on Vertebrate Paleontology,* The University of Chicago Press, Chicago and London, 312p.

Romer, A. S., B. Schaeffer, E. C. Olson, E. H. Colbert, and G. G. Simpson, 1968, Vertebrate Animals: Developments, Trends, and Outlooks in Paleontology, *Jour. Paleontology* **42:**1371-1377.

Rosen, D. E., P. L. Forey, B. G. Gardiner, and C. Patterson, 1981, Lungfishes, Tetrapods, Paleontology, and Plesiomorphy, *Am. Mus. Nat. History Bull.* **167:**163-275.

Rudwick, M. J. S., 1976, *The Meaning of Fossils,* Neale Watson Academic Publications, New York, 296p.

Sarjeant, W. A. S., 1980, *Geologists and the History of Geology: An International Bibliography from the Origins to 1978,* 5 vols., Arno Press, New York, 4530p.
Sarjeant, W. A. S., 1983, *Terrestrial Trace Fossils,* Benchmark Papers in Geology, vol. 76, Hutchinson Ross Publishing Company, Stroudsburg, Pa., 424p.
Savage, D. E., and D. E. Russell, 1983, *Mammalian Paleofaunas of the World,* Addison-Wesley Publishing Company, Reading, Mass. 450p.
Schneer, C. J., ed., 1969, *Toward a History of Geology,* MIT Press, Cambridge, Mass., and London, 476p.
Schultze, H. -P., and L. Trueb, 1981, Review of E. Jarvik, "Basic Structure and Evolution of Vertebrates," *Jour. Vert. Paleontology* **1:**389-397.
Scott, W. B., 1939, Vertebrate Paleontology Since 1888, *Geol. Soc. America Bull.* **50:**375-386.
Seeley, H. G., 1866, An Epitome of the Evidence that Pterodactyles are not Reptiles, but a New Subclass of Vertebrate Animals Allied to Birds (Saurornia), *Ann. Mag. Nat. History,* ser. 3, **17:**321-331.
Shor, E. N., 1971, *Fossils and Flies: The Life of a Compleat Scientist, Samuel Wendell Williston (1851-1918),* University of Oklahoma Press, Norman, 300p.
Simpson, G. G., 1942, The Beginnings of Vertebrate Paleontology in North America, *Am. Philos. Soc. Proc.* **86:**130-188.
Simpson, G. G., 1961, Some Problems of Vertebrate Paleontology, *Science* **133:**1679-1689.
Simpson, G. G., 1980, *Splendid Isolation: The Curious History of South American Mammals,* Yale University Press, New Haven, Conn., and London, 276p.
Simpson, G. G., 1983, *Fossils and the History of Life,* W. H. Freeman & Co., New York and San Francisco, 252p.
Smith, J. M., ed., 1982, *Evolution Now: A Century after Darwin,* W. H. Freeman & Co., San Francisco, 244p.
Steel, R., and A. P. Harvey, 1979, *The Encyclopedia of Prehistoric Life,* McGraw-Hill Book Company, New York, 218p.
Stock, C., 1946, Progress in Paleontological Research on the Pacific Coast, 1917-1944, *Geol. Soc. America Bull.* **57:**319-354.
Szalay, F. S., 1977, Phylogenetic Relationships and a Classification of the Eutherian Mammalia, in *Major Patterns in Vertebrate Evolution,* M. K. Hecht, P. C. Goody, and B. M. Hecht, eds., Plenum Press, New York and London, pp. 315-374.
Szalay, F. S., 1981, Functional Analysis and Practice of the Phylogenetic Method as Reflected by Some Mammalian Studies, *Amer. Zool.* **21:**37-45.
Szalay, F. S., 1982, A New Appraisal of Marsupial Phylogeny and Classification, in *Carnivorous Marsupials,* M. Archer, ed., Royal Zoological Society, New South Wales, Sydney, pp. 621-640.
Szalay, F. S., and E. Delson, 1979, *Evolutionary History of the Primates,* Academic Press, New York, 594p.
Thenius, E., 1959, *Tertiär,* vol. 3, *Wirbeltierfaunen, Handbuch der Stratigraphischen Geologie,* Ferdinand Enke Verlag, Stuttgart, Germany, 328p.
Ubaghs, G., 1975, Early Paleozoic Echinoderms, *Ann. Rev. Earth Planetary Sci.* **3:**79-98.
Walker, C. A., 1981, New Subclass of Birds from the Cretaceous of South America, *Nature* **292:**51-53.

Webb, S. D., 1979, Vertebrate Paleontology, *Geotimes* **24:**49–50.
Webb, S. D., 1980, Vertebrate Paleontology in 1979, *Geotimes* **25:**22–23.
Wheeler, W. H., 1960, The Uintatheres and the Cope-Marsh War, *Science* **131:**1171–1176.
Young, J. Z., 1981, *The Life of Vertebrates,* Clarendon Press, Oxford, 660p.
Zittel, K. A. von, 1900–1925, *Text-book of Palaeontology,* translated and edited by C. R. Eastman (vol. 3 revised by M. Schlosser), Macmillan and Co., London, vol. 1 (1900), 716p.; vol. 2 (1902), 292p.; vol. 3 (1925), 324p.
Zittel, K. A., 1901, *History of Geology and Palaeontology to the End of the Nineteenth Century,* Walter Scott, London, 578p.

Part I

THE SCIENCE OF VERTEBRATE PALEONTOLOGY

Editor's Comments on Papers 1 and 2

Baron Georges Léopold Chrétien Frédéric Dagobert Cuvier (1769–1832), more than any other man, can be considered the founder of vertebrate paleontology as a distinct discipline. Born in French Montbéliard, which at that time was a dependency of the Duchy of Württemberg, Cuvier early became interested in natural history and learned to dissect. He received formal training at the Caroline Academy in Stuttgart (1784–1788) and then spent seven years as a tutor in Normandy, during which time he continued his studies. He then went to Paris and received the following appointments, held simultaneously: professor of Anatomy at the Museum National d'Histoire Naturelle (1795), professor of Natural History at the College of France (1800), and professor of Anatomy at the College of France (1802).

Cuvier's greatest work is *Recherches sur les Ossemens Fossiles* (Paris, 1812, and many subsequent editions; Romer et al., 1962), which is primarily a collection of his previously published papers. This collection contains a "preliminary discourse" that was also published separately as *Discours sur la Théorie de la Terre* (Paris, 1821; Romer et al., 1962). This discourse had a number of editions and revisions. Papers 1A and 1B are excerpts from an early English translation of this work.

In this selection, Cuvier first justifies the importance of studying vertebrate fossils and then sets out his methodology for pursuing such studies. Throughout his career, two great principles guided the researches of Cuvier: (1) the facts come first and scientific theories are at best secondary; (2) God created each species with a definite role and place in nature, and species vary little through time (Bourdier, 1969). In accord with his second principle, Cuvier propounded a series of catastrophic extinctions of entire faunas, which God then replaced by new faunas. Cuvier was very much opposed to evolutionary notions involving the transmutation of species. His second principle also helped to justify his famous reconstruction of animals based on extremely fragmentary remains.

Othniel Charles Marsh (1831-1899) was one of the three founders of vertebrate paleontology in North America. (The others were Joseph Leidy [1823-1891] and Edward Drinker Cope [1840-1897].) The nephew of the philanthropist George Peabody (1795-1869), Marsh founded the Peabody Museum of Natural History at Yale University, New Haven, Connecticut, with his uncle's money. There Marsh held the position of professor of Paleontology from 1866 to the end of his life. He was a prodigious collector and writer, sending out numerous fossil collecting expeditions to the western United States (Schuchert and LeVene, 1940).

Paper 2 is Marsh's vice-presidential address to the American Association for the Advancement of Science in 1877. The address generally expresses the state of the science of vertebrate paleontology a hundred years ago. With uncanny accuracy, Marsh (Paper 2, p. 490) predicted the subsequent discovery of mammals in the Cretaceous strata of North America, many of which he described himself (see Archibald, 1982; Lillegraven, Kielan-Jaworowska, and Clemens, 1979). He also predicted the discovery of higher anthropoids and the ancestors of man in Africa (see Paper 16; Howell, 1978; Simons, Andrews, and Pilbeam, 1978; Simons and Pilbeam, 1978). As stated in the Preface, it would be impossible to here correct and update in their entirety papers reprinted in this volume. However, one important addition to Marsh's address should be pointed out. Vertebrates (in the form of supposed heterostracean fish dermal fragments) are now reported from the late Cambrian of Wyoming and are known from the Ordovician of several continents (Repetski, 1978). The "first of anatomists" mentioned by Marsh (Paper 2, last paragraph on page 472) is Thomas Henry Huxley (1825-1895), who visited Marsh for a week in 1876 (Schuchert and LeVene, 1940).

REFERENCES

Archibald, J. D., 1982, *A study of Mammalia and Geology across the Cretaceous-Tertiary Boundary in Garfield County, Montana,* University of California Publications in Geological Science, No. 122, University of California Press, Berkeley, 286p.

Bourdier, F., 1969, Geoffroy Saint-Hilaire Versus Cuvier: The Campaign for Paleontological Evolution (1825–1838), in *Toward a History of Geology,* C. J. Schneer, ed., MIT Press, Cambridge, Mass., and London, pp. 36–61.

Howell, F. C., 1978, Hominidae, in *Evolution of African Mammals,* V. J. Maglio and H. B. S. Cooke, eds., Harvard University Press, Cambridge, Mass., and London, pp. 154–248.

Lillegraven, J. A., Z. Kielan-Jaworowska, and W. A. Clemens, eds., 1979, *Mesozoic Mammals: The First Two-thirds of Mammalian History,* University of California Press, Berkeley, 322p.

Repetski, J. E., 1978, A Fish from the Upper Cambrian of North America, *Science,* **200:**529–531.

Romer, A. S., N. E. Wright, T. Edinger, and R. van Frank, 1962, *Bibliography of Fossil Vertebrates Exclusive of North America, 1509–1927,* 2 vols., Geological Society of America Memoir 87, Geological Society of America, Boulder, Colo., 1634p.

Schuchert, C., and C. M. LeVene, 1940, *O. C. Marsh—Pioneer in Paleontology,* Yale University Press, New Haven, Conn., 562p.

Simons, E. L., P. Andrews, and D. R. Pilbeam, 1978, Cenozoic Apes, in *Evolution of African Mammals,* V. J. Maglio and H. B. S. Cooke, eds., Harvard University Press, Cambridge, Mass., and London, pp. 120–146.

Simons, E. L., and D. R. Pilbeam, 1978, *Ramapithecus* (Hominidae, Hominoidea), in *Evolution of African Mammals,* V. J. Maglio and H. B. S. Cooke, eds., Harvard University Press, Cambridge, Mass., and London, pp. 147–153.

1A

Reprinted from pages 57–60 of *Essay on the Theory of the Earth,* 2nd ed., Robert Jameson, trans., William Blackwood, John Murray, and Robert Baldwin, London, 1815

HIGH IMPORTANCE OF INVESTIGATING THE FOSSIL REMAINS OF QUADRUPEDS

G. Cuvier

It is obvious that the fossil remains of the bones of quadrupeds must lead to more rigorous conclusions than any other remains of organized bodies, and that for several reasons.

In the first place, they indicate much more clearly the nature of the revolutions to which they have been subjected. The remains of shells certainly indicate that the sea has once existed in the places where these collections have been formed : But the changes which have taken place in their species, when rigorously enquired into, may possibly have been occasioned by slight changes in the nature of the fluid in which they were formed, or only in its temperature, and may even have arisen from other accidental causes. We can never be perfectly assured that certain species, and even genera, inhabiting the bottom of the sea, and occupying certain fixed spaces for a longer or shorter time, may not have been driven away from these by other species or genera.

In regard to quadrupeds, on the contrary, every thing is precise. The appearance of their bones in strata, and still more of their entire carcases, clearly establishes that the bed in which they are found must have been previously laid dry, or at least that dry land must have existed in its immediate neighbourhood. Their disappearance as certainly announces that this stratum must have been inundated, or that the dry land had ceased to exist in that state. It is from them, therefore, that we learn with perfect certainty the important fact of the repeated irruptions of the sea upon the land, which the extraneous fossils and other productions of marine origin could not of themselves have proved ; and, by a careful investigation of them, we may hope to ascertain the number and the epochs of those irruptions of the sea.

Secondly, the nature of the revolutions which have changed the surface of our earth, must have exerted a more powerful action upon terrestrial quadrupeds than upon marine animals. As these revolutions have consisted chiefly in changes of the bed of the sea, and as the waters must have destroyed all the quadrupeds which they reached, if their irruption over the land was general, they must have destroyed the entire class, or, if confined only to certain continents at one time, they must have destroyed at least all the species inhabiting these continents, without having the same effect upon the marine animals. On the other hand, millions of aquatic animals may have been left quite dry, or buried in newly-formed strata, or thrown violently on the coasts, while their races may have been still preserved in more peaceful parts of the sea, whence they might again propagate and spread after the agitation of the water had ceased.

Thirdly, this more complete action is also more easily ascertained and demonstrated; because, as the number of terrestrial quadrupeds is limited, and as most of their species, at least the large ones, are well known, we can more easily determine whether fossil bones belong to a species which still exists, or to one that is now lost. As, on the other hand, we are still very far from being acquainted with all the testaceous animals and fishes belonging to the sea, and as we probably still remain ignorant of the greater part of those which live in the extensive deeps of the ocean, it is impossible to know, with any certainty, whether a species found in a fossil state may not still exist somewhere alive. Hence some naturalists persist in giving the name of oceanic or pelagic shells to *belemnites* and *cornua-ammonis*, and some other genera, which have not hitherto been found, except in the fossil state, in ancient strata; meaning by this, that although these have not as yet been found in a living or recent state, it is because they inhabit the bottom of the ocean, far beyond the reach of our nets.

1B

Reprinted from pages 88–103 of *Essay on the Theory of the Earth,* 2nd ed., Robert Jameson, trans., William Blackwood, John Murray, and Robert Baldwin, London, 1815

OF THE DIFFICULTY OF DISTINGUISHING THE FOSSIL BONES OF QUADRUPEDS

G. Cuvier

While the study of the fossil remains of the greater quadrupeds is more satisfactory, by the clear results which it affords, than that of the remains of other animals found in a fossil state, it is also complicated with greater and more numerous difficulties. Fossil shells are usually found quite entire, and retaining all the characters requisite for comparing them with the specimens contained in collections of natural history, or represented in the works of naturalists. Even the skeletons of fishes are found more or less entire, so that the general forms of their bodies can, for the most part, be ascertained, and usually at least their generic and specific characters are determinable, as these are mostly drawn from their solid parts. In quadrupeds, on the contrary, even when their entire skeletons are found, there is great difficulty in discovering their distinguishing characters, as these are chiefly founded upon their hair and colours, and other marks which have disappeared previous to their incrustation. It is also very rare to find any fossil skeletons of quadrupeds in any degree approaching to a complete state, as the strata for the most part only contain separate bones, scattered confusedly, and almost always broken and reduced to fragments, which are the only means left to naturalists for ascertaining the species or genera to which they have belonged.

It may be stated also, that most observers, alarmed by these formidable difficulties, have passed slightly over the fossil remains of quadrupeds, and have satisfied themselves with classing them vaguely, by means of slight resemblances, or have not even pretended to give them names. Hence this portion of the history of extraneous fossils, though the most important and most instructive, has been investigated with less care than any other. *

* As I have already remarked on a former occasion, it is not my intention, by these observations, to detract from the merits

Fortunately, comparative anatomy, when thoroughly understood, enables us to surmount all these difficulties, as a careful application of its principles instructs us in the correspondence and dissimilarity of the forms of organized bodies of different kinds, by which each may be rigorously ascertained, from almost every fragment of its various parts and organs.

Every organized individual forms an entire system of its own, all the parts of which mutually correspond, and concur to produce a certain definite purpose, by reciprocal reaction, or by combining towards the same end. Hence none of these separate parts can change their forms without a corresponding change on the other parts of the same animal, and consequently each of these parts taken separately, indicates all the other parts to which it has belonged. Thus, as I have elsewhere shewn, if the viscera of an animal are so organized as only to be fitted for

---

of Camper, Pallas, Blumenbach, Sœmmering, Merk, Faujas, Rosenmuller, and other naturalists, in regard to extraneous fossils: But, though their observations have been of great value in my researches, and are quoted by me in every step, they are in general very incomplete.

the digestion of recent flesh, it is also requisite that the jaws should be so constructed as to fit them for devouring prey; the claws must be constructed for seizing and tearing it to pieces; the teeth for cutting and dividing its flesh; the entire system of the limbs, or organs of motion, for pursuing and overtaking it; and the organs of sense, for discovering it at a distance. Nature also must have endowed the brain of the animal with instincts sufficient for concealing itself, and for laying plans to catch its necessary victims.

Such are the universal conditions that are indispensable in the structure of carnivorous animals; and every individual of that description must necessarily possess them combined together, as the species could not otherwise subsist. Under this general rule, however, there are several particular modifications, depending upon the size, the manners, and the haunts of the prey for which each species of carnivorous animal is destined or fitted by nature; and, from each of these particular modifications, there result certain differences in the more minute conformations of particular parts, all, however, comformable to the general principles of structure already mentioned. Hence it follows, that in every one of their parts we discover distinct indications, not only of the classes and orders of animals, but also of their genera, and even of their species.

In fact, in order that the jaw may be well adapted for laying hold of objects, it is necessary that its condyle should have a certain form; that the resistance, the moving power, and the fulcrum, should have a certain relative position with respect to each other; and that the temporal muscles should be of a certain size: The hollow or depression, too, in which these muscles are lodged, must have a certain depth; and the zygomatic arch under which they pass, must not only have a certain degree of convexity, but it must be sufficiently strong to support the action of the masseter.

To enable the animal to carry off its prey when seized, a correspondent force is requisite in the muscles which elevate the head; and this necessarily gives rise to a determinate form of the vertebræ to which these muscles are attached, and of the occiput into which they are inserted.

In order that the teeth of a carnivorous animal may be able to cut the flesh, they require to be sharp, more or less so in proportion to the greater or less quantity of flesh that they have to cut. It is requisite that their roots shonld be solid and strong, in proportion to the quantity and the size of the bones which they have to break in pieces. The whole of these circumstances must necessarily influence the developement and form of all the parts which contribute to move the jaws.

To enable the claws of a carnivorous animal to seize its prey, a considerable degree of mobility is necessary in their paws and toes, and a considerable strength in the claws themselves. From these circumstances, there necessarily result certain determinate forms in all the bones of their paws, and in the distribution of the muscles and tendons by which they are moved. The fore-arm must possess a certain facility of moving in various directions, and consequently requires certain determinate forms in the bones of which it is composed. As the bones of the fore-arm are articulated with the arm-bone or humerus, no change can take place in the form and structure of the former, without occasioning correspondent changes in the form of the latter. The shoulder blade also, or scapula, requires a correspondent degree of strength in all animals destined for catching prey, by which it likewise must necessarily have an appropriate form. The play and action of all these parts require certain proportions in the muscles which set them in motion, and the impressions formed by these muscles must still farther determine the forms of all these bones.

After these observations, it will be easily seen that similar conclusions may be drawn with respect to the hinder limbs of carnivorous animals, which require particular conformations to fit them for rapidity of motion in general; and that similar considerations must influence the forms and connections of the vertebræ and other bones constituting the trunk of the body, to fit them for flexibility and readiness of motion in all directions. The bones also of the nose, of the orbit, and of the ears, require certain forms and structures to fit them for giving perfection to the senses of smell, sight, and hearing, so necessary to animals of prey. In short, the shape and structure of the teeth regulate the forms of the condyle, of the shoulder-blade, and of the claws.

in the same manner as the equation of a curve regulates all its other properties; and, as in regard to any particular curve, all its properties may be ascertained by assuming each separate property as the foundation of a particular equation; in the same manner, a claw, a shoulder-blade, a condyle, a leg or arm bone, or any other bone, separately considered, enables us to discover the description of teeth to which they have belonged; and so also reciprocally we may determine the forms of the other bones from the teeth. Thus, commencing our investigation by a careful survey of any one bone by itself, a person who is sufficiently master of the laws of organic structure, may, as it were, reconstruct the whole animal to which that bone had belonged.

This principle is sufficiently evident, in its general acceptation, not to require any more minute demonstration; but, when it comes to be applied in practice, there is a great number of cases in which our theoretical knowledge of these relations of forms is not sufficient to guide us, unless assisted by observation and experience.

For example, we are well aware that all hoofed animals must necessarily be herbivorous, because they are possessed of no means of seizing upon prey. It is also evident, having no other use for their fore-legs than to support their bodies, that they have no occasion for a shoulder so vigorously organized as that of carnivorous animals; owing to which, they have no clavicles or accromion processes, and their shoulder-blades are proportionally narrow. Having also no occasion to turn their fore-arms, their radius is joined by ossification to the ulna, or is at least articulated by gynglymus with the humerus. Their food being entirely herbaceous, requires teeth with flat surfaces, on purpose to bruise the seeds and plants on which they feed. For this purpose also, these surfaces require to be unequal, and are consequently composed of alternate perpendicular layers of hard enamel and softer bone. Teeth of this structure necessarily require horizontal motions, to enable them to triturate or grind down the herbaceous food; and, accordingly, the condyles of the jaw could not be formed into such confined joints as in the carnivorous animals, but must have a flattened form, correspondent to sockets in the temporal bones, which also are more or less flat for their reception. The hollows likewise of the temporal bones, having smaller muscles to contain, are narrower, and not so deep, &c. All these circumstances are deducible from each other, according to their greater or less generality, and in such manner that some are essentially and exclusively appropriated to hoofed quadrupeds, while other circumstances, though equally necessary to that description of animals, are not exclusively so, but may be found in animals of other descriptions, where other conditions permit or require their existence.

When we proceed to consider the different orders or subdivisions of the class of hoofed animals, and examine the modifications to which the general conditions are liable, or rather the particular conditions which are conjoined, according to the respective characters of the several subdivisions, the reasons upon which these particular conditions or rules of confirmation are founded become less evident. We can easily conceive, in general, the necessity of a more complicated system of digestive organs in those species which have less perfect masticatory systems; and hence we may presume that these latter animals require especially to be ruminant, which are in want of such or such kinds of teeth; and may also deduce, from the same considerations, the necessity of a certain confirmation of the esophagus, and of corresponding forms in the vertebræ of the neck, &c. But I doubt whether it would have been discovered, independently of actual observation, that ruminant animals should all have cloven hoofs, and that they should be the only animals having that particular conformation; that the ruminant animals only should be provided with horns on their foreheads; that those among them which have sharp tusks, or canine teeth, should want horns, &c.

As all these relative conformations are constant and regular, we may be assured that they depend upon some sufficient cause; and, since we are not acquainted with that cause, we must here supply the defect of theory by observation, and in this way lay down empirical rules on the subject, which are almost as certain as those deduced from rational principles, especially if established upon careful and repeated observation. Hence, any one who observes merely the print of a cloven hoof, may conclude that it has been left by a ruminant animal, and regard the conclusion as equally certain with any other in physics or in morals. Consequently, this single foot-mark clearly indicates to the observer the forms of the teeth, of the jaws, of the vertebræ, of all the leg-bones, thighs, shoulders, and of the trunk of the body of the animal which left the mark. It is much surer than all the marks of Zadig. Observation alone, independent entirely of general principles of philosophy, is sufficient to shew that there certainly are secret reasons for all these relations of which I have been speaking.

When we have established a general system of these relative conformations of animals, we not only discover specific constancy, if the expression may be allowed, between certain forms of certain organs, and certain other forms of different organs; we can also perceive a classified constancy of conformation, and a correspondent gradation between these two sets of organs, which demonstrate their mutual influence upon each other, almost as certainly as the most perfect deduction of reason. For example, the masticatory system is generally more perfect in the non-ruminant hoofed quadrupeds than it is in the cloven-hoofed or ruminant quadrupeds; as the former pcssess incisive teeth, or tusks, or almost always both of these, in both jaws. The structure also of their feet is in general more complicated, having a greater number of toes, or their phalanges less enveloped in the hoof, or a greater number of distinct metacarpal and metatarsal bones, or more numerous tarsal bones, or the fibula more completely distinct from the tibia: or, finally, that all these enumerated circumstances are often united in the same species of animal.

It is quite impossible to assign reasons for these relations; but we are certain that they are not produced by mere chance, because, whenever a cloven-hoofed animal has any resemblance in the arrangement of its teeth to the animals we now speak of, it has the resemblance to them also in the arrangement of its feet. Thus camels, which have tusks, and also two or four incisive teeth in the upper jaw, have one additional bone in the tarsus, their scaphoid and cuboid bones not being united into one; and have also very small hoofs, with corresponding phalanges, or toe-bones. The musk animals, whose tusks are remarkably conspicuous, have a distinct fibula as long as the tibia; while the other cloven-footed animals have only a small bone articulated at the lower end of the tibia, in place of a fibula. We have thus a constant mutual relation between the organs or conformations, which appear to have no kind of connection with each other; and the gradations of their forms invariably correspond, even in those cases in which we cannot give the rationale of their relations.

By thus employing the method of observation, where theory is no longer able to direct our views, we procure astonishing results. The smallest fragment of bone, even the most apparently insignificant apophysis, possesses a fixed and determinate character, relative to the class, order, genus, and species of the animal to which it belonged; insomuch, that, when we find merely the extremity of a well-preserved bone, we are able, by careful examination, assisted by analogy and exact comparison, to determine the species to which it once belonged, as certainly as if we had the entire animal before us. Before venturing to put entire confidence in this method of investigation, in regard to fossil bones, I have very frequently tried it with portions of bones belonging to well-known animals, and always with such complete success, that I now entertain no doubt with regard to the results which it affords. I must acknowledge that I enjoy every kind of advantage for such investigations that could possibly be of use, by my fortunate situation in the Museum of Natural History; and, by assiduous researches for nearly fifteen years, I have collected skeletons of all the genera and sub-genera of quadrupeds, with those of many species in some of the genera and even of several varieties of some species. With these aids, I have found it easy to multiply comparisons, and to verify, in every point of view, the application of the foregoing rules.

We cannot, in the present Essay, enter into a more lengthened detail of this method, and must refer for its entire explanation to the large work on Comparative Anatomy, which we propose to publish very soon, and in which all its laws will be explained and illustrated. In the meantime, the intelligent reader may gather a great number of these from the work now laid before him, if he will take the trouble of attending to all the applications which we have made of them. He will there find that it is by this method alone that we have been guided, and that it has almost always been sufficient for the purpose of referring every fossil bone to its peculiar species, if belonging to one that still exists; to its genus, if belonging to an unknown species; to its order, if belonging to a new genus; and, finally, to its class, if belonging to an unknown order: And, in these three latter predicaments, to assign to it the proper characters for distinguishing it from the nearest resembling orders, genera, and species. Before the commencement of these researches, naturalists had done no more than this in regard even to such animal remains as were found in an entire state.

ERRATUM

Page 93, line 5 from the top of the page should read "It is requisite that their roots should . . . ."

2

Reprinted by permission from *Nature* **16:**448–450, 470–472, 489–491 (1877)

INTRODUCTION AND SUCCESSION OF VERTEBRATE LIFE IN AMERICA[1]

O. C. Marsh

THE origin of life, and the order of succession in which its various forms have appeared upon the earth, offer to science its most inviting and most difficult field of research. Although the primal origin of life is unknown, and may perhaps never be known, yet no one has a right to say how much of the mystery now surrounding it science cannot remove. It is certainly within the domain of science to determine when the earth was first fitted to receive life, and in what form the earliest life began. To trace that life in its manifold changes through past ages to the present is a more difficult task, but one from which modern science does not shrink. In this wide field every earnest effort will meet some degree of success; every year will add new and important facts; and every generation will bring to light some law, in accordance with which ancient life has been changed into life as we see it around us to-day. That such a development has taken place no one will doubt who has carefully traced any single group of animals through its past history, as recorded in the crust of the earth. The evidence will be especially conclusive if the group selected belongs to the higher forms of life, which are sensitive to every change in their surroundings. But I am sure I need offer here no argument for evolution; since to doubt evolution to-day is to doubt science, and science is only another name for truth.

Taking, then, evolution as a key to the mysteries of past life on the earth, I invite your attention to the subject I have chosen: The Introduction and Succession of Vertebrate Life in America.

In the brief hour allotted to me I could hardly hope to give more than a very incomplete sketch of what is now known on this subject. I shall therefore pass rapidly over the lower groups, and speak more particularly of the higher vertebrates, which have an especial interest to us all, in so far as they approach man in structure, and thus indicate his probable origin. These higher vertebrates, moreover, are most important witnesses of the past, since their superior organisation made them ready victims to slight climatic changes, which would otherwise have remained unrecorded.

In considering the ancient life of America it is important to bear in mind that I can only offer you a brief record of a few of the countless forms that once occupied this continent. The review I can bring before you will not be like that of a great army, when regiment after regiment with full ranks moves by in orderly succession, until the entire host has passed. My review must be more like the roll-call after a battle, when only a few scarred and crippled veterans remain to answer to their names. Or rather, it must resemble an array of relics, dug from the field of some old Trojan combat, long after the contest, when no survivor remains to tell the tale of the strife. From such an ancient battle-field a Schliemann might unearth together the bronze shield, lance-head, and gilded helmet of a prehistoric leader, and learn from them with certainty his race and rank. Perhaps the skull might still retain the barbaric stone weapon by which his northern foe had slain him. Near by the explorer might bring to light the commingled coat of mail and trappings of a horse and rider, so strangely different from the equipment of the chief, as to suggest a foreign ally. From these, and from the more common implements of war that fill the soil, the antiquary could determine, by patient study, what nations fought, and perhaps when and why.

By this same method of research the more ancient strata of the earth have been explored, and in our western wilds, veritable battle-fields, strewn with the fossil skeletons of the slain, and guarded faithfully by savage superstition, have been despoiled, yielding to science treasures more rare than bronze or gold. Without such spoils, from many fields, I could not have chosen the present theme for my address to-night.

According to present knowledge, no vertebrate life is known to have existed on this continent in the Archæan, Cambrian, or Silurian periods; yet during this time more than half of the thickness of American stratified rocks was deposited. It by no means follows that vertebrate animals of some kind did not exist here in those remote ages. Fishes are known from the upper Silurian of Europe, and there is every probability that they will yet be discovered in our strata of the same age, if not at a still lower horizon.

In the shore deposits of the early Devonian sea, known as the Schoharie grit, characteristic remains of fishes were preserved, and in the deeper sea that followed, in which the corniferous limestone was laid down, this class was well represented. During the remainder of the Devonian fishes continue abundant in the shallower seas, and, so far as now known, were the only type of vertebrate life. These fishes were mainly ganoids, a group represented in our present waters by the gar-pike (*Lepidosteus*) and sturgeon (*Acipenser*), but, in the Devonian sea, chiefly by the placoderms, the exact affinities of which are somewhat in doubt. With these were elasmobranchs, or the shark tribe, and among them a few chimæroids, a peculiar type, of which one or two members still survive. The placoderms were the monarchs of the ocean. All were well protected by a massive coat of armour, and some of them attained huge dimensions. The American Devonian fishes now known are not as numerous as those of Europe, but they were larger in size, and mostly inhabitants of the open sea. Some twenty genera and forty species have been described.

The more important genera of placoderms are *Dinichthys*, *Aspidichthys*, and *Diplognathus*, our largest palæozoic fishes. Others are, *Acanthaspis*, *Acantholepis*, *Coccosteus*, *Macropetalichthys*, and *Onychodus*. Among the elasmobranchs were, *Cladodus*, *Ctenacanthus*, *Machæracanthus*, *Rhynchodus*, and *Ptyctodus*, the last two being regarded as chimeroids. In the

[1] Lecture delivered at the Nashville meeting of the American Association, August 30,

Chemung epoch the great dipterian family was introduced with *Dipterus*, *Heliodus*, and possibly *Ceratodus*. Species of the European genera, *Bothriolepis* and *Holoptychius*, have likewise been found in our Devonian deposits.

With the close of the Devonian came the almost total extinction of the great group of placoderms, while the elasmobranchs, which had hitherto occupied a subordinate position, increase in numbers and size, and appear to be represented by sharks, rays, and chimeras. Among the members of this group from the carboniferous were numerous cestracionts, species of *Cochliodus* of large size, with others of the genera *Deltodus*, *Helodus*, *Psammodus*, and *Sandalodus*. Of the Petalodonts there were *Antliodus*, *Chomatodus*, *Ctenoptychius*, *Petalodus*, and *Petalorhynchus*; and of the hybodonts, the genera *Cladodus*, *Carcharopsis*, and *Diplodus*. These elasmobranchs were the rulers of the carboniferous open sea, and more than one hundred species have been found in the lower part of this formation alone. The ganoids, although still abundant, were of smaller size, and denizens of the more shallow and confined waters. The latter group of fishes was represented by true lepidostidæ, of the genera *Palæoniscus*, *Amblypterus*, *Platysomus*, and *Eurylepis*. Other genera are, *Rhizodus*, *Megalichthys*, *Ctenodus*, *Edestus*, *Orodus*, *Ctenacanthus*, *Gyracanthus*, and *Cœlacanthus*. Most of these genera occur also in Europe.

From the permian rocks of America no vertebrate remains are known, although in the same formation of Europe ganoids are abundant, and with them are remains of sharks, and some other fishes, the affinities of which are doubtful. The palæozoic fishes at present known from this country are quite as numerous as those found in Europe.

In the mesozoic age the fishes of America begin to show a decided approach to those of our present waters. From the triassic rocks ganoids only are known, and they are all more or less closely related to the modern gar-pike, or *Lepidosteus*. They are of small size, and the number of individuals preserved is very large. The characteristic genera are *Catopterus*, *Ischypterus*, *Ptycholepis*, *Rhabdolepis*, and *Turseodus*. From the Jurassic deposits no remains of fishes are known, but in the cretaceous ichthyic life assumed many and various forms; and the first representatives of the teleosts, or bony fishes, the characteristic fishes of to-day, make their appearance. In the deep open sea of this age elasmobranchs were the prevailing forms, sharks, and chimæroids being most numerous. In the great inland cretaceous sea of North America true osseous fishes were most abundant, and among them were some of carnivorous habits and immense size. The more sheltered bays and rivers were shared by the ganoids and teleosts, as their remains testify. The more common genera of cretaceous elasmobranchs were *Otodus*, *Oxyrhina*, *Galeocerdo*, *Lamna*, and *Ptychodus*. Among the osseous fishes, *Beryx*, *Enchodus*, *Portheus* and *Saurocephalus* were especially common, while the most important genus of ganoids was *Lepidotus*.

The tertiary fishes are nearly all of modern types, and from the beginning of this period there was comparatively little change. In the marine beds sharks, rays, and chimæroids maintained their supremacy, although teleosts were abundant, and many of them of large size. The ganoids were comparatively few in number. In the earliest eocene fresh-water deposits it is interesting to find that the modern gar-pike, and *Amia*, the dog-fish of our western lakes, which by their structure are seen to be remnants of a very early type, are well represented by species so closely allied to them that only an anatomist could separate the ancient from the modern. In the succeeding beds these fishes are still abundant, and with them are siluroids nearly related to the modern cat-fish (*Pimelodus*). Many small fishes allied apparently to the modern herring (*Clupea*), left their remains in great numbers in the same deposits, and with them has been recently found a land-locked ray (*Heliobatis*).

The almost total absence of remains of fishes from the miocene lake-basins of the west is a remarkable fact, and perhaps may best be explained by the theory that these inland waters, like many of the smaller lakes in the same region to-day, were so impregnated with mineral matters as to render the existence of vertebrate life in them impossible. No one who has tasted such waters or has attempted to ford one of the modern alkaline lakes which are often met with on the present surface of the same deposits, will doubt the efficiency of this cause, or the easy entombment of the higher vertebrates that ventured within their borders. In the pliocene lake-basins of the same region remains of fishes were not uncommon, and in some of them are very numerous. These are all of modern types and most of them are cyprinoids related to the modern carp. The post-pliocene fishes are essentially those of to-day.

In this brief synopsis of the past ichthyic life of this continent I have mentioned only a few of the more important facts, but sufficient, I trust, to give an outline of its history. Of this history it is evident that we have as yet only a very imperfect record. We have seen that the earliest remains of fishes known in this country are from the lower Devonian; but these old fishes show so great a diversity of form and structure as to clearly indicate for the class a much earlier origin. In this connection we must bear in mind that the two lowest groups of existing fishes are entirely without osseous skeletons, and hence, however abundant, would leave no permanent record in the deposits in which remains of fishes are usually preserved. It is safe to infer from the knowledge which we now possess of the simpler forms of life, that even more of the early fishes were cartilaginous, or so destitute of hard parts as to leave no enduring traces of their existence. Without positive knowledge of such forms, and considering the great diversity of those we have, it would seem a hopeless task at present to attempt to trace successfully the genealogy of this class. One line, however, appears to be direct, from our modern gar-pike, through the lower eocene *Lepidosteus* to the *Lepidotus* of the cretaceous, and perhaps on through the triassic *Ischypterus* and carboniferous *Palæoniscus*; but beyond this, in our rocks, it is lost. The living chimæra of our Pacific coast has nearly allied forms in the tertiary and cretaceous, more distant relatives in the carboniferous, and a possible ancestor in the Devonian *Rhynchodus*. Our sharks likewise can be traced with some certainty back to the palæozoic; and even the *Lepidosiren*, of South America, although its immediate predecessors are unknown, has some peculiar characters which strongly point to a Devonian ancestry. These suggestive lines indicate a rich field for investigation in the ancient life-history of American fishes.

The amphibians, the next higher class of vertebrates, are so closely related to the fishes in structure, that some peculiar forms of the latter have been considered by anatomists as belonging to this group. The earliest evidence of amphibian existence, on this continent, is in the sub-carboniferous, where foot-prints have been found which were probably made by labyrinthodonts, the most ancient representatives of the class. Well preserved remains are abundant in the coal-measures, and show that the labyrinthodonts differed in important particulars from all modern amphibians, the group which includes our frogs and salamanders. Some of these ancient animals resembled a salamander in shape, while others were serpent-like in form. None of those yet discovered were frog-like, or without a tail, although the restored labyrinthodont of the text-books is thus represented. All were protected by large pectoral bony plates, and an armour of small scutes on the ventral surface of the body. The walls of their teeth were more or less folded, whence the name labyrinthodont. The American amphibians known from osseous remains are all of moderate size, but the foot-prints attributed to this group indicate animals larger than any of the class yet found in the Old World. The carboniferous amphibians were abundant in the swampy tropical forests of that period, and their remains have been found imbedded in the coal then deposited, as well as in hollow stumps of the trees left standing.

The principal genera of this group from American carboniferous rocks, are, *Sauropus*, known only from foot-prints, *Baphetes*, *Dendrerpeton*, *Hylonomus*, *Hylerpeton*, *Raniceps*, *Pelion*, *Leptophractus*, *Molgophis*, *Ptyonius*, *Amphibamus*, *Cocytinus*, and *Ceraterpeton*. The last genus occurs also in Europe. Certain of these genera have been considered by some writers to be more nearly related to the lizards (*Lacertilia*) among true reptiles. Some other genera known from fragmentary remains or foot-prints in this formation have likewise been referred to the true reptiles, but this question can perhaps be settled only by future discoveries.

No amphibia are known from American permian strata, but in the triassic, a few characteristic remains have been found. The three genera, *Dictyocephalus*, *Dispelor*, and *Pariostegus*, have been described, but, although apparently all labyrinthodonts, the remains preserved are not sufficient to add much to our knowledge of the group. The triassic foot-prints which have been attributed to amphibians are still more unsatisfactory, and at present no important conclusions in regard to this class can be based upon them. From the Jurassic and cretaceous beds of this continent, no remains of amphibians are known. A few only have been found in the tertiary, and these are all of modern types.

The amphibia are so nearly allied to the ganoid fishes, that we can hardly doubt their descent from some member of that group. With our present limited knowledge of the extinct forms, however, it would be unprofitable to attempt to trace in detail their probable genealogy.

The authors to whom especial credit is due for our knowledge of American fossil fishes and amphibians, are Newberry, Leidy, Cope, Dawson, Agassiz, St. John, Gibbes, Wyman, Redfield, and Emmons, and the principal literature of the subject will be found in their publications.

Reptiles and birds form the next great division of vertebrates, the sauropsida, and of these the reptiles are the older type, and may be first considered. While it may be stated with certainty that there is at present no evidence of the existence of this group in American rocks older than the carboniferous, there is some doubt in regard to their appearance even in this period. Various foot-prints which strongly resemble those made by lizards, a few well preserved remains similar to the corresponding bones in that group, and a few characteristic specimens, nearly identical with those from another order of this class, are known from American coal measures. These facts, and some others which point in the same direction, render it probable that we may soon have conclusive evidence of the presence of true reptiles in this formation, and in our overlying permian, which is essentially a part of the same series. In the permian rocks of Europe, true reptiles have been found.

The mesozoic period has been called the age of reptiles, and during its continuance some of the strangest forms of reptilian life made their appearance, and became extinct. Near its commencement, while the triassic shales and sandstones were being deposited, true reptiles were abundant. Among the most characteristic remains discovered are those of the genus *Belodon*, which is well known also in the trias of Europe. It belongs to the thecodont division of reptiles, which have teeth in distinct sockets, and its nearest affinities are with the crocodilia, of which order it may be considered the oldest known representative. In the same strata in which the belodonts occur, remains of dinosaurs are found, and it is a most interesting fact that these highest of reptiles should make their appearance, even in a generalised form, at this stage of the earth's history. The dinosaurs, although true reptiles in all their more important characters, show certain well marked points of resemblance to existing birds of the order *Ratitæ*, a group which includes the ostriches ; and it is not improbable that they were the parent stock from which birds originated.

During triassic time, the dinosaurs attained in America an enormous development both in variety of forms and in size. Although comparatively few of their bones have as yet been discovered in the rocks of this country, they have left unmistakable evidence of their presence in the foot-prints and other impressions upon the shores of the waters which they frequented. The triassic sandstone of the Connecticut Valley has long been famous for its fossil foot-prints, especially the so-called "bird-tracks," which are generally supposed to have been made by birds, the tracks of which many of them closely resemble. A careful investigation, however, of nearly all the specimens yet discovered, has convinced me that there is not a particle of evidence that any of these fossil impressions were made by birds. Most of these three-toed tracks were certainly not made by birds : but by quadrupeds, which usually walked upon their hind feet alone, and only occasionally put to the ground their smaller anterior extremities. I have myself detected the impressions of these anterior limbs in connection with the posterior foot-prints of nearly all the supposed "bird-tracks" described, and have little doubt that they will eventually be found with all. These double impressions are precisely the kind which dinosaurian reptiles would make, and as the only characteristic bones yet found in the same rocks belong to animals of this group, it is but fair to attribute all these foot-prints to dinosaurs, even where no impressions of fore-feet have been detected, until some evidence appears that they were made by birds. I have no doubt that birds existed at this time, although at present the proof is wanting.

The principal genera of triassic reptiles known from osseous remains in this country are, *Amphisaurus* (*Megadactylus*), from the Connecticut Valley, *Bathygnathus*, from Prince Edward's Island, *Belodon* and *Clepsysaurus*. Other generic names which have been applied to foot-prints and to fragmentary remains, need not be here enumerated. A few remains of reptiles have been found in undoubted Jurassic rocks of America, but they are not sufficiently well determined to be of service in this connection. Others have been reported from supposed Jurassic strata, which are now known to be cretaceous. It will thus be seen that, although reptilian life was especially abundant during the triassic and Jurassic periods, but few bones have been found. This is owing in part to the character of most of the rocks then formed, which were not well fitted for preserving such remains, although admirably adapted to retain foot-prints.

(*To be continued.*)

INTRODUCTION AND SUCCESSION OF VERTEBRATE LIFE IN AMERICA

II.

THE reptiles most characteristic of our American cretaceous strata are the *Mosasauria*, a group with very few representatives in other parts of the world. In our cretaceous seas they ruled supreme, as their numbers, size, and carnivorous habits enabled them to easily vanquish all rivals. Some were at least sixty feet in length, and the smallest ten or twelve. In the inland cretaceous sea, from which the Rocky Mountains were beginning to emerge, these ancient "sea serpents" abounded; and many were entombed in its muddy bottom. On one occasion, as I rode through a valley washed out of this old ocean bed, I saw no less than seven different skeletons of these monsters in sight at once. The mosasaurs were essentially swimming lizards, with four well-developed paddles, and they had little affinity with modern serpents, to which they have been compared.

The *Crocodilia* are abundant in rocks of cretaceous age in America, and two distinct types are represented. The tertiary marine beds of the Atlantic coast contain comparatively few crocodilian remains, and all are of modern types, the genus *Gavialis* having one eocene species, and the alligator being represented only in the latest deposits.

It is worthy of special mention in this connection that no true *Lacertilia*, or lizards, and no *Ophidia*, or serpents, have yet been detected in American cretaceous beds; although their remains, if present, would hardly have escaped observation in the regions explored. The former will doubtless be found, as several species occur in the mesozoic of Europe, and perhaps the latter, although the ophidians are apparently a more modern type. In the eocene lake-basins of Western America, remains of lizards are very numerous, and indicate species much larger than any existing to-day.

The first American serpents, so far as now known, appear in the eocene, which contains also the oldest European species.

The *Pterosauria*, or flying lizards, are among the most interesting reptiles of mesozoic time, and many of them left their remains in the soft sediments of our inland cretaceous sea. These were veritable dragons, having a spread of wings of from ten to twenty-five feet.

The strange reptiles known as *Dinosauria*, which, as we have seen, were numerous during the deposition of our triassic shales and sandstones, have not yet been found in American Jurassic, but were well represented here throughout the cretaceous, and at its close became extinct. These animals possess a peculiar interest to the anatomist, since, although reptilian in all their main characters, they show clear affinities with the birds, and have some features which may point to mammals. The cretaceous dinosaurs were all of large size, and most of them walked on the hind feet alone, like modern struthious birds. Near the base of our cretaceous formation in beds which I regard as the equivalent of the European Wealden, the most gigantic forms of this order yet discovered have recently been brought to light One of these monsters (*Titanosaurus montanus*) from Colorado, is by far the largest land animal yet discovered, its dimensions being greater than was supposed possible in an animal that lived and moved upon the land. It was some fifty or sixty feet in length, and, when erect, at least thirty feet in height. It doubtless fed upon the foliage of the mountain forests, portions of which are preserved with its remains. With *Titanosaurus* the bones of smaller dinosaurs, one (*Nanosaurus*) not larger than a cat, as well as those of crocodiles and turtles, are not uncommon. The recent discovery of these interesting remains, many and various, in strata that had long been pronounced by professional explorers barren of vertebrate fossils, should teach caution to those who decline to accept the imperfection of our knowledge to-day as a fair plea for the supposed absence of intermediate forms.

In the marine cretaceous beds of the west only a single dinosaur (*Hadrosaurus agilis*) has been found, but in the higher fresh-water beds which mark the close of this formation their remains are numerous, and indicate several well-marked species, if not genera.

The first appearance of birds in America, according to our present knowledge, was during the cretaceous period, although many announcements have been made of their existence in preceding epochs. The evidence of their presence in the trias, based on footprints and other impressions is at present, as we have seen, without value, although we may confidently await their discovery there if not in older formations. *Archæopteryx*, from the European Jura, the oldest bird known, and now fortunately represented by more than a single specimen, clearly indicates a much higher antiquity for the class. The earliest American forms at present known are the *Odontornithes*, or birds with teeth, which have been exhumed within the last few years from the chalk of Kansas. The two genera *Hesperornis* and *Ichthyornis* are types of distinct orders, and differ from each other and from *Archæopteryx* much more than do any existing birds among themselves, thus showing that birds are now a closed type, and that the key to the history of the class must be sought for in the distant past.

In *Hesperornis* we have a large aquatic bird, nearly six feet in length, with a strange combination of characters. The jaws are provided with teeth, set in grooves ; the wings were rudimentary, and useless, while the legs were very similar to those of modern diving birds. This last feature was merely an adaptation, as the more-important characters are struthious, showing that *Hesperornis* was essentially a carnivorous swimming ostrich. *Ichthyornis*, a small flying bird, was stranger still, as the teeth were in sockets, and the vertebræ biconcave, as in fishes and a few reptiles. *Apatornis* and all other allied forms occur in the same beds, and probably all were provided with teeth. It is strange that the companions of these ancient toothed birds should have been pterodactyls without teeth. In the later cretaceous beds of the Atlantic coast various remains of aquatic birds have been found, but all are apparently distinct from those of the west.

During the tertiary period birds were numerous in this country, and all yet discovered appear to have belonged to modern types.

It is now generally admitted by biologists who have made a study of the vertebrates, that birds have come down to us through the dinosaurs, and the close affinity of the latter with recent struthious birds will hardly be questioned. The case amounts almost to a demonstration, if we compare, with dinosaurs, their contemporaries, the mesozoic birds. The classes of birds and reptiles, as now living, are separated by a gulf so profound, that a few years since it was cited by the opponents of evolution as the most important break in the animal series, and one which that doctrine could not bridge over. Since then, as Huxley has clearly shown, this gap has been virtually filled by the discovery of bird-like reptiles and reptilian birds. *Compsognathus* and *Archæopteryx* of the old world, and *Ichthyornis* and *Hesperornis* of the new, are the stepping-stones by which the evolutionist of to-day leads the doubting brother across the shallow remnant of the gulf once thought impassable.

It remains now to consider the highest group of the animal kingdom, the class *Mammalia*, which includes Man. Of the existence of this class before the trias we have no evidence, either in this country or in the old world, and it is a significant fact that at essentially the same horizon in each hemisphere, similar low forms of mammals make their appearance. Although only a few incomplete specimens have been discovered, they are characteristic and well preserved, and all are apparently marsupials, the lowest mammalian group which we know in this country, living or fossil. The American triassic mammals are known at present only from two small lower jaws, on which is based the genus *Dromotherium*, supposed to be related to the insect-eating *Myrmecobius*, now living in Australia.

Although the Jura of Europe has yielded other similar mammals, we have as yet none of this class from that formation ; while, from rocks of cretaceous age, no mammals are known in any part of the world.

In the lowest tertiary beds of this country a rich mammalian fauna suddenly makes its appearance, and from that time through the age of mammals to the present, America has been constantly occupied by this type of life in the greatest diversity of form. Fortunately, a nearly continuous record of this life, as preserved, is now accessible to us, and ensures great additions to our knowledge of the genealogy of mammals, and perhaps the solution of more profound problems.

The boundary line between the cretaceous and tertiary in the region of the Rocky Mountains has been much in dispute during the last few years, mainly in consequence of the uncertain geological bearings of the fossil plants found near this horizon. The accompanying invertebrate fossils have thrown little light on the question, which is essentially whether the great lignite series of the West is uppermost cretaceous, or lowest eocene. The evidence of the numerous vertebrate remains is, in my judgment, decisive, and in favour of the former view.

This brings up an important point in palæontology, one to which my attention was drawn several years since, namely, the comparative value of different groups of fossils in marking geological time. In examining the subject with some care, I found that for this purpose plants, as their nature indicates, are most unsatisfactory witnesses ; that invertebrate animals are much better ; and that vertebrates afford the most reliable evidence of climatic and other geological changes. The sub-divisions of the latter group, moreover, and in fact all forms of animal life, are of value in this respect, mainly according to the perfection of their organisation, or zoological rank. Fishes, for example, are but slightly affected by changes that would destroy reptiles or birds, and the higher mammals succumb under influences that the lower forms pass through in safety. The more special applications of this general law, and its value in geology, will readily suggest themselves.

The evidence offered by fossil remains is, in the light of this law, conclusive, that the line, if line there be, separating our cretaceous from the tertiary, must at present be drawn where the dinosaurs and other mesozoic vertebrates disappear, and are replaced by the mammals, henceforth the dominant type.

It is frequently asserted, and very generally believed, that the large number of huge *Edentata* which lived in North America during the post-pliocene, were the results of an extensive migration from South America soon after the elevation of the Isthmus of Panama, near the close of the tertiary. No conclusive proof of such migration has been offered, and the evidence it seems to me, so far as we now have it, is directly opposed to this view. No undoubted tertiary edentates have yet been discovered in South America, while we have at least two species in our miocene, and during the deposition of our lower pliocene large individuals of this group were not uncommon as far north as the forty-third parallel of latitude, on both sides of the Rocky Mountains. In view of these facts and others which I shall lay before you, it seems more natural to conclude from our present knowledge that the migration which no doubt took place was from north to south. The edentates finding thus in South America a congenial home flourished greatly for a time, and, although the larger forms are now all extinct, diminutive representatives of the group still inhabit the same region.

The ungulates are the most abundant mammals in the tertiary, and the most important, since they include a great variety of types, some of which we can trace through their various changes down to the modified forms that represent them to-day. Of the various divisions in this comprehensive group, the perissodactyle, or odd-toed ungulates, are evidently the oldest, and throughout the eocene are the prevailing forms. Although all of the perissodactyles of the earlier tertiary are more or less generalised, they are still quite distinct from the artiodactyles, even at the base of the eocene. One family, however, the *Coryphodontidæ*, which is well represented at this horizon, both in America and Europe, although essentially *Perissodactyle*, possesses some characters which point to a primitive ungulate type from which the present orders have been evolved. Among these characters are the diminutive brain, which in size and form approaches that of the reptiles, and also the five-toed feet from which all the various forms of the mammalian foot have been derived. Of this family, only a single genus, *Coryphodon* (*Bathmodon*), is known, but there were several distinct species. They were the largest mammals of the lower eocene, some exceeding in size the existing tapirs.

In the middle eocene, west of the Rocky Mountains, a remarkable group of ungulates makes its appearance. These animals nearly equalled the elephant in size, but had shorter limbs. The skull was armed with two or three pairs of horn-cores, and with enormous canine tusks. The brain was proportionally smaller than in any other land mammal. The feet had five toes, and resembled in their general structure those of *Coryphodon*, thus indicating some affinity with that genus. These mammals resemble in some respects the perissodactyles, and in others the proboscidians, yet differ so widely from any known ungulates, recent or fossil, that they must be regarded as forming a distinct order, the *Dinocerata*.

Besides these peculiar mammals which are extinct, and mainly of interest to the biologist, there were others in the early tertiary which remind us of those at present living around us. When a student in Germany some twelve years ago, I heard a world-renowned professor of zoology gravely inform his pupils that the horse was a gift of the old world to the new, and was entirely unknown in America until introduced by the Spaniards. After the lecture I asked him whether no earlier remains of horses had been found on this continent, and was told in reply that the reports to that effect were too unsatisfactory to be presented as facts in science. This remark led me, on my return, to examine the subject myself, and I have since unearthed, with my own hands, not less than thirty distinct species of the horse tribe, in the tertiary deposits of the west alone ; and it is now, I think, generally admitted that America is, after all, the true home of the horse.

I can offer you no better illustration than this of the advance vertebrate palæontology has made during the last decade, or of the important contributions to this progress which our Rocky Mountain region has supplied.

The oldest representative of the horse at present known is the diminutive *Eohippus* from the lower eocene. Several species have been found, all about the size of a fox. Like most of the early mammals, these ungulates had fourty-four teeth, the molars with short crowns, and quite distinct in form from the premolars. The ulna and the fibula were entire and distinct, and there were four well-developed toes and a rudiment of another on the fore-feet, and three toes behind. In the structure of the feet and in the teeth, the *Eohippus* indicates unmistakably that the direct ancestral line to the modern horse has already separated from the other perissodactyles. In the next higher division of the eocene another genus (*Orohippus*) makes its appearance, replacing *Eohippus*, and showing a greater, although still distant, resemblance to the equine type. The rudimentary first digit of the fore-foot has disappeared, and the last premolar has gone over to the molar series. *Orohippus* was but little larger than *Eohippus*, and in most other respects very similar. Several species have been found in the same horizon with *Dinoceras*, and others lived during the upper eocene with *Diplacodon*, but none later.

Near the base of the miocene, in the brontotherium beds, we find a third closely-allied genus, *Mesohippus*, which is about as large as a sheep, and one stage nearer the horse. There are only three toes and a rudimentary splint bone on the fore-feet, and three toes behind. Two of the premolar teeth are quite like the molars. The ulna is no longer distinct, or the fibula entire, and other characters show clearly that the transition is advancing. In the upper miocene *Mesohippus* is not found, but in its place a fourth form, *Miohippus*, continues the line. This genus is near the *Anchitherium* of Europe, but presents several important differences. The three toes in each foot are more nearly of a size, and a rudiment of the fifth metacarpal bone is retained. All the known species of this genus are larger than those of *Mesohippus*, and none pass above the miocene.

The genus, *Protohippus* of the lower pliocene, is yet more equine, and some of its species equalled the ass in size. There are still three toes on each foot, but only the middle one, corresponding to the single toe of the horse, comes to the ground. This genus resembles most nearly the *Hipparion* of Europe. In the pliocene we have the last stage of the series before reaching the horse, in the genus *Pliohippus*, which has lost the small hooflets, and in other respects is very equine. Only in the upper pliocene does the true *Equus* appear and complete the genealogy of the horse, which in the post-tertiary roamed over the whole of North and South America and soon after became extinct. This occurred long before the discovery of the Continent by Europeans, and no satisfactory reason for the extinction has yet been given. Besides the characters I have mentioned there are many others in the skeleton, skull, teeth, and brain of the forty or more intermediate species, which show that the transition from the eocene *Eohippus* to the modern *Equus* has taken place in the order indicated, and I believe the specimens now at New-Haven will demonstrate the fact to any anatomist. They certainly carried prompt conviction to the first of anatomists who was the honoured guest of the Association a year ago, whose genius had already indicated the later genealogy of the horse in Europe, and whose own researches so well qualified him to appreciate the evidence here laid before him. Did time permit I might give you at least a probable explanation of this marvellous change, but justice to the comrades of the horse in his long struggle for existence demands that some notice of their efforts should be placed on record.

(*To be continued.*)

INTRODUCTION AND SUCCESSION OF VERTEBRATE LIFE IN AMERICA

III.

THE artiodactyles, or even-toed ungulates, are the most abundant of the larger mammals now living, and the group dates back at least to the lowest eocene. In every vigorous primitive type which was destined to survive many geological changes, there seems to have been a tendency to throw off lateral branches which became highly specialised and soon died out because they are unable to adapt themselves to new conditions. The narrow path of the persistent suilline type throughout the whole tertiary is strewn with the remains of such ambitious offshoots, while the typical pig, with an obstinacy never lost, his held on in spite of catastrophes and evolution, and still lives in America to-day. The genus *Platygonus* is represented by several species, one of which was very abundant in the post-tertiary of North America, and is apparently the last example of a side branch, before the American suillines culminate in existing peccaries. The feet in this species are more specialised than in the living forms, and approach some of the peculiar features of the ruminants ; as, for example, a strong tendency to coalesce in the metapodial bones. The genus *Platygonus* became extinct in the post-tertiary, and the later and existing species are all true peccaries. No authenticated remains of the genera *Sus*, *Porcus*, *Phacochœrus*, or the allied *Hippotamus*, the Old World suillines, have been found in America, although several announcements to that effect have been made.

In the series of generic forms between the lower eocene *Eohyus* and the existing *Dicotyles*, which I have very briefly discussed, we have apparently the ancestral line ending in the typical American suillines. Although the demonstration is not yet as complete as in the lineage of the horse, this is not owing to want of material, but rather to the fact that the actual changes which transformed the early tertiary pig into the modern peccary were comparatively slight, so far as they are indicated in the skeletons preserved, while the lateral branches were so numerous as to confuse the line. It is clear, however, that from the close of the cretaceous to the post-tertiary the bunodont artiodactyles were especially abundant on this Continent, and only recently have approached extinction.

The selenodont division of the artiodactyles is a more interesting group, and so far as we now know, makes its first appearance in the upper eocene of the west, although forms apparently transitional between it and the bunodonts occur in the dinoceras beds, or middle eocene. The most pronounced selenodont in the upper eocene is the *Oromeryx*, which genus appears to be allied to the existing deer family, or *Cervidæ*, and if so is the oldest known representative of the group. These facts are important, as it has been supposed, until very recently, that our eocene contained no even-hoofed mammals.

A most interesting line, that leading to the camels and llamas, separates from the primitive selenodont branch in the eocene, probably through the genus *Parameryx*. In the miocene, we find in *Pœbotherium* and some nearly allied forms unmistakable indications that the cameloid type of ruminant had already become partially specialised, although there is a complete series of incisor teeth, and the metapodial bones are distinct. In the pliocene the camel tribe was, next to the horses, the most abundant of the larger mammals. The line is continued through the genus *Procamelus*, and perhaps others, and in this formation the incisors first begin to diminish, and the metapodials to unite. In the post-tertiary we have a true *Auchenia*, represented by several species, and others in South America, where the alpacas and llamas still survive. From the eocene almost to the present time North America has been the home of vast numbers of the *Camelidæ*, and there can be little doubt that they originated here and migrated to the Old World.

The deer family has representatives in the upper miocene of Europe, which contains fossils strongly resembling the fauna of our lower pliocene, a fact always to be borne in mind in comparing the horizon of any group in the two continents. Several species of *Cervidæ*, belonging to the genus *Cosoryx*, are known from the lower pliocene of the west, and all have very small antlers, divided into a single pair of tynes.

The proboscidians, which are now separated from the typical ungulates as a distinct order, make their first appearance in North America in the lower pliocene, where several species of mastodon have been found. This genus occurs also in the upper pliocene and in the post-tertiary, although some of the remains attributed to the latter are undoubtedly older. The pliocene species all have a band of enamel on the tusks, and some other peculiarities observed in the oldest mastodons of Europe, which are from essentially the same horizon. Two species of this genus have been found in South America, in connection with the remains of extinct llamas and horses. The genus *Elephas* is a later form, and has not yet been identified in this country below the upper pliocene, where one gigantic species was abundant. In the post-pliocene remains of this genus are numerous. The hairy mammoth of the Old World (*Elephas primigenius*) was once abundant in Alaska, and great numbers of its bones are now preserved in the frozen cliffs of that region. This species does not appear to have extended east of the Rocky Mountains, or South of Columbia River, but was replaced there by the American elephant, which preferred a milder climate. Remains of the latter have been met with in Canada, throughout the United States, and in Mexico. The last of the American mastodons and elephants became extinct in the post-tertiary.

Perhaps the most remarkable mammals yet found in America are the *Tillodontia*, which are comparatively abundant in the lower and middle eocene. These animals seem to combine the characters of several different groups, viz., the carnivores, ungulates, and rodents. In the genus *Tillotherium*, the type of the order, and of the family *Tillotheridæ*, the skull resembles that of the bears ; the molar teeth are of the ungulate type, while the large incisors are very similar to those of rodents. The skeleton resembles that of the carnivores.

We now come to the highest group of mammals, the primates, which includes the lemurs, the apes, and man. This order has a great antiquity, and even at the base of the eocene we find it represented by several genera belonging to the lower forms of the group. In considering these interesting fossils it is important to have in mind that the lemurs, which are usually regarded as primates, although at the bottom of the scale, are only found at the present day in Madagascar and the adjacent regions of the globe. All the American monkeys, moreover, belong to one group, much above the lemurs, while the Old World apes are higher still, and most nearly approach man.

In the lower eocene of New Mexico we find a few representatives of the earliest known primates, and among them are the genera *Lemuravus* and *Limnotherium*, each the type of a distinct family. These genera became very abundant in the middle eocene of the West, and with them are found many others, all, however, included in the two families *Lemuravidæ* and *Limnotheridæ*.

In the miocene lake basins of the West only a single species of the *Primates* has been identified with certainty. This was found in the oredoon beds of Nebraska and belongs to the genus *Laopithecus*, apparently related both to *Limnotheridæ* and to some existing South American monkeys. In the pliocene and post-pliocene of North America no remains of primates have yet been found.

In the post-pliocene deposits of the Brazilian caves remains of monkeys are numerous, and mainly belong to extinct species of *Callithrix*, *Cebus*, and *Jacchus*, all living South American genera. Only one extinct genus, *Protopithecus*, which embraced animals of large size, has been found in this particular fauna.

It is a noteworthy fact that no traces of any anthropoid apes, or indeed of any Old World monkeys, have yet been detected in America. Man, however, the highest of the primates, has left his bones and his works from the Arctic Circle to Patagonia. Most of these specimens are clearly post-tertiary, although there is considerable evidence pointing to the existence of man in our pliocene. All the remains yet discovered belong to the well-marked genus *Homo*, and apparently to a single species, at present represented by the American Indian.

In this rapid review of mammalian life in America, from its first known appearance in the trias down to the present time, I have endeavoured to state briefly the introduction and succession of the principal forms in each natural group. If time permitted, I might attempt the more difficult task of trying to indicate what relations these various groups may possibly bear to each other; what connection the ancient mammals of this continent have with the corresponding forms of the Old World; and, most important of all, what real progress mammalian life has here made since the beginning of the eocene. As it is, I can only say in summing up, that the marsupials are clearly the remnants of a very ancient fauna, which occupied this continent millions of years ago, and from which the other mammals were doubtless all derived, although the direct evidence of the transformation is wanting.

The relations of the American primates, extinct and recent, to those of the other hemisphere, offer an inviting topic, but it is not in my present province to discuss them in their most suggestive phases. As we have here the oldest and most generalised members of the group, so far as now known, we may justly claim America for the birth-place of the order. That the development did not continue here until it culminated in man, was due to causes which at present we can only surmise, although the genealogy of other surviving groups gives some data towards a solution. Why the Old World apes, when differentiated, did not come to the land of their earlier ancestry, is readily explained by the then intervening oceans, which likewise were a barrier to the return of the horse and rhinoceros.

Man, however, came; doubtless first across Behring's Straits; and at his advent became part of our fauna, as a mammal and primate. In these relations alone it is my purpose here to treat him. The evidence, as it stands to-day, although not conclusive, seems to place the first appearance of man in this country in the pliocene, and the best proof of this has been found on the Pacific coast. During several visits to that region many facts were brought to my knowledge which render this more than probable. Man at this time was a savage, and was doubtless forced by the great volcanic outbreaks to continue his migration. This was at first to the south, since mountain chains were barriers on the east. As the native horses of America were now all extinct, and as the early man did not bring the Old World animal with him, his migrations were slow. I believe, moreover, that his slow progress towards civilisation was in no small degree due to this same cause, the absence of the horse.

It is far from my intention to add to the many theories extant in regard to the early civilisations in this country, and their connections with the primitive inhabitants or the later Indians, but two or three facts have lately come to my knowledge which I think worth mentioning in this connection. On the Columbia River, I have found evidence of the former existence of inhabitants much superior to the Indians at present there, and of which no tradition remains. Among many stone carvings which I saw there were a number of heads which so strongly resemble those of apes that the likeness at once suggests itself. Whence came these sculptures, and by whom were they made? Another fact that has interested me very much is the strong resemblance between the skulls of the typical mound-builders of the Mississippi Valley and those of the Pueblo Indians. I had long been familiar with the former, and when I recently saw the latter it required the positive assurance of a friend who had himself collected them in New Mexico to convince me that they were not from the mounds. A third fact, and I leave man to the archæologists, on whose province I am even now trenching. In a large collection of mound-builders' pottery, over a thousand specimens, which I have recently examined with some care, I found many pieces of elaborate workmanship so nearly like the ancient water-jars from Peru that no one could fairly doubt that some intercourse had taken place between the widely-separated people that made them.

The oldest known remains of man on this continent differ in no important characters from the bones of the typical Indian, although in some minor details they indicate a much more primitive race. These early remains, some of which are true fossils, resemble much more closely the corresponding parts of the highest Old World apes, than do the latter our tertiary primates, or even the recent American monkeys. Various living and fossil forms of Old World primates fill up essentially the latter gap. The lesser gap between the primitive man of America and the anthropoid apes is partially closed by still lower forms of men, and doubtless also by higher apes, now extinct. Analogy, and many facts as well, indicate that this gap was smaller in the past. It certainly is becoming wider now with every generation, for the lowest races of men will soon become extinct, like the Tasmanians, and the highest apes cannot long survive. Hence the intermediate forms of the past, if any there were, become of still greater importance. For such missing links, we must look to the caves and later tertiary of Africa, which I regard as now the most promising field for exploration in the Old World. America, even in the tropics, can promise no such inducements to ambitious explorers. We have, however, an equally important field, if less attractive, in the cretaceous mammals, which must have left their remains somewhere on this continent. In these two directions, as I believe, lie the most important future discoveries in palæontology.

As a cause for many changes of structure in mammals during the tertiary and post-tertiary, I regard as the most potent, *natural selection*, in the broad sense in which that term is now used by American evolutionists. Under this head, I include not merely a Malthusian struggle for life among the animals themselves, but the equally important contest with the elements, and all surrounding nature. By changes in the environment, migrations are enforced, slowly in some cases, rapidly in others, and with change of locality must come adaptation to new conditions, or extinction. The life history of tertiary mammals illustrates this principle at every stage, and no other explanation meets the facts.

The real progress of mammalian life in America, from the beginning of the tertiary to the present, is well illustrated by the brain-growth, in which we have the key to many other changes. The earliest known tertiary mammals all had very small brains, and in some forms this organ was proportionally less than in certain reptiles. There was a gradual increase in the size of the brain during this period, and it is interesting to find that this growth was mainly confined to the cerebral hemispheres, or higher portion of the brain. In most groups of mammals the brain has gradually become more convoluted, and thus increased in quality as well as quantity. In some, also, the cerebellum and olfactory lobes, the lower parts of the brain, have even diminished in size. In the long struggle for existence during tertiary time the big brains won, then as now; and the increasing power thus gained rendered useless many structures inherited from primitive ancestors, but no longer adapted to new conditions.

Another of the interesting changes in mammals during tertiary time was in the teeth, which were gradually modified with other parts of the structure. The primitive form of tooth was clearly a cone, and all others are derived from this. All classes of vertebrates below mammals, namely, fishes, amphibians, reptiles, and birds, have conical teeth, if any, or some simple modification of this form. The edentates and cetaceans with teeth retain this type, except the zeuglodonts, which approach the dentition of aquatic carnivores. In the higher mammals the incisors and canines retain the conical shape, and the premolars have only in part been transformed. The latter gradually change to the more complicated molar pattern, and hence are not reduced molars, but transition forms from the cone to more complex types. Most of the early tertiary mammals had forty-four teeth, and in the oldest forms the premolars were all unlike the molars, while the crowns were short, covered with enamel, and without cement. Each stage of progress in the differentiation of the animal was, as a rule, marked by a change in the teeth, one of the most common being the transfer, in form at least, of a premolar to the molar series, and a gradual lengthening of the crown. Hence it is often easy to decide from a fragment of a jaw, to what horizon of the tertiary it belongs. The fossil horses of this period, for example, gained a grinding tooth for each toe they lost, one in each epoch. In the single-toed existing horses all the premolars are like the molars, and the process is at an end. Other dental transformations are of equal interest, but this illustration must suffice.

The changes in the limbs and feet of mammals during the same period were quite as marked. The foot of the primitive mammal was doubtless plantigrade, and certainly five-toed. Many of the early tertiary forms show this feature, which is still seen in some existing forms. This generalised foot became modified by a gradual loss of the outer toes and increase in size of the central ones, the reduction proceeding according to systematic methods, differing in each group. Corresponding changes took place in the limb bones. One result was a great increase in speed, as the power was applied so as to act only in the plane of motion. The best effect of this specialisation is seen to-day in the horse and antelope, each representing a distinct group of ungulates with five-toed ancestors.

If the history of American mammals, as I have briefly sketched it, seems as a whole incomplete and unsatisfactory, we must remember that the genealogical tree of this class has its trunk and larger limbs concealed beneath the *débris* of mesozoic time, while its roots doubtless strike so deeply into the palæozoic that for the present they are lost. A decade or two hence we shall probably know something of the mammalian fauna of the cretaceous, and the earlier lineage of our existing mammals can then be traced with more certainty.

The results I have presented to you are mainly derived from personal observation, and since a large part of the higher vertebrate remains found in this country have passed through my hands, I am willing to assume full responsibility for my presentation of the subject.

For our present knowledge of the extinct mammals, birds, and reptiles of North America, science is especially indebted to Leidy, whose careful, conscientious work has laid a secure foundation for our vertebrate palæontology. The energy of Cope has brought to notice many strange forms, and greatly enlarged our literature. Agassiz, Owen, Wyman, Baird, Hitchcock, Deane, Emmons, Lea, Allen, Gibbes, Jefferson, DeKay, and Harlan deserve honourable mention in the history of this branch of science. The South American extinct vertebrates have been described by Lund, Owen, Burmeister, Gervais, Huxley, Flower, Desmarest, Aymard, Pictet, and Nodot. Darwin and Wallace have likewise contributed valuable information on this subject, as they have on nearly all forms of life.

In this long history of ancient life I have said nothing of what life itself really is. And for the best of reasons, because I know nothing. Here at present our ignorance is dense, and yet we need not despair. Light, heat, electricity and magnetism, chemical affinity, and motion are now considered different forms of the same force; and the opinion is rapidly gaining ground that life, or vital force, is only another phase of the same power. Possibly the great mystery of life may thus be solved, but whether it be or not, a true faith in science admits no limit to its search for truth.

Part II

BIOSTRATIGRAPHY AND BIOGEOGRAPHY

Editor's Comments on Papers 3 Through 7

3 COPE
The Horizontal Relations of the North American Tertiaries with Those of Europe

4 AMEGHINO
New Discoveries of Fossil Mammalia of Southern Patagonia

5 TANG and ZHOU
The Vertebrate-bearing Early Tertiary of South China: A Review

6 McKENNA
Sweepstakes, Filters, Corridors, Noah's Arks, and Beached Viking Funeral Ships in Palaeogeography

7 KURTÉN
Continental Drift and the Palaeogeography of Reptiles and Mammals

It is through biostratigraphy and biogeography that vertebrate paleontology contributes most directly to mainstream geology. The papers reprinted in Part II highlight these contributions. The most important and influential aspects of this work concern Cenozoic land-mammal faunas and intercontinental correlation of the Cenozoic land-mammal ages (both formal age/stage units and informal "ages" and "faunas"). This is a subject of continued confusion. To help guide the reader, I present a quick reference correlation chart for the major Cenozoic land-mammal age terms currently in common use (Fig. 1). For brief and up-to-date reviews (with numerous references) of the principles, methods, and terminology of biostratigraphic correlation utilizing fossil vertebrates, see Savage (1977), Savage and Russell (1983, pp. 2-9), Tedford (1970), and Woodburne (1977). On the basis of vertebrate paleontology, Lehman (1981) has recently proposed five new eras and erathems, replacing the traditional three (Paleozoic, Mesozoic, and Cenozoic), to encompass Phanerozoic time and rock systems. These are the Aspondylian Era (encompassing the Cambrian and Ordovician Periods/Systems), Ichthyan Era (Silurian

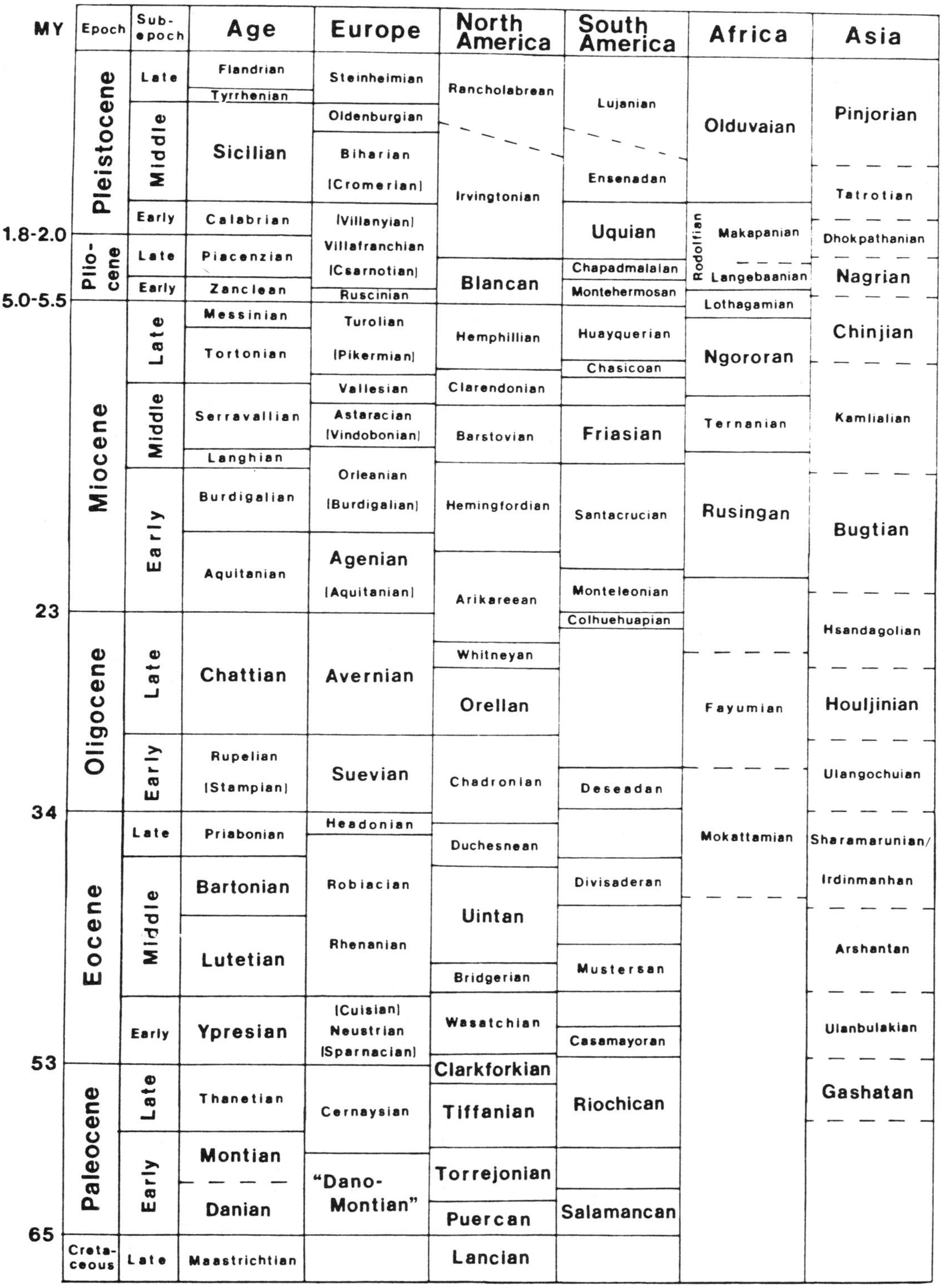

Figure 1. Correlation chart of major Cenozoic land-mammal age terms currently in common use (absolute dating of epoch boundaries is based on Odin et al., 1982). Correlations of Asian faunas are extremely uncertain. The late Clarkforkian, early Wasatchian, and early Sparnacian (early Ypresian) should possibly be referred to the late Paleocene rather than to the early Eocene. (Data is primarily from Berggren et al., 1978; Curry and Odin, 1982; Faulbusch, 1976; Savage and Russell, 1983; Szalay and Delson, 1979; Romer, 1966; Wood et al., 1941.)

and Devonian), Therapsian Era (Carboniferous, Permian, and Triassic), Sauropsidian Era (Jurassic and Cretaceous), and the Placentalian Era (Tertiary or Cenozoic).

Edward Drinker Cope (1840–1897), mentioned previously as one of the founders of vertebrate paleontology in North America, was also a renowned geologist, not just a paleontologist in the strict sense. Born and educated in Philadelphia, Cope was, successively, professor of Comparative Anatomy at Haverford College in Pennsylvania (1863–1867), an employee of the U.S. Geological Survey of the Territories (during which time he took part in the Hayden and Wheeler Expeditions to Kansas, Colorado, Wyoming, New Mexico, and Texas), and professor of Geology and Mineralogy at the Academy of Natural Sciences in Philadelphia (appointed in 1889). One of Cope's many outstanding contributions was his attempt to correlate the European with the North American vertebrate faunas (Paper 3). In Paper 3 (p. 44), Cope observed that a major discrepancy existed between the age determinations for the American Cenozoic formations as derived from mammalian faunas and as derived from paleofloras. Today the age determinations made by Cope, and other paleomammalogists, are generally accepted. (But, for a discussion of a possible late Cretaceous and early Tertiary age disparity between Arctic and midlatitudinal biotas, see Hickey et al., 1983.) Another of Cope's seminal papers is reprinted as Paper 8.

Paper 4 is a typical work by the brilliant, if eccentric, Argentine paleontologist Florentino Ameghino (1854–1911). Ameghino, the founder of vertebrate paleontology in South America, had little formal education and was helped in his collecting of fossils by his brothers Juan and Carlos, especially the latter. Ameghino specialized in fossil mammals (although he also worked on other vertebrate groups such as dinosaurs), the general geology and stratigraphy of South America, and the origin and history of man in South America. Many of his theories were very controversial (most have since been repudiated). In particular, Ameghino believed that the South American strata and their contained fossils were much older than dated by most workers, and he traced the origins of most mammal lineages to his native Argentina. In Paper 4 Ameghino discusses some important fossil finds from southern Patagonia and attempts to correlate the South American Cenozoic strata with the strata of North America.

Paper 5, by the contemporary Chinese workers Tang Xin and Zhou Ming-zhen (better known as Minchen Chow), is representative of the detailed, high quality research that has come out of China in the last few decades. Their paper reviews in detail the Paleocene through Oligocene vertebrate faunas of South China and attempts to correlate them not only to one another, but also to faunas of Inner Mongolia, North China, Europe, and North America. Finally, Tang and Zhou attempt to correlate the distribution of fossil occurrences in South China with the climatic, paleobiogeographic, and tectonic history of the area. Such considerations lead directly into the topics considered in Papers 6 and 7.

The principles and methodology of historical biogeography were developed in the first part of the twentieth century primarily by the Canadian-American vertebrate paleontologist William Diller Matthew (1871-1930) and the American vertebrate paleontologist George Gaylord Simpson, born in 1902 (see Matthew, 1974; Simpson, 1980, and review and references in Paper 6; also Ross, 1976). In hindsight, however, the major fault to be found with this early work is that it was based on a stabilist model of fixed continents that do not move through geologic time. Of course, Matthew's (1974) Holarctic dispersal plan was not entirely false; for the most part, as drifting continents moved north, the fauna tended to migrate south. The same trend was caused by the increasing northern cold of the Quaternary. In Paper 6, the contemporary worker Malcolm Carnegie McKenna utilizes the best of Matthew's and Simpson's methodology, reinterpreting it and applying it to a plate tectonic framework in which landmasses (continents) are not stable in configuration over geologic time, but collide, separate, and move across the face of the Earth. In this way, McKenna is able to reach a new understanding of the distribution and dispersal of land vertebrates. Björn Kurtén goes one step further (Paper 7) and attempts to correlate differing continental configurations and paleogeographical settings with the taxonomic variety and macroevolutionary trends produced during the reptilian and mammalian radiations in the late Paleozoic-Mesozoic and Cretaceous-Tertiary respectively (cf. similar studies by Valentine, 1973, on marine invertebrates).

REFERENCES

Berggren, W. A., M. C. McKenna, J. Hardenbol, and J. D. Obradovich, 1978, Revised Paleogene Polarity Time Scale, *Jour. Geology* **86:**67-81.

Curry, D., and G. S. Odin, 1982, Dating of the Palaeogene, in *Numerical Dating in Stratigraphy*, G. S. Odin, ed., John Wiley & Sons, Chichester, England, pp. 607-630.

Faulbusch, V., 1976, Report on the International Symposium on Mammalian Stratigraphy of the European Tertiary, *Newsl. Stratigr.* **5:**160-167.

Hickey, L. J., R. M. West, M. R. Dawson, and D. K. Choi, 1983, Arctic Terrestrial Biota: Paleomagnetic Evidence of Age Disparity with Mid-Northern Latitudes during the Late Cretaceous and Early Tertiary, *Science* **221:**1153-1156.

Lehman, J. -P., 1981, Points de Vue d'un Paléoanatomiste des Vertébrés sur la Nomenclature Géologique, *Mus. Natl. Histoire Nat. Bull.,* ser. 4, sec. C, **3:**149–156.
Matthew, W. D., 1974, *Climate and Evolution,* Arno Press, New York, 234p. (Reprint of the 1939 edition, with foreword by E. H. Colbert, biographical sketch of William Diller Matthew by W. K. Gregory, additions by W. D. Matthew and others, and a bibliography by C. L. Camp and V. L. Vander Hoof.)
Odin, G. S., D. Curry, N. H. Gale, and W. J. Kennedy, 1982, The Phanerozoic Time Scale in 1981, in *Numerical Dating in Stratigraphy,* G. S. Odin, ed., John Wiley & Sons, Chichester, England, pp. 957–960.
Romer, A. S., 1966, *Vertebrate Paleontology,* The University of Chicago Press, Chicago and London, 478p.
Ross, C. A., ed., 1976, *Paleobiogeography,* Benchmark Papers in Geology, vol. 31, Dowden, Hutchinson & Ross, Stroudsburg, Pa., 442p.
Savage, D. E., 1977, Aspects of Vertebrate Paleontological Stratigraphy and Geochronology, in *Concepts and Methods of Biostratigraphy,* E. G. Kauffman and J. E. Hazel, eds., Dowden, Hutchinson & Ross, Stroudsburg, Pa., pp. 427–442.
Savage, D. E., and D. E. Russell, 1983, *Mammalian Paleofaunas of the World,* Addison-Wesley, Reading, Mass., 450p.
Simpson, G. G., 1980, *Why and How: Some Problems and Methods in Historical Biology,* Pergamon Press, Oxford, 272p.
Szalay, F. S., and E. Delson, 1979, *Evolutionary History of the Primates,* Academic Press, New York, 594p.
Tedford, R. H., 1970, Principles and Practices of Mammalian Geochronology in North America, *North American Paleontol. Conv., First, Proc.* **1:**666–703.
Valentine, J. W., 1973, *Evolutionary Paleoecology of the Marine Biosphere,* Prentice-Hall, Englewood Cliffs, N.J., 526p.
Wood, H. E., II, R. W. Chaney, J. Clark, E. H. Colbert, G. L. Jepsen, J. B. Reeside, Jr., and C. Stock, 1941, Nomenclature and Correlation of the North American Continental Tertiary, *Geol. Soc. America Bull.* **52:**1–48.
Woodburne, M. O., 1977, Definition and Characterization in Mammalian Chronostratigraphy, *Jour. Paleontology* **51:**220–234.

3

Reprinted from *U.S. Geol. Survey Territories Rept.* **3**:21–45 (1884)

THE HORIZONTAL RELATIONS OF THE NORTH AMERICAN TERTIARIES WITH THOSE OF EUROPE

E. D. Cope

ERRATA

Page 31, line 1 from the bottom of the page to page 32, line 1 from the top of the page should read "during the Tertiary period appears . . . ." Page 32, line 2 from the top of the page should read "level appear to have . . . ." Page 32, line 4 from the top of the page should read "irregular variation." Page 32, line 5 from the top of the page should read "restricted deposits, each of which differs . . . ." Page 35, the second footnote should read "vide Scott, Speir, and Osborn." Page 45, line 13 from the top of the page should read "in North America than in Europe."

Uniformity of system requires that an identical scale of stratigraphy be employed by all geologists. In accomplishing this object, the students of distinct regions necessarily rely on paleontology as the guide in making determinations of the relations of strata, since determination by observation of continuity is impossible.

The progress of the vertebrate paleontology in North America of late years has been such as to permit of comparisons with the extinct faunæ of Europe and other continents, which give definiteness to our knowledge of the relations between their geologic periods. Comparisons made by Morton and Leidy had nearly determined the position of some of the eastern Cretaceous strata, and those of Leidy had approximately fixed those of the White River beds. Lyell and Conrad early determined the positions of the Eastern Tertiaries. My own views as to the European equivalency of our Keuper* and Laramie† were first expressed, and I later established the ages of the Wasatch,‡ Bridger,|| Loup River,§ and Permian¶ formations in America. A more detailed comparison being very necessary, I visited Europe in 1878 for the purpose of examining the rich collections of vertebrate fossils, and read a general synopsis of results before the Congress of Geologists of Paris of that year. The present section embraces a summary of that paper, with some additional matter.**

The history of the succession of life upon any one portion of the earth's surface is replete with matter for speculation. It shows us a series of faunæ succeeding each other, each of which, in many instances, commences without previous announcement in the forms of older periods, and

* Proceedings Academy Phila., 1866, p. 249.

† Report on U. S. Geol. Surv. Fortieth Parallel, iii, p—. Transac. American Philosophical Soc., xiv, 1869, pp. 40, 98, 243.

‡ Proceedings Amer. Philos. Society, February, 1872.

|| Loc. cit., August, 1872.

§ Annual Report U. S. Geol. Surv. Terrs., 1873, p. 402; Proceed. Academy Phila., 1875, p. 257.

¶ American Naturalist, 1878, p. 327; Proceedings Amer. Philos. Society, 1878, p. 530.

** This synopsis was first published in the Bulletin U. S. Geol. Surv. Terrs., v, p. 33. See Comptes Rendus Sténographiques, Congrès Internationale de Géologie 1880, p. 144.

disappears without leaving representatives in later ones. With this basis of fact, which naturally enough has been furnished by the longest explored and best-known portion of the earth, Europe, we turn to other lands, with the hope of obtaining further light upon a subject so full of mystery. These types of life, did they originate in a single centre, from which they disseminated themselves; and, if so, did each form originate in a region of its own, or not? Or, did the same types of generic structure appear at different points on the earth's surface independently; and, if so, whether cotemporaneously or at different times?

For a solution of these and similar questions, we naturally look to a comparison of the facts first established, with those obtained more recently by exploration in other regions. In this quest, no portion of the earth offers greater promise of results than America. As the second great continent, separated from the other by the greatest possible water surface, we anticipate the widest diversity in the character of its life-history. If the types of life have originated independently, we will find evidence of it by studying American paleontology; if their origin has been through gradual modification, America should furnish us with many intermediate faunæ.

Let us first consider the nature of the evidence on which we should rely in classifying faunæ and the deposits which contain them. We are accustomed, at present, to rely for our definitions upon all the faunal peculiarities upon which we can seize: the period of appearance of certain types; the duration of certain types; and the disappearance of certain types, depending on orders, families, and genera for the major divisions, and species at a given locality for the lesser. It is, of course, evident that either of the above-mentioned three criteria are variable quantities, since discovery is constantly extending our knowledge of the distribution of types. Hence the definitions are empirical and temporary. We must, then, if we desire a stable system, examine the principles involved, and endeavor to discover definitions which stand on stronger foundations than those which we now possess.

As a matter of fact, the old definitions of epochs and periods are continually invalidated by new discoveries. As a matter of theory, this should be the case.

To the believers in the doctrine of derivation, the obliteration of faunal distinctions is not a cause of surprise. Such await with confidence the day when complete phylogenies will be possible, and at present regard the interruptions in the succession of life as local only. Will the result then, be, that paleontology will cease to be available in the definition of ages and of deposits? I answer no, on various grounds. Interruptions in the succession of life in any given locality, due to various causes, have doubtless often occurred, and have left traces in the crust of the earth which are ineffaceable by discovery. But apart from this, one fact in this history is patent both to the friends and to the opponents of the doctrine of derivation: It is known that the world has witnessed, at every stage of its history, the extinction of some important type of life. Familiar examples are the *Placodermi* of Paleozoic time, various reptilian groups of Mesozoic time, and the *Amblypoda* of the Tertiary. Each minor subdivision of time offers its own record of persistences and extinctions of particular families and genera.

Now, all departments of biology compel us to recognize the law of classification, that the order of forms is from the less to the more generalized, from the simple to the more complex, and *vice versa*, whether the lines of succession be those of descent or of creative order; and this law is true in time as well as in classification. It follows from this, that all types of life are, at the time of their appearance, less distinct and more general in their characters than they are later in their history.

It also follows, as a consequence of the principle of descent, which states that the types of one age have taken their origin from generalized types of preceding ages, that there is no descent from the most specialized types; which is to say, conversely, that the genera, families, and orders whose extinction has been a marked feature of every geologic age have been the specialized types of those ages.

We now have a clue to a basis of a definition for faunæ, and hence for epochs, which discovery can safely build upon. The successive increments of structure by which an important modification of animal type is introduced, preclude the possibility of exact determination of the time at which such type may be said to have *appeared.* Even where such a point may be arbitrarily fixed, the type must then be less characteristically represented than it is at the other limit of its existence, viz, the period of its disappearance.

For these reasons I must regard the latter criterion as the true one in the discrimination of the subdivisions of geologic time, while the point of the appearance of types must be looked upon as of provisional use only, and this quite independently of the changes which discovery will from time to time compel us to make in our knowledge of the distribution of life in time and space. It must, however, be borne in mind that disappearance may be due to two causes: first, to extinction; and secondly, to modification; a distinction which is entirely essential. The case of disappearance by modification is identical with that of appearance by modification, and cannot be used otherwise in classification. It is, then, the period of extinction of types to which I have reference.

With these principles in view, we attempt the comparison of the extinct faunæ of Europe and North America, employing principally the nomenclature of D'Orbigny for the former, and Hayden for the latter, in the Mesozoic and Tertiary series.

It is well known that no remains of *Vertebrata* have yet been discovered in North America in strata of Silurian age, while several species are known from the Upper Silurian of Europe. The latter are *Placodermi* and sharks, and are not very numerous in species. They have been derived from England, Germany, and Russia. In America, the first fishes appear in the Corniferous limestone at the base of the Devonian. Professor Newberry, who has devoted much attention to this department, points out important distinctions as existing between the Devonian fish faunæ of North America and Europe, and also to important coincidences. The first of these is the occurrence of the genus *Macropetalichthys* in both continents; in Germany in the Eifel limestone, and in America in the Corniferous limestone of Ohio. The other examples are furnished by the Catskill beds of New York and Pennsylvania, which contain a part of the fauna of the old Red Sandstone of Scotland, including the genera *Holoptychius* and *Bothriolepis.**

The structure of the *Batrachia* of the Coal-Measures is not yet sufficiently well known to enable the most exact comparisons to be made, but

* Geological Survey of Ohio, i, pp. 264, 271.

close parallels, if not identities, of genera exist. Such are the *Oëstocephalus* and *Ceraterpeton* of Ohio as compared with the *Urocordylus* and *Ceraterpeton* of Great Britain.

The Permian vertebrate fauna which I discovered in Illinois and Texas exhibits close parallels, but not yet generic identity, in the two continents. Thus, the American *Clepsydrops* and *Dimetrodon* are near to the *Deuterosaurus* of Perm, in Russia, and the *Lycosaurus* of the mountains of South Africa. The Texan genus *Pariotichus* may, with further information, prove to be identical with *Procolophon* Ow. from the Tafelberg. Humeri of the type discovered by Kutorga in Russia, and by Owen in South Africa, are found in North America, and the same remarkable type has been recently discovered by Gaudry in France. The peculiar type of *Ganocephalous* vertebræ described by me under the genus *Rhachitomus* from Texas, has been discovered by Gaudry in France. The even more remarkable *Cricotus* (Cope), type of the *Embolomera*, is paralleled by the *Diplovertebron* (Fritsch) of Bohemia. *Ectosteorhachis* represents in Texas the genus *Megalichthys*. The present indications are that close similarity between the faunæ of this period in Europe and America will be discovered. Nevertheless, up to the present time no representatives of the striking American forms *Bolosaurus*, *Diadectes*, and *Empedocles* have yet been found in any other continent.

As regards the Triassic fauna, it differs from that of the Permian in being better known in Europe than America. As marine Trias is little developed in North America, so the vertebrate fauna of the Muschelkalk has not been discovered in the latter country. It is otherwise with the Keuper. The characteristic genus of that epoch, *Belodon*, existed in America, and parallels, if not identity, are seen in the genera *Thecodontosaurus* and *Palæosaurus*. These are known in America from teeth only. The reptiles are accompanied in North America, as in Europe, by Stegocephalous *Batrachia*, mostly Labyrinthodonts, but their generic affinities are yet unknown.

The great Jurassic faunæ are as yet but sparsely represented in North American paleontology. The marine *Vertebrata* of the Lias are either unknown or are represented by a few provisional identifications of unsatisfactory fragments. We do not yet know any deposits in North America which contain the typical reptilian genera *Plesiosaurus*, *Ichthyosaurus*, *Plio-*

saurus and *Dimorphodon*, or the fishes of the *Dapediidæ*. This formation, so important in Europe, is almost omitted from the North American series. Several characteristic fossils of the Rocky Mountain region represent the Oölite, particularly the Upper Oölite. Such is a genus not yet distinguishable from *Megalosaurus*. This genus has not been identified beyond doubt from above the Oölite in England. *Teleosaurus* and *Steneosaurus*, and their allies, are not yet known from North American beds. From the same beds in the Rocky Mountain region come genera which nearly resemble the one from the English Oölite (Forest Marble) called by Phillips *Cetiosaurus*, and the genus from the Oxfordian of Honfleur, called by von Meyer *Streptospondylus*. Beyond this no comparisons can be made, and we therefore pass to the rich fauna of the Kimmeridge. North America cannot show such records of this epoch as have been found in Europe. There are no *Archæopteryx*, *Rhamphorhynchus*, nor *Pterodactylus*; no *Leptolepis*, *Thrissops*, nor other of the numerous fishes of Solenhofen. The *Omosaurus* has, however, some very close relatives in the *Camarasaurus* beds of the Rocky Mountains. Remains of the primitive Marsupial fauna which occurs in the Purbeck have been recently detected in the Western Continent. A partial representation of the Wealden fauna of Europe is found in the beds of the Rocky Mountains mingled with the types of the Oölite and Kimmeridge already mentioned. The important genus *Camarasaurus* represents the *Ornithopsis* of Europe, and with *Amphicœlias* included the most gigantic of land animals. The relationships of this fauna to those of the European Jurassic series may be thus exhibited:

American.	**European.**
CAMARASAURUS BEDS.	WEALDEN.
?	*Iguanodon.*
? *Hypsilophodon.*	*Hypsilophodon.*
	Hylæosaurus.
? *Cetiosaurus.*	*Cetiosaurus.*
Camarasaurus.	*Ornithopsis.**
Amphicœlias.	
? *Goniopholis.*	*Goniopholis.*

* *Chondrosteosaurus* Owen.

American.	European.
	KIMMERIDGE.
Hypsirhophus.	*Omosaurus.*
Caulodon.	? *Caulodon.**
	OXFORD.
Epanterias.	*Streptospondylus.*
	OÖLITE.
	" *Cetiosaurus.*"
? *Megalosaurus.*	*Megalosaurus.*

From the above table it will be seen how difficult it is to parallelize the related beds of the Jurassic periods of the two continents at the present time. All that can be said is, that many types resembling† nearly those of different horizons of the European Jurassic are found to have lived together or near together in the Rocky Mountain region of North America.

That the Cretaceous fauna of North America was the richest in the cold-blooded *Vertebrata* is indicated by the present state of discovery. The ocean of the interior of the continent deepened from the beginning of the period until the epoch of the Niobrara, and then gradually shallowed until the elevations of the bottom began to divide the waters. The closing scenes of this great period were enacted amid a labyrinth of lagoons and lakes of brackish and fresh water, whose deposits form the beds of the Laramie epoch.

The fauna of the deep-sea epoch, the Niobrara, is the best known. Here the remains of *Pythonomorpha* constitute its prevailing characteristic, while *Elasmosaurus* and *Polycotylus*, with but few species, represent the numerous *Sauropterygia* of Europe. Crocodiles were apparently wanting, while turtles and a peculiar group of *Pterosauria* were only moderately abundant. The fish fauna was very rich and varied. Here the *Saurodontidæ*, like the molluscous family of the *Rudistes*, appeared and as soon dis-

* *Iguanodon præcursor* Sauv.

† A near affinity has been shown by Professor Owen to exist between *Eucamerotus* and *Camarasaurus*. Professor Owen believes these genera to be identical; but the neural spines of the anterior dorsal vertebra are very different, being single in the former and double in the latter.

appeared, accompanied by the peculiar form *Erisichthe*, and the family of *Stratodontidæ*. The genera of Mount Lebanon, *Leptotrachelus* and *Spaniodon*, occur in this bed in Dakota; but the closest parallelism is exhibited with the Lower Chalk or Turonian of Western Europe. The general *facies* of the reptilian fauna is that of the Lower Chalk, and there is little doubt that several genera are identical in the two continents, *e. g. Elasmosaurus*. The apparent peculiarity of the Chalk in America is the abundance of forms (four genera) of *Pythonomorpha*, with numerous species, while but two genera have yet been found in Europe, and the presence of birds with *biconcave* vertebræ and teeth. This interesting type, which was first discovered by Seeley in the genus named by him *Enaliornis*, and afterwards found by Marsh to possess teeth, has been found at a lower horizon in England, the Upper Greensand. But in England, France, and Westphalia occur the genera of fishes above mentioned, as *Portheus*, *Ichthyodectes*, *Saurodon*, *Saurocephalus*, *Erisichthe*, *Empo*, *Pachyrhizodus*, *Enchodus*, *Leptotrachelus*, etc. This close relationship of the horizons permits an identification, and it is the first instance which appears to me to be susceptible of satisfactory demonstration.

The next horizon of the Cretaceous which has yielded many vertebrate remains in North America is the Fox Hills formation (including the Fort Pierre bed). Here the genus *Mosasaurus* appears in America, and is accompanied by the earliest crocodiles with procœlous vertebræ, and by numerous marine turtles which partake of the characters of both *Chelydridæ* and *Cheloniidæ*, which I have called the *Propleuridæ*. *Beryx* appears first here in America. The predominant genus of fishes is *Enchodus*, and the principal *Dinosauria* are *Lælaps* and *Hadrosaurus*. This horizon has been parallelized with the Maestricht of Europe, and several genera are common to the two beds; such are *Mosasaurus* and *Enchodus*. The genus *Hadrosaurus*, and the family of turtles I have called the *Adocidæ*, remain undiscovered in Europe; hence the identity of faunæ cannot be established.

The lacustrine beds, or summit of the American Cretaceous series, the Laramie of Hayden, present the remains of a populous fauna and a rich flora. The students of the palæobotany have declared this flora to be of Eocene, and the later portions of Miocene, character, while the lacustrine constitution of the strata has influenced the stratigraphic geologists to concur in the view that the formation should be arranged with the Tertiary epochs. That the fauna was of a mixed character is the result of a study of its vertebrate fossils. The predominant type in North America was the *Dinosauria*, which were abundant in species and individuals, and this fact alone will suffice most paleontologists as a reason for referring the epoch to the Cretaceous series. The genera of *Dinosauria* (*Palæoscincus*, *Cionodon*, *Diclonius*, *Monoclonius*, *Dysganus*, etc) have not yet been found in any other part of the world. Mingled with them were species of crocodiles and turtles of indifferent character, while a number of other forms existed which had a limited range in time, and hence are important indicators of stratigraphic position. Such are the genera of fishes, *Myledaphus* Cope and *Clastes* Cope, which have been found also near Reims, France, by Dr. Lemoine, in the Sables de Bracheux, which are regarded as the lowest Tertiary. Such is the curious Saurian type *Champsosaurus* (Cope) (*Simœdosaurus* Gerv.), and the turtle genus *Compsemys* Leidy, which Lemoine finds a little higher up in the series, in the conglomerate of Cerny, which is in the lower part of the Suessonian. In France, a genus of the Laramie, *Polythorax*, extends into the Lignite or Upper *Coryphodon* bed of the Suessonian Thus the Laramie is intercalated by its characters between the Cretaceous period on the one hand and the Tertiary on the other, and its fauna includes genera and orders of both great series. These relations may be exhibited in tabular form as follows. I here include the faunæ of the Sables de Bracheux and of the conglomerate of Cerny as one, since both possess the types of the Laramie, while the horizon of the Lignite of Meudon, or the Suessonian, does not.

Sables de Bracheux and Conglomerat de Cerny.	Laramie.
α. Tertiary.	
Lophiochœrus.	
Plesiadapis.	
Pleuraspidotherium.	
Arctocyon.	
Clastes.	*Clastes.*

β. PECULIAR.

Champsosaurus.	*Champsosaurus.*
Compsemys.	*Compsemys.*
Myledaphus.	*Myledaphus.*
	Scapherpeton.
SABLES DE BRACHEUX AND CONGLOMERAT DE CERNY.	LARAMIE.

γ. CRETACEOUS.

	Palæoscincus.
	Dysganus.
	Monoclonius.
	Diclonius.
	Cionodon.
	Lœlaps.
	Aublysodon.

If the Conglomerate of Cerny be the same horizon as the Conglomerate of Meudon, we must add *Coryphodon* to the upper left-hand column, and probably *Gastornis* also. The result is clear that the French and American formations together bridge most completely the interval between the Cretaceous and Tertiary series, as has been anticipated by Hayden, in America, on geological grounds. It is also evident that another formation must be added to the series already recognized in France, viz, the Laramie or Post-Cretaceous. This will be defined as the beds of the genera *Champsosaurus* and *Myledaphus.* In France, the presence of mammalia will characterize the formation as a subdivision, for which it is probable that the name Thanetian must be retained; while to the American division, which is characterized by the presence of *Dinosauria,* the name of Laramie beds has been given.

In arranging the Laramie Group, its necessary position is between Tertiary and Cretaceous, but on the Cretaceous side of the boundary, if we retain those grand divisions, which it appears to me to be desirable to do. The reasons for retaining it in the Cretaceous are two, viz: (1) because

Dinosauria are a Mesozic type, not known elsewhere from the Tertiary; (2) because *Mammalia* (should they be found in the future in the Fort Union) are not equal as evidence of Tertiary age, since they have been also found in Jurassic and Triassic beds.

The Eocene fauna is so varied, especially in Europe, that it is necessary to compare the divisions separately, as in the case of the Cretaceous. Thus, the fauna of the Suessonian is quite as distinct from that of the Calcaire Grossier and Gypse (Parisian and Tongrian) in France as are those of the Wasatch and Bridger epochs in North America.

I have already identified the Wasatch with the Suessonian or Orthrocene, on account of the community of the following genera in the two continents: *Coryphodon*, *Hyracotherium*, *Amblyctonus*, *Clastes*, and a form of birds close to *Gastornis*. I can now add *Phenacodus*, *Orotherium* (Cope), and very probably *Hyopsodus*, *Adapis*, *Opisthotomus*, and *Prototomus*. But, as above mentioned, in the lower beds of the Suessonian in France occur genera which are, so far as yet known, wanting in the Wasatch of America, but present in the beds of the Laramie.

The parallelism of the American Wasatch with the Upper Suessonian of France is the second identification which may be regarded as provisionally established. The only important discordant element at present known is the *Tœniodonta* of the Wasatch, which have not so far been found in Europe.

Above the Suessonian, a divergence in the characters of the European and North American faunæ commences, and continues to be marked throughout the remainder of Tertiary time. So far as the *Mammalia* are concerned, the diversity between the continents was greater during the periods of the Upper Eocene and Miocene than at the present era. During these periods, a limited number of genera, common to the two continents, was associated with numerous genera in the one which did not exist in the other. As a consequence, our paleontological means of identification of the horizons are limited to a restricted list, and the task of applying a uniform nomenclature is, under the circumstances, difficult. Another difficulty in the way of determining the place of the American beds in the European scale consists in the fact that the physical history of the two continents during the Terrary period appears to have been different. In America, the changes of evel appear to have been more uniform in character over large areas. Each deposit has a wider geographical extent, and the fauna presents less rregular variation. In Europe we have a great number of comparatively estricted deposits, each of which differs from the others in possessing more or less peculiarity of fauna. After a study of these faunæ, their natural arrangement in Europe into three series—Eocene, Miocene, and Pliocene—does not appear to rest on any solid basis. This is especially true of the distinction between the first two; and authors are at variance as to the point of demarkation between the last two. Thus, the Tongrian is the summit of the Eocene according to Renevier, while Gaudry, with Filhol and others, places it at the base of the Miocene. One opinion is as well supported by facts, as now interpreted, as the other.

As an essential aid in the estimation of the limits of the formations, I appeal to the criterion adopted at the opening of this chapter, viz, the period of extinction of animal groups.

If we take a general view of the Tertiary faunæ, we find that the following well-marked types representing families and higher groups have become extinct, and have left no living descendants or successors: Among *Bunotheria*, the American groups *Tæniodonta* and *Tillodonta ;* also the *Mesodonta* of both continents; of *Edentata*, *Macrotherium*, and *Ancylotherium* in Europe, and the *Megatheriidæ* in North America; among the *Carnivora*, the *Hyænodons* and *Proviverræ*, with the *Drepanodons ;* of *Ungulata*, the entire order of *Amblypoda*, which, however, doubtless disappeared in some of its members by modification; but its only known suborders, the *Pantodonta* and the *Dinocerata*, become absolutely extinct. Among *Perissodactyla*, both continents lost by extinction the *Chalicotheriidæ*, which terminated in a great development in North America; and the *Rhinoceridæ*. Of *Artiodactyla*, two great divisions, representative of each other in the two continents, totally disappeared, viz, the *Oreodontidæ* and the *Anoplotheriidæ;* to which must be added the *Hyopotamidæ*. Of true ruminants, the most important type which has disappeared from both continents is that of the *Camelidæ*. Of Suilline genera, *Anthracotherium* and *Elotherium* may be looked upon as having left no persistent successors. Last of all, the *Proboscidea* retreated to the continents of the south.

In view of the complexity of the European record, I first present the relations of the above-mentioned phenomena as displayed in the simpler American system. As the present essay commences with the earliest periods, I exhibit the succession in descending order on the page. The horizons of the Tertiary which present distinct terrestrial faunæ in North America have been named the Wasatch, the Bridger, the Uinta, the White River, the Loup Fork, the *Equus* beds, and the Champlain. The types which became extinct* with the close of each of these epochs are the following:

WASATCH.
- *Gastornithidæ.*
- *Pantodonta.*

BRIDGER.
- *Baënidæ.*
- *Tillodonta.*
- *Stypolophus.*
- *Dinocerata.*

UINTA.
- ?*Mesodonta.*
- *Amynodon.*

WHITE RIVER.
- *Leptictidæ.*
- *Hyænodon.*
- *Chalicotheriidæ.*
- *Hyopotamidæ.*

LOUP RIVER.
- *Rhinoceridæ.*
- *Hippotherium.*
- *Oreodontidæ.*

EQUUS BEDS.
- *Megatheriidæ.*
- *Drepanadon.*
- *Tapiridæ.*
- *Elephas.*
- *Camelidæ.*

The above table exhibits the present state of our knowledge; it will doubtless be much extended by future discovery, but not otherwise modified.

The numerous able writers on European vertebrate palæontology have more frequently recorded the appearance of types in defining their faunal

* This means, as already mentioned, the forms which left no direct successors in the Nearctic and and Palæarctic faunæ.

divisions than their disappearance. The following table is compiled from the writings of Gervais, Gaudry, Pomel, Filhol, Renevier, and others, but is not as complete as I would desire.

SUESSONIAN.
Pantodonta.
PARISIAN (Bruxellian, Bartonian, and Sestian).
Palæophis (Bruxellian).
Proviverra.
Pterodon.
Mesodonta.
Lophiodon (Bruxellian).
TONGRIAN.
Palæotheridæ.
Chalicotherinm.
Anoplotheridæ.
Elotherium.
AQUITANIAN.
Hyænodon.
Hyopotamus.

FALUNIAN.
Anchitherium.
Anthracotherium.
Palæochœrus.
Cænotherium.
OENINGIAN.
Ancylotherium.
Dinotherium.
Hippotherium.
Aceratherium.
SUBAPENNINE.
Mastodon.
Tapiridæ.
DILUVIAL.
Hyæna.
Drepanodon.
Elephas.
Rhinocerus.
Hippopotamus.

The above tables show that the history of mammalian life in the two continents presents many points of resemblance; but that there is a great difficulty in correlating the epochs represented by the known faunæ. As regards the two primary divisions, Eocene and Miocene, they have no special *raison d'être*, as such faunæ as the Tongrian and Oeningian are absolutely transitional in their character. More detailed comparisons of the European and American faunæ bring out many relationships not displayed by the above tables, and which I will now briefly consider.

In the American Bridger, various genera of *Mesodonta* represent the few *Adapidæ* of the Parisian, the genus *Adapis** Cuv. being probably common to the two continents. A near ally of the American *Anaptomorphus*, a

* *Notharctus* is undistinguishable from *Adapis* in inferior dental characters.

true Lemur, has been found by M. Filhol in the Phosphorites, and named *Necrolemur.* The characters of the numerous *Carnivora* of the Bridger are as yet unknown. The *Stypolophus* of the Bridger is perhaps the *Prototomus* of the Wasatch, and this again has been discovered by M. Filhol* in France; while a very similar genus has been discovered in the Swiss Siderolitic, and named *Proviverra. Hyænodontidæ* probably occur in the Bridger. Nowhere in Europe do we find the *Dinocerata* and *Tillodonta* of the Bridger. *Palæosyops* is also unknown in Europe, but it plays the part in America of the *Palæotherium,* from which it does not greatly differ in structure. The latter genus is most largely developed in the Parisian, but is also characteristic of the Tongrian. *Hyrachyus* is the American *Lophiodon,* the difference between them being but slight; both are found in France; the former in the Lower Parisian, the latter in the Phosphorites. *Tapirulus*† Gerv. is a genus common to the Bridger and to more than one horizon of the Parisian. The squirrel-like rodents of the Bridger are like those of the Parisian, but they are not confined to either epoch. The character which distinguishes the Parisian most widely from the Bridger, besides the absence of the *Dinocerata* and *Tillodonta,* is the presence of numerous Selenodont *Artiodactyla,* as *Xiphodon, Cænotherium, Amphimeryx, Anoplotherium,* etc. These are of primitive type, it is true; the *Anoplotheriidæ* especially having probably four toes in the very short manus (*Eurytherium*), including the pollex, and three behind. They also display the character of a fifth crescent of the superior molars, which is wanting in the higher Selenodont types. But even these genera are absent from the Bridger. The *ensemble* is then that the latter displays relationships backwards, or to the Suessonian, while the Parisian has a later *facies,* constituting an approach to the Tongrian and White River. ‡

The following table presents the relations of the Bridger fauna succinctly, but it is much less complete than we hope to make it when its numerous species are fully described. The Parisian is here regarded as including the divisions Bruxellian, Bartonian, and Sestian (Gypse).

* It is described as *Cynohyænodon* with two species.

† Gervais, 1850; *Helaletes* Marsh, 1872, vide Scott, Spier, and Osborne.

‡ See Ann. Rept. U. S. Geol. Surv. Terrs. 1873, pp. 461–462, where this view is proposed.

PARISIAN.	BRIDGER.
Didelphys.	? *Didelphys.*
Vespertilionidæ.	*Vespertilionidæ.*
Plesiarctomys.	*Plesiarctomys.*
	Tillodonta.
Hyænodontidæ.	*Hyænodontidæ.*
Adapis.	*Adapis.*
Necrolemur (Phosph).	*Anaptomorphus.*
	Dinocerata.
	Palæosyops.
Palæotherium.	
Lophiodon.	
Hyrachyus (Phosph.).	*Hyrachyus.*
Tapirulus.	*Tapirulus.*
Anthracotherium.	*Achænodon.*
Chæropotamus.	
Dichobune.	
Anoplotherium.	
Xiphodon.	
Amphimeryx.	

The rich Tongrian (Stampian) fauna is, according to authors, represented in the Sables de Fontainebleau, Puy en Velay, Ronzon, Hempstead, and Cadibona in Italy. We find here *Didelphys* in abundance, *Hyænodon, Amphicyon, Canis, Palæotherium, Paloplotherium, Chalicotherium,* and *Aceratherium.* Of *Artiodactyla,* the Suillines are *Anthracotherium* and *Elotherium;* the Selenodonts, *Hyopotamus* and *Gelocus.* This list is the nearest known counterpart of that of the fauna of the White River epoch of North America. To reproduce the latter we must omit from the above catalogue the genera of *Palæotheriidæ,* and replace them by the allied Chalicotheroid *Menodus* and *Symborodon,* subtract *Anthracotherium,* and add the great body of the *Oreodontidæ.* Then there are included in the White River fauna the higher Selenodont Artiodactyles of the *Poëbrotheriidæ* and *Hypertragulidæ,* the corresponding types of which belong to the fauna of St. Gerand le Puy in France, or the Aquitanian epoch, which directly succeeded the Stampian. In Europe, we have here *Dremotherium, Amphitragulus, Lophiomeryx,*

Dorcatherium; in America, *Leptomeryx*, *Hypertragulus*, *Hypisodus*, and *Poëbrotherium.* It is curious that while *Leptomeryx* is also European,* it has not yet been found above the Phosphorites. Among Suillines, the *Palæochærus*† of the Oregon White River beds has also not been found below the Aquitanian in Europe. But the American *Didelphys*‡ *Hyænodon*, *Amphicyon*, *Elotherium*, and *Hyopotamus*, with the numerous Chalicotheroid species, show clearly that the White River fauna may be looked upon as a mixture of those of the Stampian and Aquitanian, the former of which is sometimes referred with reason to the Upper Eocene, while the latter is always left in the lowest Miocene. And the solution of this question of position as regards the White River beds appears to me to be at present by no means easy.§ According to the system of Naumann, it should be called Oligocene.

Although Artiodactyles with Selenodont molars are far more abundant in both continents during this period than the last, a remarkable difference is to be observed between them. Those of Europe still largely consist of the types with five crescents, as represented by the numerous *Hyopotami* and *Cænotheria*, while in America the modern four-crescent-bearing molar characterizes almost the entire suborder, the only exception being two species of *Hyopotamus*.

The following table will represent the relations of the White River fauna:

Stampian and Aquitanian.	White River.
Didelphys.	*Didelphys.*
	Leptictidæ.
Protomyidæ	*Protomyidæ.*‖
	Saccomyidæ.¶
Stampian and Aquitanian.	**White River.**
Steneofiber.	*Steneofiber.*
Leporidæ.	*Leporidæ.*

* I think M. Filhol's *Prodremotherium* is identical with *Leptomeryx.*

† *Thinohyus* Marsh appears to be the same.

‡ *Herpetotherium* Cope; *Peratherium* Aym.

§ See Ann. Report U. S. Geol. Surv. Terrs. 1878, p. 462, where the White River beds are determined as Lower Miocene.

‖ *Ischyromys* Leidy.

¶ *Entoptychus* and *Pleurolicus* Cope.

STAMPIAN AND AQUITANIAN.	WHITE RIVER.
Hyænodon.	*Hyænodon.*
Amphicyon.	*Amphicyon.*
Canis.	*Canis.*
	Temnocyon.
	Enhydrocyon.
*Gulo.**	*Gulo.**
Proælurus.	
Aclurogale (Phosp.).	*Archælurus.*
	Nimravus.
	Dinictis.
Palæotheriidæ.	*Hoplophoneus.*
	Chalicotheriidæ.
	Hyracodon.
Aceratherium.	*Aceratherium.*
	Anchitherium
Elotherium.	*Elotherium.*
Palæochœrus.	*Palæochœrus.*
Anthracotherium.	
	Oreodontidæ.
Anoplotheriidæ.	
Hyopotamidæ.	*Hyopotamidæ.*
	Poëbrotherium.
Lophiomeryx.	*Hypertragulus.*
Amphitragulus.	
Leptomeryx (Phosph.).	*Leptomeryx.*
Dremotherium.	
	Hypisodus.

The Falunian epoch includes in the large sense the Langhian, Helvetian, and Tortonian divisions, embracing the rich deposits of the Orléanais, of Simorre, and of Sansan. We have here the true Miocene fauna, of which the following genera are characteristic: Edentata, *Macrotherium;* Proboscidea, *Dinotherium*, *Mastodon;* Perissodactyla, *Anchitherium*, *Listriodon;*

* *Amphictis* Pom.

† Cope 1874; *Procervulus.* Gaudry, 1878; *Dicrocerus* Cope, 1874 (not Lartet); *Merycodus* et *Cosoryx* Leidy, nomina nuda.

Artiodactyla, *Palæomeryx*, *Dicrocerus*, *Cosoryx;*† Carnivora, *Amphicyon*, *Hyænarctos*, *Drepanodon;* Quadrumana, *Pliopithecus*. The ancient genera *Anthracotherium* and *Cænotherium* continue throughout, and the existing genera *Arvicola*, *Lutra*, and *Sus* appear. The succeeding epoch, the Oeningian, including with it the horizons of Epplesheim and Pikermi, presents the additional genera *Dorcatherium*, *Helladotherium*, several genera allied to *Antilope*, with *Hippotherium*, the huge edentate *Ancylotherium*, and the monkey *Mesopithecus*.

It is from these materials that we must determine by comparison the American Loup Fork epoch, whose deposits are widely spread, and whose fauna is of well-marked character. Although called by my predecessors Pliocene in age, I have insisted that it should be referred to the Miocene series, and I think that the evidence to that effect which I have produced will be found conclusive. Nevertheless here, as in other American Tertiary horizons, the element of geographical peculiarity enters, and diminishes the number of identical types.

Falunian.	Loup Fork.
Steneofiber.	*Steneofiber.*
	Mylagaulus.
Macrotherium. *Ancylotherium.* }	*Morotherium.*
Amphicyon.	*Amphicyon.**
Dinotherium.	*Cænobasileus.*
Tetralophodon.	*Tetralophodon.*
Aceratherium.	*Aphelops.*
Anchitherium.	
Listriodon.	
Hippotherium (Oeningian).	*Hippotherium.*
	Protohippus.
	Hippidium.†
Cænotherium.	*Oreodontidæ.*
Anthracotherium.	
Palæomeryx.	*Blastomeryx.*
Dicrocerus.	

* *Canis ursinus* Cope.

† *Pliohippus* Marsh.

FALUNIAN.	LOUP FORK.
Cosoryx.	*Cosoryx.*
	Protolabis.
	Procamelus.

The existing genera mentioned as found in the Falunian fauna are paralleled by the *Dicotyles*, *Hystrix*, and *Mustela* of the Loup Fork beds. It is evident that this latter horizon retains in its *Oreodontidæ* the same traces of antiquity that the Falunian does in its *Cænotherium*, but shows a more modern aspect in the omission of *Anchitherium* and its replacement by *Hippotherium* and *Protohippus*, and in the still more modern type *Hippidium*. Although but six genera of the two continents are determined as identical in the above table, yet others, which are facing on the same line, are very nearly allied. Other differences are geographical. The *facies* of the Loup Fork horizon is then a compound of that of the Falunian and Oeningian, or Middle and Upper Miocene.

In commenting on the above-described fauna in 1874,* I remarked that "the proper discrimination of the American Pliocene remains to be accomplished." It was not long after that date that material for making the identification of this horizon on this continent first came into my hands. This was derived from the superior Tertiary of Oregon, and includes a considerable number of species of fishes, birds, and *Mammalia*. I published a list of some of the species in 1878.† The character of the fauna from that region coincides with that which has from time to time been unearthed in the caves and other Eastern deposits, to such an extent, as to lead us to suspect that the differences between them are geographical only. In Europe the Pliocene, or Subapennine, includes, according to D'Orbigny (1855) and Gaudry (1878), the Plaisancian and Astian, which are represented at the following localities:

SUBAPENNINE.

Plaisancian.—Montpellier; Casino (Tuscany).

Astian.—Perrier, near Issoir, Coupet, Vialette (Upper Loire), Chagny; English Crag; part of deposits of the Val d'Arno.

The characteristic of this fauna is the fact that the species belong

*Report Lieut. G. M. Wheeler, iv, Paleontology of New Mexico, 1874, p. 364.

†Bull. Hayden's U. S. Geol. Surv. Terrs. iv, 1878, p. 389.

mostly to existing genera, the chief exception being *Hippotherium*. The horses are chiefly represented by *Equus*. Common genera are *Arctomys*, *Lepus*, *Elephas*, *Mastodon*, *Tapirus*, *Sus*, *Cervus*, *Antilope*, *Bos*, *Canis*, *Drepanodon*, *Felis*, *Ursus*. In the *Equus* beds of Oregon a few extinct genera in like manner share the field with various recent ones, while not a few of the bones are not distinguishable from those of recent species. I give the following list, the extinct species being in italics:

Mylodon sodalis.	Canis latrans.	*Auchenia magna.*
Thomomys (nr.) clusius.	*Elephas primigenius.*	*Auchenia vitakeriana.*
Thomomys talpoides.	*Equus occidentalis.*	*Cervus fortis.*
Castor fiber.	*Equus major.*	
Lutra near *piscinaria.*	*Auchenia hesterna.*	

The species derived from the cave formations of the Eastern States are more numerous, and differ from the Oregon fauna in many respects; yet the parallelism is close in the genera with the *Equus* beds on the one hand and the Pliocene of Europe and South America on the other. The differences distinguishing it from the *Equus* beds of Oregon are, however, such as compel me to regard it as a distinct division of the Pliocene, under the name of the *Megalonyx* beds.*

Megatherium (p).	*Castoroides.*	*Drepanodon Smilodon* (vel).
Mylodon (p).	Lagomys (s).	*Mastodon* (sp).
Megalonyx (p).	Lepus (s).	Equus (sp).
Sciurus (s).	*Anomodon.*	? *Hippotherium* (s).
Arctomys (s).	Scalops.	Tapirus (s).
Jaculus (s).	Arctotherium (p).	Dicotyles (p),
Arvicola (s).	Procyon.	Cariacus (p).
Erethizon.	Canis (sp).	Bos (s).
Hydrochœrus (p).	Mustela (sp).	

In the above list the extinct genera are marked in italics. There exists, as a marked feature of the North American Pliocene, to which I called attention several years ago,† a considerable representation of the fauna of

* Bull. U. S. Geol. Surv. Terrs. v, p. 52, 1879.

† Proc. Acad. Phila. 1857, 156; Proc. Am. Philos. Soc. 1869, 178.

the Pampean formation of South America; such are twelve genera, of which six are extinct genera, and four are peculiar to that formation and fauna. The genera found in the Pampean are marked (*p*), and those of the Subapennine (*s*). In the list from the Oregon localities, *Mylodon* and *Auchenia* were observed to be the only distinctively Pampean genera. As a conclusion of the comparison of the American *Equus* beds in general with those of Europe, it may be stated that the number of identical genera is so large, that we may not hesitate to parallelize them as stratigraphically the same. On the other hand, the agreement with the South American Pampean formation is so marked in some respects as to induce us to believe that the distinction is geographic rather than stratigraphic. Believing that the Pampean formation contains too large a percentage of extinct genera to be properly regarded, as it has been, as Post-Pliocene or Quaternary, its characters, both essentially and as a result of the comparison which I have been able to make, refer it properly to the Pliocene. It appears, then, that the term Pliocene or Subapennine is applicable to the horizon of this fauna in Europe and North and South America.

RÉSUMÉ OF COMPARISONS.

The conclusions to be derived from the facts enumerated in the preceding pages are as follows:

I. Portions of all the faunæ of all the primary divisions of geologic time have been recognized on both the European and North American continents.

II. Parallels requiring general identification of principal divisions of these faunæ may be detected. These are: the Coal Measures; the Permian; the Laramie; the Maestrichtian; the Eocene; the Miocene.

III. Exact identifications of restricted divisions may be made in a few instances only; such are the Turonian and the Niobrara; the Suessonian and the Wasatch; the *Equus beds* and the Pliocene.

It is not impossible that some of the relations mentioned in II will be by the accession of further information, referrible to the list of exact comparisons in III. In all cases of identification it will be necessary to employ the name first proposed with definition for the horizon, other names taking

WEST EUROPE.			NORTH AMERICA.	
Astian.	Pliocene.	Tertiary.	Pliocene.	Megalonyx beds.
Plaisancian.				Equus beds.
Oeningian.	Oeningian.		Loup Fork.	Procamelus beds.
Tortonian.	Falunian.			Ticholeptus beds.
Langhian.				
Aquitanian.	Aquitanian.		White River.	Truckee.
Stampian.	Tongrian.			White River.
Sestian.	Parisian.		Uinta.	Uinta.
Bartonian.			Bridger.	Amyzon beds.
Bruxellian.				Bridger.
Suessonian.	Suessonian.		Wasatch.	Green River.
				Wasatch.
Thanetian.	Thanetian.	Post-Cretaceous.	?Puerco.	Puerco.
			Laramie.	Fort Union.
				Bear River.

Maestrichtian.		**Cretaceous.**		Fox Hills.
Campanian.	Senonian.		Fox Hills.	
Santonian.				Fort Pierre.
Turonian.			Colorado.	Niobrara.
Carentonian.	Cenomanian.			Fort Benton.
Rhotomagian.			Dakota.	Dakota.
Vraconian.	Gault.			
Albian.				
Aptian.				
Rhodanian.	Urgaptian.			
Urgonian.				
Hauterivian.	Wealden.	**Jurassic.**	Camarasaurus beds.	
Valangian.				
Purbeckian.				
Portlandian.	Portland.			
Kimmeridgian.				
	Corallian.			
	Oxfordian.			
	Bathian.			
	Upper Lias.			
	Lower Lias.			

	Rhaetic.	**Trias.**		
Karnian.	Keuper.		Keuper.	
Norian.				
	Muschelkalk.			
Thuringian.	Permian.	**Carboniferous.**	Permian.	Clepsydrops shales.
Lodevian.				Eryops beds.
Coal Measures.	Coal period.		Coal period.	Coal Measures.
Conglomerate.				Conglomerate.
Mountain limestone.				Mountain limestone.
Fammenian.	Upper Devonian.		Upper Devonian.	Catskill.
				Chemung.
Eifelian.	Middle Devonian.			Hamilton.
Coblenzian.	Lower Devonian.		Lower Devonian.	Corniferous.
				Oriskany.

Ledburian.				Lower Helderberg.
Ludlovian.	Upper Silurian.		Upper Silurian.	Salina.
Wenlockian.				Niagara.
Llandoverian.				Hudson.
Caradocian.				
	Lower Silurian.	**Silurian.**	Lower Silurian.	
Llandeilian.				Trenton.
Tremadocian.				
				Calciferous.
	Primordial.		Primordial.	
				Potsdam.
		Archæan.		

places as synonymes But in the majority of strata it will be necessary to preserve the special names: thus those of Bear River, Bridger, White River, and Loup Fork, applicable to beds having no exact equivalents in Europe, cannot be set aside for older ones, but must themselves be applied to corresponding faunal horizons elsewhere, should any such be found in future. And it will rarely happen that the minor subdivisions of such faunæ will be found to have an extent sufficient to warrant their having other than special names.

In the accompanying diagram the series of strata of Europe and North America, as determined by their paleontology, are placed side by side for the purpose of comparison. Complete parallelism can only be predicated of divisions of the first order, separated by horizontal lines. Such relation is indicated by exact opposition of the areas representing the epochs in question. In giving the minor divisions of the European epochs I have generally restricted myself to those of the epochs which have American equivalents. Where there is no equivalent on one side or the other, the vacancy is represented by a diagonal line. In employing names for epochs and their divisions, I have adhered to the law of priority as far as my knowledge of the literature allows.* I have given a few names to American formations, but only in instances where such had not been previously given. In such cases I have preferred employing the name of some characteristic genus of fossils, rather than one of local origin.

COMPARISON WITH THE SCALE DERIVED FROM PALEOBOTANY.

I now consider another kind of relation presented by the American and European horizons. I allude to the floræ, for my knowledge of which I am necessarily dependent on the labors of others. I first exhibit the determinations of the ages of the American formations, already discussed, made by Mr. Lesquereux on the basis of the vegetable remains which they contain. I place by the side of these my own determinations of the ages of the same beds, as already related. The former are derived from the full memoir of Mr. Lesquereux in the Annual Report of the United States Geological Sur-

*In the European system I have been much aided by the writings of Woodward, Gervais, Hébert, Pomel, Gaudry, etc., and by the atlas of Professor Renevier, of Lausanne.

vey of the Territories for 1872, pp. 410–417. It will be observed that there is a constant discrepancy between the two tables.

Lesquereux.	Formation.	Cope.
....	Loup Fork..............................	Miocene.
....	White River	Oligocene.
Upper Miocene..................	Bridger	Middle Eocene.
Miocene	{ Wasatch .... { Green River................. }	Lower Eocene.
Lower Eocene.....................	Laramie	Upper Cretaceous.

If the determinations of Mr. Lesquereux be correct,* it is evident from the above that the vegetable life of North America reached its present condition one epoch or period earlier than the higher *Vertebrata*, and that the nomenclature is thus thrown back by so much. It would appear that the recent flora of North America is a period older than the fauna, *i. e.*, has persisted longer than the latter by a certain length of geologic time. Applying the same reasoning to the past, I embodied the idea in reference to the Laramie period ("Fort Union") in the statement that "a Cretaceous fauna was then contemporary with a Tertiary flora";† and, later, that "an Eocene fauna was contemporary with a Miocene flora." It may have to be added that a Miocene fauna was contemporary with a Pliocene flora. Since Mr. Lesquereux has the support of the best paleobotonists of Europe in his conclusions, it is useless to take the ground assumed by a few of my colleagues, that the former gentleman has simply erred in his determinations. He gives us grounds for believing that he has not done so, by giving us the European standard by which his identifications are governed.‡ It is as follows :

Pliocene...	**Lower limits not positively fixed; largely developed in Italy. †(Subapennine, E.D.C.)**
Miocene ...	**Oeningian; Mayencian; Aquitanian.**
Oligocene..	**Tongrian.**
Eocene	**Gypse of Aix; Alum Bay; Mt. Bolca; London Clay; Sheppey; Grès of the Sarthe.**
Paleocene..	**Upper Landenian; Sezanne (=Paniselian).** **Suessonian (Lignitic Soissonais; Sables de Bracheux); Lower Landenian.** **Hersian; Gelinden.** **Limestone of Mons, overlying unconformably the Maestrichtian.**

*** The above parallels are well presented by Dr. Peale in his report to Dr. Hayden, Ann. Rept. U. S. Geol. Surv. Terrs. 1874, p. 141 *et seq.*** **† Bulletin U. S. Geol. Surv. Terrs. I, art. 2. p. 16, April, 1874**
‡ Ann. Report U. S. Geol. Surv. Terrs. 1874, p. 285.

This system, it will be observed, is almost exactly identical with that employed in the preceding pages as the standard of comparison for the *Vertebrata*. Yet it has resulted, from a most careful comparison of both faunæ and floræ of America with this standard scale, that two distinct paleontological series have to be adopted, the one for the vertebrate life and the other for the plants, of the Western Continent. If this result be accurate, and there appears to be no avoiding it, an explanation must be sought. There are only two possible ones; either the animal life of North America has lagged behind that of Europe by one period during past geologic time; or, secondly, the vegetable life of America has been equally in advance of that of Europe during the same period. In other words, if the plant-life of the continents was contemporaneous, ancient types of animals remained a period longer in North America than in Erope. If animal life was contemporaneous, plant-life had advanced by one period in Europe beyond that which it had attained in North America. In any case, either the faunal or the floral standard of estimation of geologic age of strata for North America is a false one, since there can be but one standard of comparison for anything. But this great fact being understood, the evidence of each of the great departments of life possesses its own intrinsic value.

In conclusion, it may be observed that the lacunæ in the series as presented by one continent, render us dependent on the other for the evidence necessary for the complete elucidation of the laws of the creation of animal life. Phylogenies can be thus constructed which would otherwise be impossible, and the results of researches into the earliest types of *Vertebrata* become intelligible. Thus I have been able to prove, in support of a thesis published in 1874, that the earliest Ungulate *Mammalia* were pentadactyle and plantigrade. I have also shown that the ankle-joint had not, in the primitive *Mammalia*, the hinge-like character that it has in the later ones, but that it is without the interlocking superior articulation. The small size of the brain of early *Mammalia*, already pointed out by Lartet, has received extensive confirmation by the researches of Marsh, who has also shown the progressive increase in size of the whole body in various mammalian lines. To these results I have added another, which is derived from the study of numerous Permian *Vertebrata*, viz, that the earliest land vertebrates had a persistent chorda dorsalis.

4

Reprinted from *Am. Naturalist* **27**:439–449 (1893)

NEW DISCOVERIES OF FOSSIL MAMMALIA OF SOUTHERN PATAGONIA.

By Florentino Ameghino.[1]

I.

At this time, the most important discovery in South American paleontology is the presence of true monkeys in the Eocene of Patagonia. Certain paleontologists have expressed doubts as to the age of the remains. I believe that these doubts have no foundation from whatever point of view one approaches the subject.

From the geological standpoint, the latest researches of my brother, Carlos Ameghino, show that the marine formation, which I have named *Subpatagonian*, and the formation of sub-aerial or terrestrial origin that I have designated *Santacruzian* (the formation which contains the mammals in question), are conformable without a break. The characteristic fossil of Subpatagonian formation is *Ostrea bourgeoisii* (R. de C.). In the east, toward the Atlantic, the lower part of the Santacruzian formation, is found to be intersected with beds of *Ostrea bourgeoisii*, representing the upper part of the Subpatagonian with all the beds of the two formations resting upon each other conformably. In the west, toward the Andes, in the lower reaches of the Rio Sehuen, and in the vicinity of the lakes Viedrua and Saint-Martin, the marine beds of the Subpatagonian formation rest on the Dinosaurian beds of terrestrial origin, contemporary with the North American Laramie (Pehuenche formation); the transition from one formation to the other is almost imperceptible. Throughout this region no mammalian remains have been found in any formation more recent than the Santacruzian. It is then certain that the fossil monkeys belong to this formation. The *Homunculus* specimens have always been found in place, and in the lower Santacruzian beds. It is beyond a doubt that the fossil monkeys belong to the Santacruzian fauna.

[1] From the Revue Scientifique, Jan. 7, 1893.

It remains now to ascertain if the Santacruzian fauna is really Eocene, or if it belongs to a more recent epoch. The affinities of the lower part of this formation with the Dinosaur beds, the presence of Creodonts and Plagiaulacidae, the absence of the modern groups of Ungulates, seem to me to demonstrate plainly that we have to do here with a fauna which dates at least from the commencement of the Tertiary epoch, and which represents the development upon the spot of an autochthonic fauna dating from an anterior geological epoch. Another proof of its antiquity is that a large number of Laramie mammals from the United States described by Marsh, have their nearest allies in the Eocene of Patagonia.

On the other hand, some of the forms characteristic of the Puerco fauna of North America are found in beds much more recent (Lower Oligocene and Upper Eocene) of the Argentine Republic in the neighborhood of Paranà (*Periptychus* Cope). It must also be said that all the authors who have studied the conchological fauna of the rocks of Paranà (d'Orbigny, Darwin, Bravard, Doering, Philippi, etc.), have referred this fauna to the Upper Eocene or to the Lower Oligocene. But, between the Santacruzian fauna and the fossil fauna of the Paranà, there is, when we consider the development of forms, a considerable hiatus, almost an abyss, that can only be filled by the aid of a half-dozen intermediate faunas yet to be discovered. Moreover, the greater part of the types of the Santacruzian fauna, considered from the point of view of an evolutionist, represent forms less advanced than do the totality of similar forms found in other continents.

Thus all the evidences—geological, paleontological and evolutional—are in favor of the great age of the Santacruzian fauna. I place this fauna at the beginning of the Tertiary, and consider it to be contemporary with the fauna of the Puerco of North America. I believe that the more we seek to place this fauna at a later date, the harder it will be to explain the affinities it presents with the faunas of other continents.

Returning now to the fossil monkeys of this formation, I can say that my brother, in his last exploration, found some new specimens, among them, a skeleton, almost complete, of *Ho-*

munculus patagonicus, in the lowest bed of the Santacruzian formation. This skeleton is imbedded in a large block of stone, and, as yet, only a few bones have been exposed. The lower jaw has been taken out almost intact, with its entire dentition. This new specimen, the preservation of which leaves nothing to be desired, shows that the jaw is not so narrow nor so compressed as I have figured recently in drawing from a specimen whose symphysis was evidently distorted by pressure. In the new specimen now at hand, the distance between the internal posterior border and the second true molar on each side is 10 mm., 5. It thus results that the two dental series are more widely separated at the back, and that the whole jaw is much more elevated.

The exact dental formula is I $\frac{2}{2}$, C $\frac{1}{1}$, P M. $\frac{3}{3}$, M $\frac{3}{3}$. The first two true inferior molars are very nearly of the same size, but the third is smaller. The internal incisors are weaker than the external ones. The canine is a little more developed than in the first specimen, and is separated from the premolar by a small diastema. It is probable that the two specimens are respectively male and female. Unfortunately, the individual represented by the new specimen, was very old, so that one can no longer detect the details of the surface of the crown of the molars. The dental series measure, in a straight line, 31 millimeters. The femur is 11 centimeters, and the radius 95 millimeters long. The humerus has a large epitrochlear perforation, but it lacks the intercondylar foramen. Each of these bones, in form, are those of a man in miniature.

Among the other forms belonging to the Santacruzian formation, and which ought to be better known, I will select only the most important, beginning with the Ungulates which are exceedingly interesting.

II.

I have been able to reconstruct the foot of the Mesorhinidæ (the genera Mesorhinus, Theosodon, Pseudocoelosoma), which are the predecessors of the Macraucheniidæ. The genus Theosodon had five toes in front as well as behind; the three in the middle were well-developed, very similar to those of *Macrau-*

chenia, but the internal and external on each foot were very small.

The Homalodontheridæ also had five toes on each foot, but with them all the toes are well-developed. The feet are strong and robust, and it is worthy of note that the hoof-bearing phalanges are split up, and have the same shape as those of the Chalicotherium of Europe. The carpal and tarsal bones have the alternate arrangement. The legs are equally robust, and the humerus has an epitrochlear foramen. The bones of the feet and of the legs bear a singular resemblance to those of the Edentates, but there are also more characters of specialization altogether peculiar, without which they might be taken for the direct ancestors of the Edentates. The Homalodontheridæ are the ancestors of the more modern Chalicotheridæ of the Northern Hemisphere. It is a mistake to look for the stock of the Chalicotheridæ in the Meniscotheridæ. The latter represent a type allied to the Proterotheridæ, with which they ought to have a common ancestor as yet unknown.

The Proterotheridæ present us with peculiarities still more surprising. The discovery of the leg bones of the Thoatherium prove that it was a monodactyl like the horse. Moreover, in the *Thoatherium minusculum*, the reduction of the parts was still more advanced than in the horses; the posterior feet are very slender, provided with a single toe, the third, while the second and the fourth are represented only by vestiges of the metatarsals, much more atrophied than are the same bones in the horse. One sees, in this case, an example of parallel evolution very remarkable. A complete reduction of the bones of the foot among the ungulates has taken place in two different families, at two different epochs, a fact perhaps without precedent in the Vertebrate series.

It is, however, certain that there exists a certain ancestral relation between the Proterotheridæ and the Equidæ, for the latter have descended from a form approaching the Protere-therium, but with complete dentition. I have elsewhere stated that in Europe as in North America, one may be able to trace the genealogy of the horse beyond Anchitherium (Mesohippus included), and that one ought to separate definitely from the ancestral line of the horses, the genera Palaeotherium and Hyracotherium, as well as the different types of the sub-order of Condylarthra. In my opinion, the horses originate from a group of Litopterna, from which sprang, at the same time, the Paleotheriidæ and the Hyracotheriidæ, and this ancestral type would be closely related to the Proterotheriidæ.

It would seem that I am right, for the splendid monograph of the genus Mesohippus, recently published by M. Scott, shows that the calcaneum of this genus still possessed a very small articular facet for the fibula, the last vestige of the litoptern organization, however, the shape of the astragalus, as a whole, is already like that of a Perissodactyl.

We now have evidence to warrant the statement that the horses are descended from a form allied to the Proterotheridæ, and which ought to differ from it above all by its orbits open behind, and its complete dentition, in a continuous series. This ancestral is, perhaps, the *Notohippus* or a kindred genus, but, be that as it may, it is undoubtedly in the Northern Hemisphere that the group of horses has acquired the characters which distinguish it at the present time.

The genus *Astrapotherium* is now known, not only by a skull almost entire, but by many of the bones of the skeleton. The skull, in its anterior part, agrees in a number of points with those of the Proboscidians, and could have carried a proboscis as fully developed as that of the elephants. The bones of the limbs also greatly resemble those of the elephants, and there is no doubt that this type represents the nearest relative of the Proboscidians discovered up to this time in the older formations. The genus *Astrapotherium* is certainly not the direct ancestor of the Proboscidians, but only a collateral branch of the trunk from which the latter have sprung. However, some of the other genera of the family of the Astrapotheridæ, Astrapodon for example, may well be considered the ancestor of the living Proboscidians.

In any case, the latter have acquired their characters upon some other continent, as it is certain that during the epoch of the formation of the fossiliferous beds of the Paranà, the Proboscidians had no representatives in our country. These animals did not make their appearance in the Argentine Republic until the end of the Miocene.

The new specimens of fossil rodents taken from the Santacruzian formation afford unexpected opportunities for comparison. The fossil Cercolabinæ of Patagonia are the ancestors of all the hystricomorphous rodents. The Stiromys have, while young, five superior molars on each side, but the next to the last premolar is soon lost. From the primitive Cercolabinæ are derived the Eocardiinæ, the Eryomyinæ, and the Echinomyinæ. The Eocardiinæ, in their turn, have given rise to the Caviidæ and to the Dasyproctinæ. The Octodontidæ, which now seem to be isolated, have come from certain forms of primitive Echinomyinæ. The discovery of complete skulls of the genus Scieromys, enables me to recognize in this genus the true origin of the Octodontidæ. There are still some intermediate forms wanting, but, no doubt, they will be found soon in the fossiliferous formation of the Paranà.

Not only are the Cercolabinæ of the Eocene of Patagonia the primitive source of all the Hystricomorpha, but they are also the ancestors of another group of rodents, the Myomorpha, which seem also, for the present at least, entirely isolated in that order. The origin of the rats is found in the genera Acaremys and Sciamys, from the lower Eocene of Patagonia. The Myomorphs are not derived from the Acaramyinæ, they having lost their premolars and undergone various modifications in the conformation of the skull. If fossil rats have not yet been found in the Paranà beds, it is on account of their fragile bones, but, in all probability, they will be found there some day.

I now know a part of the dentition of the animal that I have named Tidaeus—this name being preoccupied, I substitute for it, Mannodon. The *Mannodon trisulcatus* is the first genus of Plagiaulacidæ of Patagonia which has the inferior molars constructed on the same type as those of the Multituberculata. It is a genus closely related to Neoplagiaulax and Ptilodus.

Many of the types of the Plagiaulacidæ present a feature that is strange to be found among mammals. The Epanorthidæ, for example, have the inferior incisors beveled on their *internal* border, and not on their upper face as is the case among rodents and all other known mammals which have chisel-like incisors the mandibular branches of the same animals have no symphyseal suture, the surface of the bone, at this point, being nearly as smooth as the rest of the jaw. It thus results that the two mandibular rami of the Epanorthidæ were susceptible of lateral movements, and the singular incisors could perform, up to a certain point, the function of horizontal pincers.

Also, although I am quite convinced that the Plagiaulacidæ and the allied fossil forms have been derived from the Didelphian Diprotodontia, the relation with the living Diprotodontia is not so close as I had thought. The discovery I made showing that the large furrowed tooth of the mandible of the Plagiaulacidæ is not the last premolar, as had been believed, but is the first true molar, separates the latter from the living Diprotodontidæ. This character justifies a sub-order in the Diprotodontia, to which can be given the name Plagiaulacoidea, a name which is as appropriate for the northern forms known as Multituberculata as it is for those of the Eocene of Patagonia.

The recent studies that I have made of the Microbiotheriidæ have confirmed me in my opinion that they represent the direct ancestors of the living Didelphyidæ. The latter have certainly originated in South America, and although their remains have not yet been found in the Paranà formations, they will be found there some day.

One of the fossil groups of Patagonia which presents the greatest interest is that of the primitive Carnivora, allied both to the Dasyuridæ and to the Creodonts. It seems to me impossible to find a line of demarcation between them. In studying the material in my possession, I arrive at the following conclusions: The Dasyuridæ are descended from the Microbiotheriidæ and are then modified into Creodonts; these, in their turn, are transformed into the more modern Carnivores. The dental formulæ of these primitive carnivora are varied—take, for example, the superior incisors, of which the Microbiotheridæ have five pairs like the Didelphydæ; the Dasyuridæ have only four pairs, and the same number is found among other forms allied at once to the Dasyuridæ and to the Creodonts. The larger number of the Creodonts have only three pairs, and, in some genera, but two, and even a single pair. There are some specimens in which one can trace the evolution from the primitive formula. Thus, in the genus Amphiproviverra, the internal pair of superior incisors are entirely atrophied and useless, while the other three pairs are well-developed. This proves that it is the internal pair of upper incisors that is lost in the passage from the Dasyurian form to the creodont and carnivorous forms.

Perhaps the most remarkable specimen discovered during the last exploration is a skull almost entire of *Borhyæna tuberata*, an animal which has hitherto been known only by some insignificant fragments. It is certainly one of the most singular carnivores that is known. It must have been rather large since its skull is 205 millimeters long, and nearly 15 centimeters in transverse diameter at the level of the zygomatic arches. The upper arch of the skull is in an almost horizontal plane with the cerebral cavity excessively reduced, but the sagittal crest is very long and well-developed; the frontals have no postorbital apophyses, and the nasals are widened behind in an extraordinary manner; the anterior nasal opening has the characteristic form of placental Carnivora. The lachrymal canals open entirely inside of the orbits. The malar bone is prolonged behind to the glenoid cavity, of whose border it forms a part of the boundary as in the Marsupials. The palate has no palatine foramina. The upper incisors are reduced to two pairs, of which the internal one is atrophied. Between the canine and the external incisor on each side, is a large groove intended to accommodate the lower canine; this groove replaces the deep fossa which is found in the same place in Dasyurus and Amphiproviverra. These grooves and the contraction of the anterior part of the intermaxillary bone, followed by the reduction in the number of the incisors, gives the anterior part of the skull the appearance of a rodent. The canines, premolars, and molars, of each side, form a continuous series. There are seven molars, of which three are premolars and four are back molars. The first have but a single cusp, but they have a posterior heel, well-developed in the last. The first three true molars increase regularly in size from the first to the third, and have the anterior internal cusp atrophied, which gives them a cutting form, especially noticeable in in the third. The fourth is much reduced, and occupies the same position as in the genera Thylacynus, Prothylacynus and Amphiproviverra. Perhaps Dynamictis, of which only the lower jaw is known, is a synonym for Borhyæna. In any case, the latter animal was a formidable Carnivore, as dangerous as our tigers, and which it is hard to classify. It may belong with the placental Carnivores, or among the marsupial flesh-eaters (Dasyures), or with the ambiguous forms known as Creodonts.

It would take too long to speak of the numerous fossil Edentates of the Santacruzian formation and of the many peculiarities they present. However, I cannot resist mentioning one of the greatest novelties of the new family of the Metopotheriidæ, which comprises the genera Metopotherium, Pelecyodon, and Zamicrus. In this group, each of the horizontal branches of the mandibula is formed of two distinct bony parts, one in front, the other behind. These two pieces are united by a suture which commences at the alveolar border, very nearly in the middle of the dental series, and is directed obliquely back and down, terminating in the lower border, almost just below the last tooth. This is evidently a character inherited from the reptiles. The same suture can be seen, but not quite so plainly, in the mandible of some of the Orthotheriidæ.

I remarked, in the beginning of this article, that I had found some forms characteristic of the Puerco of North America in the beds much more recent than those of the Santacruzian epoch. In fact, I have received from the Paranà, a fragment of a upper maxillary, which shows several tooth sockets and one tooth in place. The latter corresponds to a third premolar. It is impossible to separate this specimen generically from Periptychus (Cope). The tooth resembles perfectly the third upper premolar of *Periptychus rhabdodon* and shows the same grooves or wrinkles in the enamel which distinguishes the North American species. However, the South American species (*Periptychus argentinus* Amegh.) is easily distinguished by its size—almost a third smaller—and by the insertion of the molar teeth which are closer together, forming a continuous series. This specimen was found in the Lower Oligocene or the Upper Eocene in the neighborhood of Paranà.

While I am speaking of the fossils of that region, I must add that the doubts which I had concerning the real nature of Ribodon are now set at rest. The new material at hand shows that *Ribodon limbatus* belongs to the Sirenia, and is one of the family of Halitheriidæ.

III.

I would like, in closing, to reply to the objections that have been made to my classification of the Tertiary beds of the Argentine Republic, and particularly to the assertion of Professor Steinmann, who maintains that the beds I have designated Miocene and Pliocene are only Plistocene (or Quaternary).

I have not the complete work of Professor Steinmann, and therefore will refer only to the extracts that he has published.[1]

I am more and more convinced that the Pampean formation is tertiary, and not quaternary. The Araucanian, of which Professor Steinmann speaks, is entirely distinct from the Pampean formation. The vast difference which exists between the Paranà fauna and that of Monte-Hermoso is easily explained by taking into account the fact that the first is Lower Oligocene or Upper Eocene of the old geologists, while Monte-Hermoso is Upper Miocene. Some day, when the Lower Miocene of Catamarca (an Araucanian bed) is explored, and the mammalian fauna of that formation is satisfactorily known, the great break that is now held to exist between the fauna of the Paranà and that of Monte-Hermoso will largely disappear.

I am astonished that any one can still speak of a glacial epoch and of traces of glacial action in connection with the Pampean, and still more so in regard to the pre-Pampean formations. The Araucanian bed is formed by the accumu-

[1]American Naturalist, 1891, p. 855.

lations of volcanic detritus. The Pehuelche is of fluviatile origin. Is it necessary to repeat for the hundredth time, that there is not the least vestige of glacial action in the Pampean formation?

It is impossible to name single fossil species from the Pampean formation which is characteristic of a colder climate than now prevails in the same country. On the contrary, not only does the mammalian fauna indicate a warmer climate than at present, but the fresh water conchological fauna also furnishes similar evidence. The many fossil plants of the Pampean formation are the same that are seen to-day in the province of Buenos-Ayres, or in the northern part of the Argentine Republic—not a single form belonging to a cold climate. I am almost certain that an examination of the marine shells of the same formation would lead to the same result. I am about to make a collection of them to send to M. von Jhering, whom I will request to study them from this point of view.

ERRATA

Page 447, lines 5 and 6 from the top of the page should read "noticeable in the third." Page 449, line 5 from the top of the page should read "to name a single fossil species . . . ."

5

The vertebrate-bearing early Tertiary of South China: A review

Tang Xin and Zhou Ming-zhen

UDC 566:551.781(510-13)(047.1)951=20

ABSTRACT: All the early Tertiary vertebrate localities known to date in South China are systematically reviewed with regard to the characters of the sediments and vertebrate fauna. Emphases are laid on the analyses of the fauna and the geographical distribution and stratigraphical ranges of the leading piscine and mammalian forms. A tentative correlation of all the more important lower Tertiary sections based on the relationship of vertebrate fauna with those of North China and the adjacent countries in southeast Asia is also attempted. General conclusions regarding the paleogeography and tectonic history of the various regions in South China during early Tertiary times are also briefly discussed as a necessary background toward an understanding of the evolution of various sedimentary basins and their vertebrate fauna. --From English summary in Vertebrata PalAsiatica.

* * *

I. INTRODUCTION

Every province of China south of the Yangtse* has a fair number of red-bed series and similar terrigenous facies, which occur in every independent or semi-independent basin.[1] As sediments they have no distinct aspect and are lacking well-defined stratigraphic characteristics and correlation features. In addition, they are almost without fossils. Thus the difficulties involved in differentiation, correlation, and age classification have resulted in controversy.

Until recently, very few fossils have been found in the early Tertiary of South China; vertebrate fossils had been found in only a few places (Lunan and Qujing [Ch'uching] in Yunnan [Yünnan], Hengyang in Hunan, and Ichang* in Hupei*), but these indicated that the lower Tertiary does occur in southern China. Much progress has been made in research on early Tertiary vertebrate fossils and stratigraphy since the liberation. A few of the localities have already been described, and these have been included to supplement and extend the more important new evidence on the nature of the various groups of fossils. Most are newly discovered localities and are of much greater significance.

Here we consider parts of the lower Tertiary known to contain vertebrate fauna (the fossil evidence for the age assignment is fairly reliable) in order to draw up a general classification in accordance with the succession of strata. After a detailed study of the stratigraphy and geography of the distribution of these T_1 vertebrate fauna, we deal with some aspects of the characteristics of the fossil groups, of age assignment, and of the associated paleogeography, paleoclimatology, and structure.

II. EARLY TERTIARY VERTEBRATE-FAUNA HORIZONS IN SOUTH CHINA AND NATURE OF THE FAUNA GROUPS

1. Lunan Basin

Lunan Basin lies in East Yunnan*; the beds are mainly red gritstones and fine-grained sandstones with thin bands of mudstone, oölites, and bedded calcareous strata. Total thickness up to 380 m. Unconformity with underlying Carboniferous-Permian limestone.

The Lunan red beds were at first (Deprat) thought to be Upper Permian; later (1937) Bian Mei-nian [Pian Mei-nien] discovered remains of Mammalia near Shaoji [Shaochi] village about 5 li north of Lunan City. He and Yang Zhong-jian [Yang Chung'-chien] (1939) found there the following: Carnivora indet., Amynodon cf. mongoliensis, Caenolophus minimus, Teleolophus sp., Deperetella sp., ? Protitanotherium sp., Anthracotheriidae indet., and Chelonia indet. Of these, A. mongoliensis and C. minimus are major fossils of the Shara Muren**

Translated by Scripta Technica, Inc. from Vertebrata PalAsiatica (in Chinese), 1964, v.8, no.2, p.119-133. The system of transliteration used here is Pin-Yin (the latinized form now approved by the Chinese government). Proper names as well as certain place names, usually on first mention only, have been further transliterated - in brackets - by the editors of International Geology Review into the older Wade-Giles system, more familiar to American readers. Series and formation designations and those of fossils have not been retransliterated. Author Tang Xin of the Petroleum Institute, Canton, becomes T'ang Hsin in the Wade-Giles system; coauthor Zhou Ming-zhen of the Institute of Vertebrate Paleontology and Paleoanthropology, Academia Sinica, becomes Chou Minchen. Zhou is one of many Chinese notables who have further adapted their names for English readers - a strict Wade-Giles transliteration would have been spelled Chou Ming-chen.

[1] Names given with one asterisk* have been transcribed as traditionally written, not in the Pin-Yin system. Names given with two asterisks** are not Han names and have not been transcribed in the Pin-Yin system. Most of them are Mongolian.-- transl.

series of the upper Eocene of Inner Mongolia. They therefore considered these Lunan red beds to be of Eocene age (32).

Zhou Ming-zhen [Chou Minchen] reexamined the these fossils in 1957 and gave a new list, as follows: Carnivora indet., Deperetella sp., Lunania youngi Chow, Caenolophus medius Chow, Amynodon sp., Titanotherium sp., Anthracotheriidae indet., and Chelonia indet.; he also went some way towards demonstrating the previous assumption as to the age.

In 1959-1960 the Paleovertebrate and Paleoanthropological Research Institute collaborated with the Geology Department of the Museum of Geology in a study of a fairly rich find of mammalian fossils near the footbridge south of Lunan; details have been published over a number of years (16, 18, 26). Zhou used this new evidence to suggest a new division of these red beds into two parts(10), the upper belonging to the lower Oligocene and the lower to the upper Eocene. Later, Shu Yu-xuan [Shu Yu-hsuan] and Qiu Zhan-xiang [Ch'iu Chan-hsiang'] (26) published a list of the Lunan fossils, in which the lower part was termed the Lunan series and the upper the Anrencuen series. The fossils were given as follows:

Lunan series:
- Carnivora: Creodont indet.
- Perissodactyla: Titanotherium sp. (Rhinotitan?), Lunania youngi, Deperetella sp., Teleolophus sp., Caenolophus medius, Amynodon sp. (cf. sinensis Zdansky).
- Artiodactyla: Anthracotheriidae indet.

Anrencuen series:
- Carnivora: Carnivora indet.
- Perissodactyla: Rhinotitan quadridens Xu et Chiu, Parabrontops lunanensis Chow et Hu, Eomoropus ulterior Chow, Telelophus medius Matthew et Granger, Amynodon altidens Xu et Chiu, Cf. Metamynodon sp., Prohyracodon meridionale Chow et Xu, P. progressa C. et X., Ilianodon lunanensis C. et X., Indricotherium parvum Chow, Rhinoceratidae indet.
- Artiodactyla: Eoentelodon yunnanense Chow, Probrachyodus panchiaoensis X. et C., Brachyodus hui (Chow), Anthracotheriidae indet.

The identification and horizon assignment are relatively reliable for the fossils of the lower (Lunan) series; most of them can be correlated with species in the Shara Muren formation of Inner Mongolia; i.e., Titanotherium sp. (Rhinotitan?), Deperetella sp., Caenolophus medius, Caenolophus sp. correlate with Rhinotitan, Deperetella cristata, Caenolophus obligus, and C. minimus, only Lunania youngi and Teleolophus being relatively primitive. The Lunan series may thus be dated as late Eocene, with the correlated Shara Muren species of North China and Inner Mongolia a little earlier in date, slightly earlier than previously assumed (10).

The composition of the fossil groups in the Anrencuen series is rather complicated. Shu and Qui [Ch'iu] considered that the groups include representatives of three distinct stages of evolution, that these fossils all came from the same set of horizons (Anrencuen series) and that they were of early Oligocene age. They further believed the groups to be equivalent to those of the Aertung Obo** series of Inner Mongolia, or even of slightly earlier date (26). But many of these fossils are, in fact, merely found at the site where they were ultimately deposited; their original horizon and locality are by no means certain. Hu Zheng-zhi [Hu Cheng-chih] et al. collected fossils from the center part; their records, together with identifications by one of us (ZMZ), indicate that the Anrencuen series has two or three fossil beds (zones). Parabrontops was found in the upper part of the white marls near Anrencuen; Prohyracodon progressa and Eomoropus ulterior were found on the Xiaosha He [Hsiaosha Ho] (Little Sand river) in a section containing concretions in the lower part of the white beds that yielded Parabrontops. Prohyracodon meridionale occurs in a zone in red silty sandstone lower down. The fossils described by Shu and Qiu had no locality, and most, apart from Metamynodon (?), are typical of the lower Eocene; only Indricotherium parvum and Prochyodus hui of their list belong to the Oligocene.

Current evidence indicates that the Anrencuen series (namely, the fossil-bearing strata above the Lunan series) can be taken as a relatively broad group (the Anrencuen formation) containing two or three fossil zones. The lower Tertiary of the Lunan basin can thus be arranged tentatively as shown below:

Anrencuen series	Parabrontops lunanensis zone (Oligocene)	
	Prohyracodon progressa zone	Upper Eocene
	P. meridionale zone	Upper Eocene
locally with unconformity or discontinuous sedimentation:		
Lunan series	Lunania youngi zone	Upper Eocene

The lower Tertiary is also found southwards in Yunnan towards Milei and Kaiyuan [K'aiyüan]. The section at Milei is well-developed, the structure and nature of the sedimentation being much as at Lunan. It may be that the two were once parts of the same large sedimentation unit, the upper part being later lost by erosion.

2. Qujing Caijia Chong [Ch'uching Ts'aichia Ch'ung'] (Yuezhou) Basin

The lower Tertiary occurs as the Caijia Chong series at that locality at Yangjia Chong [Yangchia Ch'ung] in the northern part of the Qujing Dongnan [Ch'üching Dungnan] (Yuezhou) basin in Yunnan. The base consists of a limestone breccia and an angular conglomerate; it has iron concretions. Upwards this passes into massive dark-red fine-grained sandstones with occasional broken limestone lenses. Above the sandstones is a set of beds of lacustrine facies, the base having a dark-green marl containing vertebrate fossils, above which is a thick band of fresh-water limestone containing fresh-water Spirogyera and Chara braunii Gmel. Finally there is arenaceous friable sandstone. Total thickness about 450 m.

In 1932 Yang Zhong-jian first reported fossils Caijia Chong series, which were: Carnivora indet., Rhinoceratidae indet., Merycopotamus sp., the last being a major fossil of Indian Middle Siwalik time. Yang then assigned this series to the early Pliocene (Pengti time) (41). Later (Colbert, 1940) the distribution of Merycopotamus in India and Burma (lower Pliocene to upper Pliocene or lower Pleistocene) was used to conclude that the Caijia Chong series could belong after the middle Pliocene or Pengti time.

In 1939 Bien Mei-nian discovered numerous vertebrate fossils at Yangjia Chong (42 sites)(32), and in the same year Yang Zhong-jian published the following list: Cadurcotherium cf. ardynensis (Osborn), Tragulidae or Cervidae indet., Anthracotheriidae indet., Chelonia, Crocodilus sp. The C. cf. ardynensis indicated that the Caijia Chong series was of early Oligocene age, which would make it equivalent to the Aertung Obo series of Inner Mongolia (49).

In 1957 one of us (ZMZ) made a fresh study of the fossils in this series and concluded that one identified as Merycopotamus sp. should rather be Bothriodon sp.; this and the C. cf. ardynensis then define the age of the series as early Oligocene (33, 34). Later (1961) Shu Yuxuan added fresh material and published the following list (24):

Perissodactyla: Brontotheriidae indet., Cadurcodon ardynensis (Osborn), Gigantamynodon giganteus Xu, cf. Metamynodon sp., Caenopinae indet., Indricotherium sp.

Artiodactyla: Bothriodon chowi Xu, cf. Miomeryx sp.

The [bones] in the Oligocene Caijia Chong series have a distinct color; [those] of C. ardynensis are precisely the color of those from the Aertung Obo series. Gigantamynodon giganteus is slightly larger than the G. cessator of Aertung Obo, but it is clearly from the same horizon. Cf. Metamynodon sp. was formerly unrecorded for China; the characteristics of the teeth make it closely resemble the Cadurcodon of the Oligocene of Western Europe and the Metamynodon of the lower Oligocene of North America, especially in the latter case. Indricotherium sp. is fairly similar to the I. parvum of the Arencuen formation. The basic structure of Bothriodon chowi is very close to that the B. velaunum of the lower Oligocene of Western Europe, but few specimens are available. We may say that the fossil zones of the Caijia Chong series can be correlated with the Aertung Obo series of Inner Mongolia and with the upper fossil zones of the Anrencuen formation of Lunan; all are early Oligocene.

3. Luoping (Shizong) [Lop'ing (Shihtsung)] Basin

No very detailed information is available for the lower Tertiary of the Luoping (Shizong) basin in East Yunnan. In 1959 Zhou Ming-zhen and Shu Yu-zuan [Shu Yü-ksüan] reported a large rhinoceros found at Datongziang (Tat'ung'hsiang) in Shizong (17). (Indricotherium sp.), whose age was determined as middle or late Oligocene. In 1960 Qui Zhan-xiang [Ch'iu Chan-hsiang] found a few such fossils at Luoping (I. intermedium Chiu) and another (as yet unidentified) large rhinoceros fossil. The characteristics of the former make it more advanced than the I. parvum of the upper part of the Anrencuen formation at Lunan but more primitive than the I. grangeri of Inner Mongolia (middle Oligocene or a little later). Qui therefore concluded that the sediments at Luoping could possibly include the base of the middle Oligocene or the top of the lower Oligocene (20).

4. Baise (Tiandong, Tianyang) [T'ientung', T'ienyang'] Basin

This lies along the banks of the Zuojiang [Tsochiang'] and consists mainly of gray sandstones, brown mudstones, and carbonaceous shales intercalated with lignite and oil shales (total thickness > 2000 m). The upper part forms the Yong'ning [Yung'ning] formation (upper Tertiary) and the lower the Yongfu [Yungfu] series, which is largely lacking in fossils and has been taken to be lower Tertiary.

In 1954 Yu Qi-ping [Yü Ch'i-p'ing] and Zhu Long-kuei [Chu Lung-k'uei] found some vertebrate fossils at Tiandong and Tianyang, in which Zhou Ming-zhen (7, 9, 36) identified: Creodont indet., Metatalmatherium cf. browni Colbert, Deperetella sp., Amynodontidae indet. Anthracothema rubrica (Pilgrim et Cotter), Anthracokeryx cf. birmanicus (Pilgrim et Cotter, A. cf. moritus Pilgrim, Tragulidae indet., Emys sp., Amyda sp., Crocodilus sp.

These are all typically Eocene and are precisely the groups found at Pondaung (North Burma), which have been studied in detail by Pilgrim and Cotter (1916), Pilgrim (1926, 1928), and Colbert (1938), who assigned them to the latter part of the Eocene. From this it is concluded that the Tiandong and Tianyang fossils are from the lower part of the Yongfu series and correspond broadly to the Yuanqu [Yüanch'ü] series of Shanxi [Shanhsi] and Henan [Honan].

5. Maoming Basin

This basin in West Kwangtung contains the Youganwo series (yellow-brown mudstones, siltstones, and sandstones, with some oil shales; total thickness about 70 m). Fossils reported (4, 8, 13, 14, 40) at various times are:

Pisces: Cyprinus maomingensis Liu
Chelonia: Anosteira maomingensis Chow et Liu, Isometremys lacuna Chow et Yeh, Aspideretes impressus Yeh
Crocodilia: Tomistoma petrolica Yeh

Differences of opinion occur over the age of the series. The earliest Cyprinus horizon in Europe is of Miocene (middle Cenozoic) age, or upper Cenozoic in Asia; moreover, this series and the Shangcuen series about it have a gap against the Huangniuling series. The unconformity is not very pronounced, and upper Cenozoic Spirogyra d'Orb. have been found in the Shangcuen series, from which Liu Xian-ting [Liu Hsien-t'ing] concluded that the Youganwo series was middle Cenozoic or lower Miocene; yet other results on the fossils almost all point to an upper Eocene age. The series may thus be assigned to the early Tertiary, very likely the late Eocene.

6. Sanshui Basin

The red beds of Guangzhou [Kuangchou] and Qilingqiu [Ch'iling'ch'iu] are fairly fully exposed in the Sanshui area. The lower part Xikengbu formation) is probably Cretaceous; above this lies a sequence of intercalated red sandy conglomerates and mudstones (the Lanmentou formation), and above this a gray-green mudstone series containing oil shales and thin bands of gypsum, mirabilite, etc. (Buxin series). The top is a sandy conglomerate of true river facies (Tasimagang series).

The Lanmentou formation contains Spirogyra, which may derive from the early part of the early Tertiary. The base of the Buxin series (at Honggang Duan [Hungkang Tuan] has yielded (1962) fishes (Canton Petroleum Research Institute), which are stated to include Osteochilus linliensis Tang and Osteochilus sp. (sp. nov.). These fossils are all found at Linli and Xiangxiang [Hsienhsien] in Hunan and also at Dangyang [Tangyang] in Hupei [Hup'ei]; they are very widely distributed and date either to the late Eocene or early Oligocene. Similar oil shales and fish have been discovered near the Sanshui basin, e. g. on the highroad between Guangzhou and Foshan. The South China Normal Academy has also reported fish [Tungtingichthys sp. (cf. gracilis Liu, Liu et Tang)] from red beds at Xiqiaoshan [Hsich'iaoshan] in Nanhai. These resemble Osteochilus linliensis, and it would seem that the Xiqiaoshan fish beds may correspond to the lower part of the Buxin series of Sanshui.

7. Nanxiong [Nanhsiung] Basin

Much work has been done on the red beds of the Nanxiong-Shixing [Nanhsiung-Shihsing] area (North Kwangtung), but for lack of fossil data the age and stratigraphic assignment are in doubt.

In 1961 the Geological Survey of Canton Province reported fish fossils and the remains of a reptile in the red beds at a locality 3-4 km northwest of Wujinggu [Wuchingku] in Nanxiong [Nanhsiung] in dark purple conglomerates and mudstones. This fossil, named Anosteira lingnancia Young et Chow in 1962 by Yang Zhong-jian and Zhou Ming-zhen (30), has some features resembling those of the "Pseudotrionyx" delheidi of the Paleogene of Belgium and North America. Yang and Zhou took the view that the beds were probably of Eocene or Paleocene age. Later the Institute of Vertebrate Paleontology and Paleoanthropology found vertebrate fossils in various horizons around Nanxiong. Zhang Yu-ping [Chang Yü-p'ing] and Tong Yong-sheng [T'ung Yung-sheng] (22) reported that these included a representative of the Pantolambdidae and cf. Dissacus sp. (primitive carnivore) from the base of the beds formerly called the Nanxiong beds. They gave the beds another name (Luofuzhai series) and determined the age as Paleocene.

8. Hengyang Basin

This basin is in Xiangzhong [Hsiangchung] and has red beds up to several kilometers thick, which were at one time assigned to the Cretaceous; this was later altered to Tertiary or early Tertiary. They have also been taken as Cretaceous-Tertiary.

In 1938 Yang Zhong-jian, Bian Mei-nian, and Li Yue-yan found vertebrate fossils at Yangmeiqias [?] and Jiangjunzeng [Chiangchuntseng] (north of Hengyang), which contained remains of Crocodilia, Pisces, and a genus of Mammalia. This last was at first taken to be possibly related to the Lingchang (?) [Lingch'ang] Adapidium of the Yuanqu basin (45) Yang et al. originally assigned the Hengyang horizons containing this to the upper Eocene.

In 1944 Yang Zhong-jian made a fresh study of this fossil and concluded that it was a representative of the Palaeotheriidae (40), which he called Propalaeotherium hengyangensis Young. This occurs in middle Eocene horizons in Europe and has been found also in the Mengingzhuang series (Shandong) [Shantong]. This is earlier than the Irdingmannah** series of Inner Mongolia, being at least middle Eocene. Later (1948) Yang studied two Crocodilia which he also assigned to the middle Eocene (44), namely Lianghusuchus hengyangensis Young and Tienosuchus hsiangi Young. Recently (1962) Zhou Ming-zhen examined material from the basin of the Xinyu Jiangxi [Hsinyü (Chianghsi)] and concluded that Propalaeotherium hengyangensis could even be as early as early Eocene.[2]

In addition, Tang Xin [T'ang Hsin] (1960) discovered fossil fish ahd flora in purple-red sandy shales near sandstones and sandy shales containing gypsum and rocksalt bands at Beibaoshan [Peipaoshan] in Hengyang. The fish were poorly preserved, but they resembled Tungtingichthys gracilis Liu, Liu et Tang from Xiangxiang and the Linli-Shimen area. Later age assignments were late Eocene or early Oligocene (3).

The most recent evidence would indicate that the Hengyang red beds can be divided into three large formations, of which the middle and lower contain ostracoda and may be of Jurassic-Cretaceous age, while the upper (the Hengyang formation in the narrow sense) is early Tertiary. The Palaotheriidae and Crocodilia all occurred in the middle and lower parts of the Hengyang formation; the fish-bearing beds form the top part, the Baoshanyou series (Geology and Paleontology Research Institute, 1962.[3]

9. Xiangxiang Basin

Similar red beds are seen around Xiawanpu [Hsiawanp'u], Tanshi [T'anshi] and Zhujindu [Chuchintu] in the Xiangxiang area of northwest Hengyang. The base is called the Tanshi formation and has not yielded any fossils; the top is called the Xiawanpu series (Xiawanpu shales) and is rich in fish and other fossils. There appears to be an unconformity or gap between the two.

The Xiawanpu series was formerly not clearly distinguished from the Tanshi formation, being referred to as the "Tanshi series." Initial age assignments were Miocene (Tian Qi-xi [T'ien Ch'i-hsi], 1932), Eocene (Tian, 1936), and lower Tertiary (Yang, Bian, and Li, 1938). In 1954 Si Xing-jian [Szu Hsing'-chien] and Li Gong-mo [Li Kong'-mo] distinguished the Xiawanpu shales from the "Tanshi series" and assigned them (on the basis of plant fossils) to the late Tertiary, possible Pliocene. This assignment has usually been accepted by later workers, although some consider the age to be early Pleistocene (Peking Geological Institute, Quaternary Geology Series, 1960; Li Wen-yi [Li Wen-yi'i], 1961). It is clear that the evidence is inadequate.

Zheng Jia-jian [Cheng Chia-chien] (1962) and Tang Xin (1963) found that the Xiawanpu fossil fish consist of the following (22):

Cypriniformes: Pararutilus orientalis Tang, Osteochilus linliensis T., O. hunanensis Cheng, Aphyocypris xiangxiangensis Tang, Barbinae indet., Aoria lacus Cheng.
Perciformes: Tungtingichthys gracilis Liu, Liu et Tang, T. hsiawanpuensis Cheng.

Of these, Osteochilus occurs in parts of southeast Asia, while in China it occurs only in the provinces of Kwangtung and Kwangsi*, first appearing at the start of the Tertiary. The species found at Xiawanpu are very similar to those in the early Tertiary of Indonesia. Aoria has been found in the Eocene of Asia and Africa, while Tungtingichthys is similar to the Toxotes found in the mergelschiefer (marl shale series) of the middle parts of the Inn and to the Acanus of Switzerland (Glarus). The Xiawanpu has therefore been assigned either to the latter half of the Eocene or to the latter part of the Oligocene.

10. Linli-Changde [Ch'angte] Basin

Similar red beds occurs around Linli, Shimen [Shihmen], Changde, and Taoyuan in northwestern Hunan; they resemble those found in Hengyang and Xiangxiang in many ways.

In 1959 Tang Xin examined the Cyprinus found at Xiejiashan Sue[a?]njiaquiao [Hsiehchia shan Suanchiach'iao] and Shimen Sanbanquiao [Sanbanchiao] (Linli area) (27) Osteochilus linliensis Tang); in 1962 Liu Dong-sheng [Liu Tung-sheng] and Tang Xin [T'ang Hsin] (3) examined a perciform fish, Tungtingichthys gracilis Liu, Liu et Tang. These two have been dealt with above, and it is clear that the Linli and Shimen fish beds are also late Eocene or early Oligocene.

In 1960 and 1961 the Mineral Dressing Institute of Canton and the Geological Survey of Hunan Province found further fish fossils at sites such as Taoyuan [T'aoyüan], Changde [Chan'gte], and Lixian [Lihsien], most of which

---

[2] Zhou Ming-zhen [Chou Minchen], after studying some new material from Hengyang, has kindly told the authors that P. hengyangensis might represent forms still older, lower Eocene in age. (This footnote translated from an author's footnote found in Vertebrata PalAsiatica, v.6, no.2, p.127, to which writers of this article referred. - ed.)

[3] The series has been assigned to the early Tertiary on the basis of studies of the flora, Chara braunii Gmel., gastropods, and ostracods.

were O. linliensis. The red beds of the Changde-Taoyuan [T'aoyüan] basin can thus be correlated with those of the Linli-Shimen area; the two areas are adjacent, which would mean that they formed parts of a deposition basin in early Tertiary time.

11. Jianghan [Chienhan] Basin

The Jianghan plain covers Dongtinghu [Tungt'inghu] and the middle and lower reaches of the Han Shu (Han river); it is a vast Mesozoic-Cenozoic basin, whose inner parts are now filled with thick Cenozoic sediments, the older strata now being seen only at the margins. The red beds are most abundant around Ichang*, Yidu [Yi'itu], and Dangyang. The Jianghan red beds have been termed the "Donghu series" (Li Siguang [Li Szu-kuang], 1924); they have lately been studied in considerable detail, but there is as yet no agreed lithologic or biostratigraphic classification.

Jia Han-po [Chia-Han-p'o] (1936) found a broken pantodont skull and other scattered bone fragments in the "Donghu series"; in 1938 Yang Zhong-jian et al. (39) classified these as Eudinoceras kholobolchiensis Osborn, which dates to the late Eocene. In 1959-60 Zhang Yu-ping and Zhong Yong-sheng [Chung Yung-sheng] (Institute of Vertebrate Paleontology and Paleoanthropology) found pantodont fossils from a very similar horizon in the same locality. The new material is clearly different from the Eudinoceras of Inner Mongolia and resembles the Shandong-Xintai type more closely (Coryphodon flerowi Chow). If this is so, the Yangxi [Yanghsi] fossil beds must be a little earlier and so nearer to the Hengyang formation or Xinyu series. The beds are termed the Pailoukou series (Institute of Geology and Paleontology, 1963).

In 1963 Liu Xian-ting [Liu Hsien-t'ing](5) reported Knightia yuyanga Liu from purple-gray shale in the lower part of the "Donghu formation" near Xianghe Yan [Hsienho Yen] in Yidu Chengnan [Yi'itu Ch'engnan]. This fish is often found in the Green river shales of the Rocky Mountains, so the beds are of Eocene age, at latest the late Eocene.

These Yidu shales are thus close to the above mammal beds, but we cannot yet say whether the productive horizons belong to the Yangxi or Pailoukou series (Wang Zhen [Wang Chen], 1963; private communication).

Also, the Fangjiahe series of the Dangyang Dongyue Miao [Tangyang-Tung'yüeh Miao] area yields fish fossils together with Spirogyra, Ostracoda, Chara, spores, and so on. The fish are Osteochilus sp. together with O. linliensis, so this series clearly correlates with the Xiawanpu series.

The red beds of the Xichuan [Hsichu'an] basin, southwestern Henan [Honan] are also notable. In 1939 Bing Zhi [Ping Chih](38) reported Sinohadrianus sichuanensis Ping from Fenghuang in Xichuan, and in 1930 Derigin described Lophialetes sp., which is very similar to a mammal of the Irtingmannah series of Inner Mongolia. Vertebrate fossils have also recently been found in the Hetaoyuan series of the Nanyang basin: Sciuravus sp., Miacis sp., ? Colodon sp., Dinocerata indet., Chungchienia siahuanensis Chow, Lophialetes cf. minutus Matthew et Granger. These beds are probably equivalent to the Mongolian series or a little earlier.

In 1962 Zhou Ming-zhen and Tong Yong-shen [T'ung Yung-shen](19) reported ?Gobiatherium sp. from the freshwater limestone of the Yuhuangding series (Liguanquiao [Likuanchiao] basin, Xichuan); the horizon lay below the Hetaoyuan series, and the age is probably earlier than late Eocene.

The lower Tertiary of Xichuan (or Nanyang) basin is evidently a continuation of that of the Jianghan basin; the two basins formed parts of a larger deposition area at the time.

12. Yuanshui (Xinyu) [Yüanshui (Hsinyü)] Basin

The red beds in the Yuanshui basin (Jiangxi [Chianghsi]) can be divided into the Xinyu formation (top) and Linjiang series (base) (21). In 1954 Lao Huei-yuan [Lao Hui-yüan] reported finding fish, crocodiles, and Pantodonta in the red conglomerate of the Xinyu formation. Zhou Ming-zhen (1959) (11) identified these as Crocodilus sp., Emys spp., Anosteira sp., Trionyx sp., and Coryphodon. The historical distribution of Coryphodon and Anosteira would indicate that the Xinyu formation can be taken as not later than middle Eocene. These red beds may also be equivalent to the Hengyang horizons containing Propalaeotherium, Hengyangensis and crocodiles.

In 1962 Zhou Ming-zhen and Tong Yong-sheng (19) reported Probathyopsis sinyuensis Chow et Tung (Dinocerata) from the Ningjiashan [Ningchiashan] red beds at Xinyu; this is very similar to the P. successor of the early Eocene of North America, so these red beds were tentatively assigned to the early Eocene.

The Linjiang series is gray-green and resembles the Xiawanpu series; there are also many points of resemblance between the fish fossils here and those at Xiawanpu and Linli. The horizons are also correlated (21). The red beds of the Yuanshui basin are therefore on both grounds to be correlated with the Hengyang, Xiangxiang, and (even) Jiangshan series.

13. Qujiang (Dangshi) [Ch'üchiang (Tangshih)] Basin

The red beds in the area enclosed by Jinhua

[Chinhua], Quxian [Ch'uhsien], Longyou [Lungyu], Dangshi, and Jiangshan (West Zhejiang) [Chiangshan (West Chechiang)] are together called the "Qujiang formation" (Qujiang sandstone). In 1962 Ye Xiang-kuei [Shi Hsiang-kuei] (6) reported Amyda sp. from Lanxi, which was everywhere present, on which basis he considered that the fossil bearing horizons could vary in age from Eocene to Oligocene.

III. CONCLUSIONS

1. Correlation of Horizons

Table 1 shows the correlation of the early Tertiary vertebrate-bearing horizons and the nature of the fossils.

We do not yet have much detailed evidence on the Paleocene for this area or for China as a whole; only the horizons in northern Kwangtung bearing mammals and reptiles can be taken as certainly Paleocene. These are the lowest known horizons on the Cenozoic in China at present.

The Eocene is relatively common in South China, especially the upper Eocene, which occurs in nearly every province; the middle and lower Eocene are not common, known sites being Xinyu, Hengyang, and (possibly) Jianghan. However, the Tertiary beds in provinces such as Jiangxi, Hunan, Hupei, and Kwangtung are similar at many points, so it is highly probable that the middle and lower Eocene are also present.

The Oligocene is of more restricted (local) occurrence than the Eocene, the known horizons being restricted to a few small basins in eastern Yunnan and also at Lanshi in Zhegiang.

2. Fossil Occurrence

Early Tertiary vertebrates (fish, reptiles, mammals) are quite common in South China. The mammals are the most common and have been studied in the most detail. The numbers of genera and species (respectively) are as follows: fish, 7 and 9 (3 orders: Clupeiformes, Cypriniformes, Perciformes); reptiles, 10 and 13 (orders Chelonia and Crocodilia); and mammals, no less than 30 genera and 40 species (not including ones that cannot be classified as

TABLE 1. Simplified table showing correlation of early Tertiary vertebrate-bearing horizons in South China

Epoch / Fossil horizon / Region			Southwest: Lunan	Southwest: Caijia Chong	Southwest: Luoping	Kwangsi-Kwangtung Basin: Baise	Kwangsi-Kwangtung Basin: Mao-ming	Kwangsi-Kwangtung Basin: Sanshui	Kwangsi-Kwangtung Basin: Nan-xiong
			Series			Group	Series		
Overlying beds			Ciying	?	?	Yong-ning	Huang-niou	Tasima-gang	Hongli-ling
LOWER TERTIARY	Oligocene E_3	U	?						Danxia series (?)
		M			Inthricotherium ?				
		L	Anrencuen formation or Lunan series	Caijia Chong Series (Lunan series) ?		Yongfu series	Youganwo series	Tixin series ?	
	Eocene E_2	U							
		M						(Lan-mentou group) ?	
		L							
	Paleocene E_1	U							Luofucai
		L			?				
Underlying beds			C-P limestone	Yangxin formation	?	Nazhen formation	Cretaceous	Formation: Xikeng-b	Formation: Nan-xiong

to genus or species), which fall in 5 orders: Carnivora, Pantodonta, Dinocerata, Perissodactyla, and Artiocactyla.[4] Table 2 shows the distributions of genera and species. The small amount of evidence indicates that the proportion of Paleocene forms is very similar to that for North America. The information on the abundance of early and middle Eocene fossils is very restricted, and all we can say is that there are a few fish (Knightia), Chelonia (Anosteira, Emys, and Trionyx), and crocodiles (Lianghusuchus, Tienosuchus, and Crocodilus), as well as mammals (Propalaeotherium, ? Probathyopsis, and Coryphodon). The latter are clearly related to those of North China. Further conclusions cannot be drawn until more specimens have been found.

Fossils of the late Eocene are relatively abundant and widespread. There are clearly at least four different associations:

1) Fish found in the Xiawanpu series of Xingxiang. This association occurs in the middle and lower reaches of the Yangtse in Hupei, Hunan, and Jiangxi, as well as in the lower reaches of the Zhujiang (Pearl river) in central Kwangtung. The associations found in each locality are very similar and consist mainly of Osteochilus and Tungtingichthys, as well as of the following reference types: Pararutilus, Aphyocypris, and Aoria. With these occur ostracods, Chara, and higher plants, but very few other higher vertebrates.

2) Association of lower vertebrates, as in the Youganwo series of Maoming. The fish, Chelonia, and crocodiles here form an association quite different from that in the Xiawanpu series; it includes Chelonia and crocodiles not yet found in the latter. The main forms are Cyprinus maomingensis, Anosteira maomingensis, Isometremys, and Tomistoma; there may also be Hypophthalmichthys moritrix C. and V. and Andrias scheuchzeri. At present there is only one known locality, but similar fossils may occur in western Kwangtung (Qinfang area) and in the northern part of Hainan Island.

[4] There would be at least 50 species if the undescribed species from Nanxiong (Kwangtung) and Lunan (Yunnan) were included.

TABLE 1. Simplified table showing correlation of early Tertiary vertebrate-bearing horizons in South China (Concluded)

Middle reaches of Yantse Basin					Southeast	Inner Mongolia	North China	Division outside China	
Hengyang	Xiangxiang	Changliao	Jainghan	Yuanshui	Qujiang			Europe	North America
Series									
Baishajing			?	Ganxian (Kwangsi Province)	Foyun	Tongguer	Shanwang		
					Qujiang formation	Huerjing series		Chattian	Whitneyan
						Batasuo		Stampian	Orellan
						Aerding Ob	Huangu fm.	Sannoisian	Chadronian
Hengyan fm.: Chashanao series	Xiawanpu series	Xiejiashan series	Fangjiahe series	Linjiang series	Beds containing Amyda	(1) (2) (3)	Lushi series	Ludian Bartonian	Duchesnean Uintan
Hengyan fm.: Xialioushi series, Daijiaping	(Tanshi series)	?	Pailoukou series or Yangxi series ?	Xinyu series		Aerbutou series	Guanzhuang ?	Auversian Lutetian	Bridgerian
								Cuisian Sparnacian	Wasatchian
?	?	?	?	?	?			Thanetian	Clarkforkian Torrejonian
								Montian	Dragonian Puercan
Formation: Guanyinqiao	Formation: Tianyan Shan	Cretaceous	Jiayucai fm.	Cretaceous	Zhejiang rhyolite	Paleozoic fm.			

Table 2. Distribution of early Tertiary vertebrate fossils in South China

	Genus and Species	Paleocene E	Paleocene L	Eocene E	Eocene N	Eocene L	Oligocene E	Oligocene M	Oligocene L	Locality
MAMMALIA	*Propaplaeotherium hengyangensis* Y.			? —	—					Hengyang
	Brontotheriidae									
	Metatelmatherium cf. *birmanius* Colbert					—				Tiandong, Tianyang
	Rhinotitan quadridens Xu et Chiu						—			Lunan
	Titanotherium sp. (*Rhinotitan*?)					—				Lunan
	Parabrontops lunanensis C. et H.						—			Lunan
	Chalicotheriidae									
	Eomeropus ulterior Chow					—				Lunan
	Lunania youngi Chow					—				Lunan
	Helaletidae									
	Deperetella sp. (Y. et B.)					—				Lunan
	D. sp. (Chow)					—				Tiandong, Tianyang
	Teleolophus medius M. et G.						—			Lunan
	T. sp. (Y. et B.)					—				Lunan
	Hyracodontidae									
	Caenolophus medius Chow					—				Lunan
	C. sp. (X. et C.)					—				Lunan
	Amynodontidae									
	Amynodon cf. *mongoliensis* Osborn					—				Lunan
	A. sp. (cf. *sinensis*)					—				Lunan
	A. altidens X. et C.						—			Lunan
	A. sp. (X. et C.)						—			Lunan
	Cadurcodon ardynense (Osborn)						—			Qujing
	C. sp. (Xu)						—			Qujing
	Gigantamynodon giganteus X.						—			Qujing
	cf. *Metamynodon* sp. (Xu)						—			Qujing
	cf. *Metamynodon* sp. (Xu et Chiu)						—			Lunan
	Paramynodon cf. *birmanicus* P.					—				Tiandong, Tianyang
	Rhinocerotidae									
	Indricotherium parvum Chow						—			Lunan
	I. intermedium Chiu						?	—		Luoping
	Prohyracodon meridionle C. et X.						—			Lunan
	P. progressa C. et X.						—			Lunan
	Ilianodon lunanensis C. et X.					—				Lunan
	ARTIODACTYLA 偶蹄目									
	Entelodontidae									
	Eoentelodon yunnanense Chow					—				Lunan
	Anthracotheriidae									
	Anthracokeryx birmanicus Pilgrim et Cotter					—				Tiandong, Tianyang
	A. cf. *moriturus* P.					—				Tiandong, Tianyang
	Anthracothema rubrica (P. et C.)					—				Tiandong, Tianyang
	Probachyodus panchiaoensis X. et C.					—				Lunan
	Brachyodus hui (C.)						—			Lunan
	Bothriodon chowi Xu						—			Qujing
	Hypertragulidae									
	cf. *Miomeryx* sp. (Xu)						—			Qujing

Note: There are no Carnivora, as all fossils of indeterminate species and genus have been omitted.

Table 2. Distribution of early Tertiary vertebrate fossils in South China (Concluded)

	Genus and Species	Paleocene E	Paleocene L	Eocene E	Eocene M	Eocene L	Oligocene E	Oligocene M	Oligocene L	Locality
PISCES	CLUPEIFORMES 鲱形目									
	Clupridae									
	Knightia yuyanga Liu			? —	—					Yido
	CYPRINIFORMES 鲤形目									
	Cyprinidae									
	Cyprinus maomingensis Liu					—	?			Maoming
	Osteochilus linliensis Tang					—				Linli, Shimen, Changde, Xiangxiang, Dangyang,
	O. hunanensis Cheng					—				Xiangxiang /Sangshui
	O. sp. (Tang)					—				Sanshui
	Pararutilus orientalis Tang					—				Xiangxiang
	Aphyocypris xiangxiangensis Tang					—				Xiangxiang
	Bagridae									
	Aoria lacus Cheng					—				Xiangxiang
	A. sp. (Tang)					—				Xiangxiang
	PERCIFORMES 鲈形目									
	Tungtingichthys gracilis Liu, Liu et Tang					—				Linli, Xiangxiang, Hengyang?
	T. hsiawanpuensis Cheng					—				Xiangxiang
	T. sp. (Tang)					—				Xigiaoshan (Kwantung)
REPTILIA	CHELONIA 龟鳖目									
	Trionychilidae									
	Anosteira maomingensis Chow et Liu					—				Maoming
	A. lingnanica Young et Chow	? —	—	—						Nanxiong
	A. sp. (Chow)			—						Xinyu
	Aspideretes impressus Yeh					—				Maoming
	Amyda sp. (Chow)					—				Tiandong, Tianyang
	A. sp. (Yeh)			- -	- -	- -	- -	- -	- -	Lanxi
	Trionyx sp. (Chow)			—						Xinyu
	Emydidae									
	Isometremys lacuna Chow et Yeh					—				Maoming
	Emys spp. (Chow)			—						Nanxiong
	CROCODILIA 鳄目									
	Crocodilidae									
	Lianghusuchus hengyangensis Young			? —	—					Hengyang
	Tienosuchus hsiangi Young			?	—					Hengyang
	Tomistoma petrolica Yeh					—				Maoming
	Crocodilus sp. (Chow)			—						Xinyu
MAMMALIA	PANTODONTA 钝脚目									
	Coryphodontidae									
	Coryphodon sp. (Cf. *flerowi*) (Chow)			—	?					Xinyu
	"*Eudinoceras kholobolchiensis* Osborn"			? —	—					Ichang
	DINOCERATA 恐角目									
	Prodinoceratidae									
	?*Probathyopsis sinyuensis* Chow et Tung			—						Xinyu
	PERISSODACTYLA 奇蹄目									
	Palaeotheriidae									

3) Mammal association represented by the Baise Yongfu series, which is almost exactly the Pondaung association of Berma. Its main feature is that it contains representatives of the relatively primitive family Anthracotheriidae: Anthracokery birmanicus, A. moriturus, and Anthracothema rubrica.

4) Mammal association represented by the "Lunan series" and the middle part of the "Anrencuen formation." Most of the forms are relatively standard ones from the late Eocene of Inner Mongolia: Lunania, Deperetella, Teleolophus, Caenolophus and Amynodon; and also Rhinotitan, Eomoropus, Ilianodon, Eoentelodon and Probrachyodus, etc. This may be viewed as an extension of the late Eocene fauna of Inner Mongolia (North China).

Zhou Ming-zhen (1957) examined these associations and concluded that the deposition facies of the horizons must be the principal factor responsible for variations in composition of the associations and for geographic differences between horizons (apart from effects arising solely from the scarcity of mammal fossils for all parts of Asia). For example, the Xiawanpu series and equivalent horizons are facies of inland shrinking lakes; the Youganwo series belongs to the facies of a hot marsh near the sea; the Yongfu series of Tiandong and Tienyang is a river valley facies; and the Lunan series represents the river-lake facies of an inland basin, as do the middle and lower parts of the Anrencuen formation.

Oligocene fossil associations are at present known only for Diandong, where the forms are very close to those found in Inner Mongolia; their stage of evolution indicates that they may be taken as the southern branch of the latter association. The main types are Indricotherium, Parabrontops, Metamynodon, Gigantamynodon, and Cadurcodon.

Some species indicate that South China and adjacent areas could have been radiation or source centers for certain species. For instance, the early Tertiary Cyprinus of South China contains far more species and actual numbers than in other parts of the world. Nikol'skiy (1954) considered that southern Asia and the adjacent part of Africa may have been the site where Cyprinus originated. The current evidence would show that the radiation center was probably East Asia or South China in the early period. In 1959 Zhou Ming-zhen (39) observed for the Eoentelodon of Lunan that East Asia was the source center for the Entelodontidae. This family did not evolve from the late Eocene Eoentelodon into Archaeotherium until the Oligocene, when it crossed the Bering Straits land bridge into North America. Another branch, represented by Entelodon, migrated westward to Europe. The origin of Probrachyodus is the same.

3. Distribution of Deposition Basins

Figure 1 gives a general picture of the distribution of early Tertiary basins in South China; the fossil evidence also shows that there are at least four distinct types:

1) Eastern small basins, including Zhejiang, Fujian, and possibly some in southern Jiangxi and Eastern Kwangtung. These are essentially independent basins, whose lower Tertiary lies with unconformity on late Mesozoic volcanics (Wuzang formation, Zhejang rhyolite). This shows that the late Mesozoic was a time of rapid uplift in this area, which began to break up at the start of the Tertiary to give small isolated basins. These basins have so far yielded very few fossils, so differentiation and dating of horizons are difficult. The exact history remains in doubt.

2) Central large and medium basins. Some of these include large or medium contiguous basins. The strata start with the Cretaceous (Shanlu beds, molasse), showing that the basins were sinking rapidly at that time. This feature persisted at the start of the early Tertiary and continued into the middle Eocene. The uplifting of the surrounding zones and the sinking of the basins gradually halted in the late Eocene; large stable inland lakes or lake areas were formed, with universal deposition of gray-green oolite, oil shale, and evaporites (e.g., gypsum). The basins rose after the Oligocene and deposition ceased.

3) Southern basins near the sea. These include a few medium and small basins in southwestern Kwangtung, the northern part of Hainan Island, and southwestern Kwangsi. The early history was as for the above central basins; in the early Tertiary they continued to sink, with rapid movement persisting until the late Tertiary, when large uplifts occurred and ultrabasic intrusions were formed. The lower Tertiary here consists mainly of red and brown beds, with abundant organic and oil shales, which reflect the humid climate close to a sea. However, the Baise basin lies far inland and closely resembles an inland basin; it may represent a transitional type.

4) Small western basins. These include several small graben-type fragmented-block basins in eastern Yunnan and perhaps south-central Kweichow. The lower Tertiary here always has an unconformity with very much older strata at the bottom of the basin. The oldest strata within the basins are upper Eocene and even Oligocene. These basins ceased to develop after the Tertiary period, except in a few places, where new basins gave rise to lacustrine facies (lignite beds). These basins are now separate and small; but they may once have been contiguous.

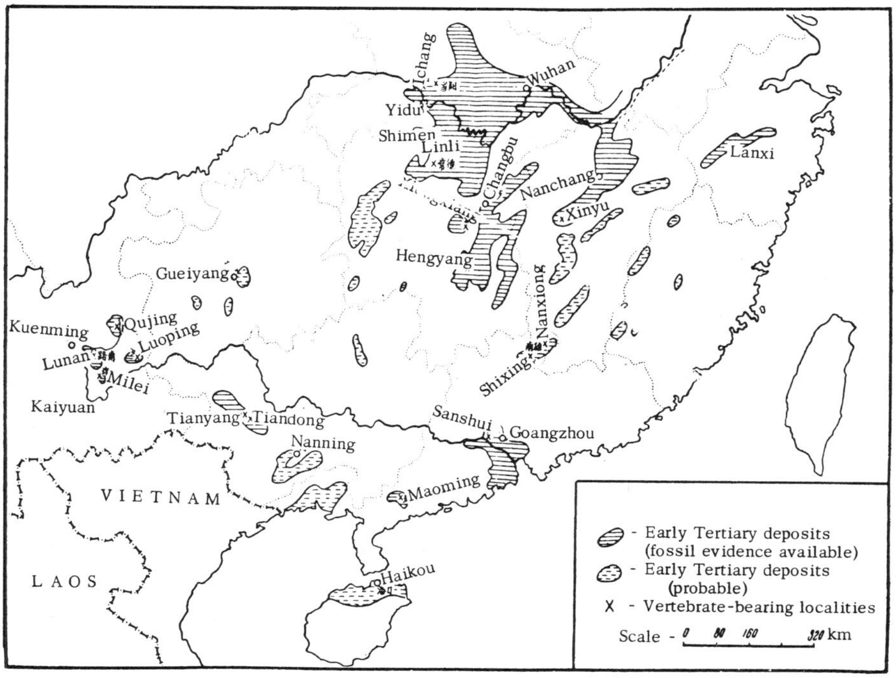

FIGURE 1. Distribution of lakes and basins in early Tertiary of South China

It is thus clear that the early Tertiary of South China was an epoch with two positive zones and one negative one. The eastern and western parts mainly rose and so gave fairly small basins, whereas the central south was a negative (depression) zone, which gave fairly large basins; in the southwest the period and amplitude of the depression were much greater.

The fossil groups of the early Tertiary of China (East Asia) were basically the same as those of North America before the middle Eocene but differences appeared after that time. The differences between the mammals of China and North America were considerable in the late Eocene and Oligocene; it is doubtful whether there was any extensive animal exchange between the two at that time. These conditions are clearly reflected in the early Tertiary mammals of South China.

4) The early Tertiary beds and fauna at many sites in South China indicate clearly that a belt with a dry climate passed over South China at that time. For instance, the fish fossils found in the middle and lower reaches of the Yantse and in Kwangtung have their present-day descendants widely distributed in lower latitudes southeast Asia, southern parts of Kwangtung and Kwangsi), which shows that the early Tertiary climate must have been warmer than the present one in South China. Species closely related to these fossil fish now live in marine or estuarine conditions; some fall between the true marine and true freshwater types and have transitional features. Studies on the fossil flora (28) (spores and pollen) indicate that the climate was hot and dry. There is also evidence that gypsum, glauberite, and other evaporites are intercalated in the early Tertiary at many places in South China; the feldspar content of the beds is high. These facts establish the climate as hot and arid, perhaps with periodic (seasonal) droughts. The dryness not pronounced at the start of the early Tertiary, developed to a maximum in the late Eocene. It may be that the climate gradually became more humid after the Oligocene.

A torried belt thus extended from central Asia via western China to central and southern China; humid conditions occurred to the north and south. Brown (rather than red) beds were formed in the humid zone, as in the Youganwo series of Maoming and the "Guangzhang series" (part) of Shandong or the Fushuen series of the northeast, all of which are similar in their fossils and are clearly of the same type.

This climate gave rise to numerous inland salt lakes during the early Tertiary, in which large amounts of salts were laid down; there were no upper through currents, and so the conditions were those of oxygen lack, which gave rise to the oil shales and salt beds so common in the early Tertiary of South China.

* * *

Our present knowledge is restricted by the lack of fossil evidence. Some horizons have not yet been firmly identified, while some are known only for small and widely separated areas. Very few fossils other than those of vertebrates have been found, and little research has been done. This makes stratigraphic differentiation and international correlation very difficult for the early Tertiary of South China; it also prevents us from drawing detailed conclusions on the history of the period. Many gaps cannot be filled in until fresh fossils are found and new studies have been made.

REFERENCES

1. Zhongguo kexue yuan gujizhui dongwu yu gurenlei yanjiou-suo dideng dongwu yanjiou-shi (Lower Animal Research Laboratory, Paleovertebrate and Paleoanthropology Research Institute, Academia Sinica). Zhongguo jizjui dongwu huashi shouce, yulei, liangshilei, paxinglei bufen (Sections Pisces, Amphibia, Reptilia, Chinese Vertebrate Fossil Handbook). Kexue chuban-she (Science Press), 1961.

2. Zhongguo kexue yuan gujizhui dongwu you gurenlei yanjiou-suo gaodeng dongwu yanjiou shi (Higher Animal Research Laboratory, Paleovertebrate and Paleonanthropology Research Institute, Academia Sinica). Zhongguo jizhui dongwu huashi shouce, buru dongwu bufen (Section Mammalia, Chinese Vertebrate Fossil Handbook): Kexue chyban-she (Science Press), 1960.

3. Liu Dong-sheng, Liu Xian-ting, and Tang Xin, Hunan Linli Lu-xing lei yi xin shu [A NEW GENUS OF PERCIFORMES AT LINLI, HUNAN]: Gujizhui dongwu yu gurenlei (Vertebrata PalAsiatica) v. 6, no. 2, p. 121-129, 1962.

4. Liu Xian-ting, Guangdong maoming de liyu huashi [CYPRINIFORMES FOSSIL FROM MAOMING, KWANGTUNG]: Gujizhui dongwa yu gurenlei (Vertebrata PalAsiatic), v. 3, no. 2, p. 151-153, 1957.

5. ___________, Habei yidu aishiyu (Knightia) de paxien ji qi yiyi DISCOVERY AND SIGNIFICANCE OF KNIGHTIA AT YIDU, HUPEI]: Gujizhui dongwu u gurenlei [Vertebrata PalAsiatica]: v. 7, no. 1, p. 31-37, 1963.

6. Ye Xiang-kuei, Ji Shandong, Zhejiang liang ge bielei huashi [NOTES ON TWO CHELONIA FROM SHANDONG AND ZHEJIANG]: Gujizhui dongwu yu gurenlei (Vertebrata PalAsiatica): v. 6, no. 4, p. 384-388, 1962.

7. Zhou Ming-zhen, Hua-nan shixintong diceng zhi xin ziliao [NEW DATA ON SOUTH CHINA EOCENE BEDS]: Dizhi zhishi (Geological Knowledge); no. 4, p. 19-20, 1956a.

8. ___________, Guanyu Maoming yu he you guan dizhi wenti de buchong cailiao FRESH DATA ON MAOMING FISH AND RELATED GEOLOGY]: Gushengwu xuebao (Acta Paleontologica Sinica): v. 4, no. 2, p. 233-238, 1956b.

9. ___________, Hua-nan di san ji he di si ji chuqi buru dongwu qun de xingzhi he duibi [FEATURES AND CORRELATION OF SOUTH CHINA TERTIARY AND EARLY QUATERNARY MAMMALIA]: Kexue Tongboa (Scientia or Science Bulletin): no. 3, p. 396, 1917.

10. ___________, Yunnan Lunan ceng de shidai wenti [PROBLEM OF DATING LUNAN BEDS IN YUNNAN]: 1958.

11. ___________, Jiangxi Xinyu shixinshi jizhui dongwu huashi de paxian [DISCOVERIES OF EOCENE VERTEBRATA AT XINGU IN JIANGXI]: Gujizhui dongwu yu gurenlei (Vertebrata PalAsiatica]: v. 3, no. 2, p. 79-80, 1959.

12. ___________, Yunnan Lunan disanji yi xin zhong zhuati shoulei [NEW ARTIODACTYLA IN TERTIARY AT LUNAN IN YUNNAN: Gujizhui dongwu yu gurenlei (Vertebrata PalAsiatica]: v. 6, no. 3, p. 219-224, 1962.

13. Zhou Ming-zhen and Liu Jia-long, Guangdong Maoming yi xin zhong shixinshi Anosteira shu yu huashi '[NEW SPECIES OF EOCENE ANOSTEIRA TURTLE FOSSIL FROM MAOMING, KWANGTUNG]: Gushengwu xuebao (Acta Paleontologica Sinica), v. 3, no. 4, p. 275-282, 1955.

14. Zhou Ming-zhen and Ye Xiang-kuei, Guangdong Maoming yuke huashi yi xin shu [NEW GENUS OF FISH FOSSIL FROM MAOMING, KWANGTUNG]: Gujizhui dongwu yu gurenlei (Vertebrata PalAsiatica), v. 6, no. 3, p. 225-231, 1962.

15. Zhou Ming-zhen and Qui Zhan-xiang, Diandong disanji he zao disiji diceng [TERTIARY AND EARLY QUATERNARY BEDS AT DIANDONG]: Zhongguo gushengwu xuehui di er jie daibiao dahui di jiu jie xueshu nianhui luenwen zhaiyao [SUMMARY OF PAPERS AT NINTH ANNUAL SCIENTIFIC CONGRESS OF CHINESE PALEONTOLOGY SOCIETY'S SECOND REPRESENTATIVE GENERAL ASSEMBLY 1962.

16. Zhou Ming-zhen and Hu Cheng-zhi, Yunnan Lunanjianxinshi leishou huashi [OLIGOCENE TITANOTHERIUM FOSSILS AT LUNAN, YUNNAN]: Gushengwu xuebao (Acta Paleontologica Sinica): v. 7, no. 2, p. 85-88, 1959.

17. Zhou Ming-zhen and Shu Yu-xuan, Xinjiang xinpaxian de juxi huashi [NEWLY DISCOVERED GIANT RHINOCEROS FOSSILS IN XINJIANG]: Gujishui dongwu yu gurenlei (Vertebrata PalAsiatica): v. 3, no. 2, p. 69-72, 1959.

18. Zhou Ming-zhen and Shu Yu-xuan, Yunnan Yiliang shixinshi zhenxilei huashi [EOCENE TRUE RHINOCEROS FOSSILS AT YILIANG, YUNNAN]: Gujizhui dongwu yu gurenlei (Vertebrata PalAsiatica): 1961 (4): p. 291-304, 1961.

19. Zhou Ming-zhen and Tong Yong-sheng, Zhongguo shixinshi kongjiao shoulei de xin ciliao [NEW DATA ON EOCENE DINOCERATA]: Gujizhui dongwu yu gurenlei (Vertebrata PalAsiatica): v. 6, no. 4, p. 368-374, 1962.

20. Qiu Zhan-xiang, Ji Yunnan Luoping de juxi huashi bing luen geshi juxi de xingghi [NOTES ON GIANT RHINOCEROS FOSSIL FROM LUOPING, YUNNAN AND COMMENTS ON GESHI VARIETY]: Gujizhui dongwu yu gurenlei (Vertebrata PalAsiatica): v. 6, no. 1, p. 57-71, 1962.

21. Zhang Yu-ping and Tong Yong-sheng, Jiangxi Yuanshui peridi "hong ceng" shidai de tantao [AGE OF RED BEDS OF YUANSHUI BASIN, JIANGXI]: Gujizhui dongwu yu gurenlei (Vertebrata PalAsiatica): v. 7, no. 2, p. 177-181, 1963.

22. Guangdong Nanxiong pendi "hong ceng" de huafen [STRATIGRAPHIC DIVISION OF RED BEDS IN NANXIONG BASIN, KWANGTUNG]: Gujizhui dongwu yu gurenlei (Vertebrata PalAsiatica): v. 7, no. 3, p. 249-260, 1963.

23. Zheng Jia-Jian, Hunan ziangziang zaodisanji yu huashi ji xiawanpu zu de shidai [EARLY TERTIARY FISH FOSSILS AT XIANGXIANG, HUNAN, AND AGE OF XIAWANPU SERIES]: Gujizhui dongwu yu gurenlei (Vertebrata PalAsiatica): v. 6, no. 4, p. 334-346, 1962.

24. Shu Yu-xuan, Yunnan qujing jianxinshi burulei huashi [OLIGOCENE MAMMAL FOSSILS AT QUJING, YUNNAN]: Gujizhui dongwu yu gurenlei (Vertebrata PalAsiatica): v. 5, no. 4, p. 315-325, 1961.

25. __________, Shanxi, Yunnan zaodisanji ji zhong shitan shou [SEVERAL SPECIES OF EARLY TERTIARY COAL FAUNA FROM SHANXI AND YUNNAN: Gujizhui dongwu yu gurenlei (Vertebrata PalAsiatica]: v. 6, no. 3, p. 232-243, 1962.

26. Shu Yu-xuan and Qiu Zhan-xiang, Yunnan Lunan zaodisanji buru dongwu [EARLY TERTIARY MAMMALIA AT LUNAN, YUNNAN]: Gujizhui dongwu yu gurenlei (Vertebrata PalAsiatica): v. 6, no. 4, p. 313-327, 1962.

27. Tang Xin, Hunan Linli like huashi yi xin zhong [NEW SPECIES OF FOSSIL AT LINLI, HUNAN]: Gujizhui dongwu yu gurenlei (Vertebrata PalAsiatica): v. 3, no. 4, p. 211-213, 1959.

28. Si Xing-jian and Li Gong-mo, Hunan disanji wangi zhiwu-qun [FLORA OF HUNAN UPPER TERTIARY]: Gushengwu xuebao (Paleontologica Sinica), v. 2, no. 2, p. 189-206, 1954.

29. Yang Zhong-jian, Jizhui dongwu de yanhua [EVOLUTION OF VERTEBRATES]: Kexue chuban-she (Science Press): 1955.

30. __________, Zhongguo gujizhui dongwu huashi yanjiu de huigu [ASSESSMENT OF ACHIEVEMENTS OF CHINESE PALEOVERTEBRATE FOSSIL RESEARCH]: Kexue tongbao (Scientia or Science Bulletin): no. 23, p. 713-716, 1957.

31. __________, Cong jizhui dongwu huashi tan jiefang yilai zai diceng duibishang de xin fazhan [NEW DEVELOPMENT IN BED CORRELATION IN LIGHT OF VERTEBRATE FOSSIL EVIDENCE SINCE LIBERATION]: Gushenywu xuebao (Paleontologica Sinica): v. 7, no. 5, p. 329-344, 1959.

32. Yang Zhong-jian and Bian Mei-nian, Lunan xinshengdai dizhi [GEOLOGY OF LUNAN CENOZOIC]: Dizhi Luenping (Geological Review): v. 4, no. 3-4, p. 165-172, 1939.

33. Yang Zhong-jian and Zhou Ming-zhen, Yue bei "hong ceng" zhong de jizhui dongwu huashi [VERTEBRATE FOSSILS IN RED BEDS IN NORTH KWANGTUNG]: Gujizhui dongwu yu gurenlei (Vertebrata PalAsiatica): v. 6, no. 2, p. 130-135, 1962.

34. Pei Wen-zhong, Zhou Ming-zhen and Zheng Jia-jian, Zhongguo de xinshengjie [CENOZOIC IN CHINA]: Quanguo diceng hui yi xueshu baogao huipian [NATIONAL (CHINA) STRATIGRAPHY CONGRESS REPORT (SUMMARY)]: Kexue chuban-she (Science Press): 1963.

35. Bien, M. N., PRELIMINARY OBSERVATIONS ON THE CENOZOIC GEOLOGY

OF YUNNAN: Bull. Geol. Soc. China, no. 20, p. 197-204, 1940.

36. Chow, M. C., ON SOME EOCENE AND OLIGOCENE MAMMALS FROM KWANG-SI AND YUNNAN: Vertebrata PalAsiatica, v. 1, no. 3, p. 204-214, 1957.

37. __________, EOENTELODON, A NEW PRIMITIVE ENTELODONT FROM LUNAN, YUNNAN: Vertebrata PalAsiatica, v. 2, no. 1, p. 30-37, 1958a.

38. __________, SOME OLIGOCENE MAMMALS FROM LUNAN, YUNNAN: Vertebrata PalAsiatica, v. 2, no. 4, p. 263-267, 1958b.

39. __________, MAMMALIAN FAUNAS AND CORRELATION OF TERTIARY AND EARLY PLEISTOCENE OF SOUTH CHINA: Jour. Palaeont. Soc. India, no. 31, p. 123-130, 1958c.

40. __________, A XENARTHRAN EDENTATA FROM THE EOCENE OF HONAN: Scientia Sinica, v. 12, no. 12, 1889, 1960.

41. Ping, C., A NEW LAND TURTLE FROM HONAN: Bull. Geol. Soc. China, v. 8, no. 1, p. 241-242, 1928.

42. Teilhard de Chardin, P., and C. C. Young: A MONGOLIAN AMBLYPOD IN THE RED BEDS OF ICHANG: Bull. Geol. Soc. China, no. 15, p. 217-224, 1938.

43. Yeh, H. K., A NEW CROCODILE FROM MAOMING, KWANGTUNG: Vertebrata PalAsiatica, v. 2, no. 4, pp. 237-242, 1958.

44. Young, C. C. ON SOME FOSSIL MAMMALS FROM YUNNAN: Bull. Geol. Soc. China, no. 11, p. 383-393, 1932.

45. __________, A REVIEW OF THE EARLY TERTIARY FORMATIONS OF CHINA: Bull. Geol. Soc. China, no. 13, 1934.

46. __________, NOTES ON THE FIRST EOCENE MAMMALS FROM SOUTH CHINA: Amer. Mus. Novitates, no. 1268, p. 1-3, 1944.

47. __________, FOSSIL CROCODILES IN CHINA, WITH NOTES ON DINOSAURIAN REMAINS ASSOCIATED WITH THE KANSU CROCODILES: Bull. Geol. Soc. China, v. 28, no. 3-4, p. 255-288, 1948.

48. Young, C. C., Bien M. N. and Y. Y. Lee, RED BEDS OF HUNAN: Bull. Geol. Soc. China, v. 18, no. 3-4, p. 259-300, 1938.

49. Young, C. C., and M. N. Bien. NEW HORIZONS OF TERTIARY MAMMALS IN SOUTH CHINA: Proc. 6th Pacific Science Congress, p. 531-534, 1939.

50. Young, C. C. and M. C. Chow. LATEST DISCOVERIES IN VERTEBRATE PALEONTOLOGY IN CHINA. Scientia Sinica. v. 5, no. 6, p. 603-610.

ERRATA

Pages 1346 and 1347 should be transposed. In the column heads on pages 1346 and 1347, E = Early, M = Middle, and L = Late. Page 1346, column 1, line 1 should read *"Propalaeotherium hengyangensis* Y." Page 1348, column 1, line 6 from the top of the page should read *"Anthracokeryx birmanicus, ..."*

6

Reprinted from pages 295-308 of *Implications of Continental Drift to the Earth Sciences*, vol. 1, D. H. Tarling and S. K. Runcorn, eds., Academic Press, London, 1973

SWEEPSTAKES, FILTERS, CORRIDORS, NOAH'S ARKS, AND BEACHED VIKING FUNERAL SHIPS IN PALAEOGEOGRAPHY

Malcolm C. McKenna

The principles of historical biogeography have been developed and shaped by W. D. Matthew (especially 1915, 1939) and G. G. Simpson (especially 1940, 1943, 1946, 1947a, 1947b, 1952, 1953, but exemplified in many other works). More than any other scientist, Simpson has brought order to the subject by codifying its principles and by investigating its processes quantitatively, developing the criteria for recognition of various types of dispersal routes: sweepstakes routes, emphasizing the operation of chance; corridors, emphasizing free movement of a balanced biota; and filters, an intermediate condition. I will not attempt to go into all aspects of Simpson's work here; his papers are clearly stated and no more than a summary is necessary for the present purpose. Simpson's work was based upon the stable continent tectonic model, however, and it is of great importance to examine, even speculatively, the possible effect of recent developments in plate tectonics on his biogeographical work.

The thesis of this paper will be that Simpson's principles of biogeographic analysis still hold, although his application of them was incomplete in certain cases and overextended in others because of his acceptance of the stabilist rationale. Viewed against a plate tectonic background, Simpson's corridors, filters, and sweepstakes actually take on more importance than ever as methods of dispersal that can be identified by biological analysis. Particular corridors, etc., have not always had the same physical configurations as at present, however, and plate motions have rearranged past dispersal routes in ways not believed possible by supporters of the stable continent rationale. In addition, time-delayed one-way dispersal of a balanced fauna is possible under the plate tectonic rationale, although such dispersal was explicitly denied by Simpson (1940).

Corridors, Sweepstakes, and Filters

Simpson set the stage for his quantitative work in a paper published in 1936, in which he compared the living mammalian faunas of the states of Florida and New Mexico (Tables 1, 2). These states of the southern Temperate Zone in North America are, of course, fully connected by land and have been so connected for a long time, but they occupy different positions within the continent and have different climates and environments. Simpson (1936) at first remarked that, '. . . the larger fauna gives a better basis for inference as regards the smaller than does the smaller for the larger . . . ,' but in later studies of faunal resemblance he came to rely on a formula of $100C/N_1$, where C is the number of faunal units at a given taxonomic level common to the two areas and N_1 is the total number of such units in the *smaller* of the two faunas (e.g. Simpson, 1943, p. 20; 1947a; 1960; Peters, 1971). Advantages of this formula are that the smaller of the two faunas is not assumed to be a filtered sample of the larger, and the effect of unequal knowledge of the faunas, especially with fossil examples, is reduced. The formula does not, however, take into account the effects of distance.

Similar figures based upon living mammals were obtained by Simpson in 1953 for the faunal resemblance of Oregon and New York. I have rearranged them here (Tables 3, 4) to agree in format with Tables 1 and 2.

Simpson (1940) utilized his comparison of the mammalian faunas of New Mexico and Florida as an example of what to expect in the analysis of the faunas of both ends of a corridor, a type of dispersal route which he defined as follows: 'If no barrier at all exists between two areas, it is to be expected that their faunas will be very similar, or as far as genera or larger groups are concerned practically identical. Such radical differences as exist will be mainly or wholly caused by the survival or development of local forms in some narrow environment, that is, will be facial and not geographic in a broader sense.' Although his definition stated that taxa as low as the generic level should be 'practically identical,' his own examples (1936, 1953) showed that as little as 29% of the total number of mammalian genera may be held in common at any one time. His values of $100C/N_1$ were 67% in the case of living mammalian genera of Florida/New Mexico and 56% for Oregon/New York. He also gave (Simpson, 1943, p. 20; 1953, p. 58) $100C/N_1$ values of 82% for Ohio/Nebraska and 64% for France/northern China. Clearly, $100C/N_1$ values above 50% or so for mammalian genera characterize corridors several thousand kilometers in length. The longer the overland distance between the two faunas compared, the lower would be the $100C/N_1$ value expected, other factors being considered equal. The France/northern China example thus stands out and, moreover, suggests that additional corrections, notably for relative altitude and latitude, could also be applied.

In 1953 Simpson amplified the corridor concept by remarking that, 'a corridor is a route along which the spread of many or most of the animals of one region to another is probable' (Simpson, 1953, p. 21). He also stated (*ibid*, p. 24), 'where a corridor or filter exists, most or all of the groups adaptively capable of following the route will do so rather promptly.' Remarking about land bridges, of which a corridor would be merely a topologically broad example with little filter action, Simpson noted that such a route 'serves for spread of balanced faunas to occupied areas, and tends to maintain full and balanced faunas on both of the areas connected.'

Sea water (for land mammals) and continuous land (for marine organisms) are certainly formidable barriers, affecting nearly all land or marine organisms alike, but a whole series of lesser barriers exist within those major environments and are the cause of differential filtering during the dispersal process. What might filter one organism within this context might still allow another to pass. Simpson (1940) was especially concerned with the filter action of isthmian links, pointing out that if organisms could not reach such filter bridges in the first place, then they surely could not cross, and that potential travellers are generally drawn from a restricted sample of diversity because the bridge is in a particular position. Filter bridges are nevertheless like corridors in that

TABLE 1 Living mammalian faunas of New Mexico and Florida: Raw data

	New Mexico	Florida	Both States	Total
Orders	6	5	5	6
Families	19	13	12	20
Genera	54	27	18	63
Species	133	33	6	160

TABLE 2 Living mammalian faunas of New Mexico and Florida: Comparisons

	Common to both States (x)	Peculiar to one or the other (100–x)	Florida fauna occurring also in New Mexico ($100C/N_1$)	New Mexico fauna occurring also in Florida ($100C/N_2$)
Orders	83%	17%	100%	83%
Families	60%	40%	92%	63%
Genera	29%	71%	67%	33%
Species	4%	96%	18%	5%

TABLE 3 Living mammalian faunas of Oregon and New York: Raw data

	Oregon	New York	Both States	Total
Orders	5	6	5	6
Families	19	15	14	20
Genera	45	37	25	57
Species	93	50	23	120

TABLE 4 Living mammalian faunas of Oregon and New York: Comparisons

	Common to both States (x)	Peculiar to one or the other (100–x)	New York fauna occurring also in Oregon ($100C/N_1$)	Oregon fauna occurring also in New York ($100C/N_2$)
Orders	83%	17%	100%	83%
Families	70%	30%	74%	93%
Genera	44%	56%	56%	68%
Species	19%	81%	25%	46%

dispersal is multiple, not completely unbalanced ecologically, two-way in action, and relatively deterministic.

A second type of dispersal discussed by Simpson is the 'sweepstakes', sometimes known as waif dispersal. This (e.g. for land mammals) is characteristically over long distances of water, but may of course be across any formidable barrier, depending on the organism. Sweepstakes dispersal is inherently improbable although not impossible at any one time, but with sufficient time sweepstakes dispersal becomes probable,

TABLE 5 Principal Cenozoic corridors, sweepstakes, and filters under the stable continent rationale

Corridors	Filters	Sweepstakes
Europe/China since about 30 m.y.	Isthmus of Panama since about 4 m.y.	Panama/Colombia prior to about 4 m.y.
Eurasia/Beringia/North America at 49–54 m.y.	Beringia with minor exceptions, whole Cenozoic	Beringia when flooded
		Africa/Madagascar, whole Cenozoic
Each continent within itself except when completely split by epicontinental seas (e.g. Turgai Straits)	Various examples within continents, such as Late Cenozoic Sahara	West Indies/Americas, whole Cenozoic
		Eurasia/Africa, Early Cenozoic
	Eurasia/Africa via Sinai (and possibly elsewhere) in Late Cenozoic	Asia/Australia, whole Cenozoic
		Epicontinental seas completely isolating sections of continents (e.g. Turgai Straits)

although random and unpredictable. Under such conditions balanced faunas do not spread together, but individual taxa of a biota may spread piecemeal and sporadically. In contrast to corridors, travel is one-way at any one time, and organisms that potentially could cross the barrier in either direction do not necessarily do so immediately, if at all.

Corridor dispersal and sweepstakes dispersal may be viewed, then, as near the opposite ends of a spectrum of dispersal probabilities. The sweepstakes is erratic; *ex post facto* description of the long term results of such dispersal is similar, in certain respects, to description of diffusion of a few atoms of gas and to the results of just a few throws of dice. In contrast, equilibria tend to be reached in corridors, though the equilibria are momentary and constantly affected by shifting physical and ecological conditions and by changes in biotic composition generated by the evolution of new kinds of organisms with new tolerances, by local extinctions, and by arrivals of new taxa to one part or another of the corridor from outside the system.

Armed with quantitative methods for estimating from biological data whether past dispersals had been by corridor, filter bridge, or sweepstakes, Simpson proceeded to show convincingly that the data from mammalian distribution were not in harmony with most proposed Tertiary transoceanic bridges nor with transoceanic continents. Developing the ideas of Matthew, he argued persuasively for northern dispersal routes coupled with sweepstakes routes as sufficient to explain all mammalian distribution.

The principal Cenozoic corridors, sweepstakes, and filters of the world under the stable continent rationale espoused by Simpson would be those shown in Table 5. Various additional, minor filters and sweepstakes could be listed. Stable continent supporters made the additional assumption that essentially these same relations would have held in pre-Cenozoic time as well, modified only by vertical crustal motion, volcanic piles, fold mountains, etc., but not by large scale horizontal crustal motions.

The final sentence in Simpson's (1943) paper on mammals and the nature of continents was 'The distribution of mammals definitely supports the hypothesis that continents were essentially stable throughout the whole time involved in mammalian history.' Alex du Toit's reaction was not long in coming: 'Simpson has merely shown that such inter-continental linkages could not have persisted as long, nor have permitted terrestrial migration as freely, as believed, but he has by no means disproved the probability of Continental Drift, which has been based on a wealth of other evidence, much of it long antedating the evolution of the mammals' (du Toit, 1944, p. 145). In retrospect, we can see that Simpson overgeneralized in attempting to deal with continental drift, extending his conclusions too far into the past.

Biogeography against a Plate Tectonic Backdrop

The revolution in geology, which has occurred mostly since 1960 and which will be taken for granted here, has not changed the basic principles developed by Simpson for biological methods of recognition of corridors, filter bridges, and sweepstakes routes, but it does offer a number of geophysically based models of ancient landmass configurations which are either actually or potentially testable by biogeographic evidence. The time scale of plate tectonics, moreover, is such that the times of final separation of certain present landmasses, even of continental dimensions (Australia/Antarctica, Europe/North America), and the collisions of others is within the time span of the Mammalia (Heirtzler *et al.*, 1968; Pitman & Talwani, 1972; McKenna, 1972b). Mammalian evidence has become more complete in recent years as well, particularly in the Cretaceous of North America, Asia, and, just recently, South America (Sigé, 1971). Although it is true that most of the now major effects of continental separation were initiated in pre-Cenozoic times, it is not true that this was prior to 'the evolution of mammals.' Mammals have been on earth since the end of the Triassic, and the boundary between the Mammalia and their ancestors, the therapsid reptiles, is merely a matter of definition. It is true, to be sure, that the usefulness of mammals in discussions of palaeogeography is still mainly confined to the Cretaceous and Cenozoic, where their record is sufficiently good, but mammals, as well as other organisms not so well represented yet in the known fossil record, certainly have been present as passengers on moving plates. It is of considerable importance to begin to investigate what effects, if any, such movements might have had upon mammalian distribution and to what degree modifications in distribution caused basically by plate separations and recombinations or by the crossing of latitude belts are now discernible. Not only must we try to mesh concepts of filter, corridor, and sweepstakes action with plate tectonic theory, but also we must be alert for other effects on dispersal that may be inherent in the plate tectonic model.

Under the plate tectonics rationale the principal Cenozoic dispersal routes would be those listed in Table 6. If the list were extended to the beginning of the Palaeozoic it would be several times longer. The landmasses listed in Table 6 have been in relative motion and the timing of final separation, culled here mostly from the geophysical literature, is shown in Fig. 1. Geophysical and geological estimates vary, but I have

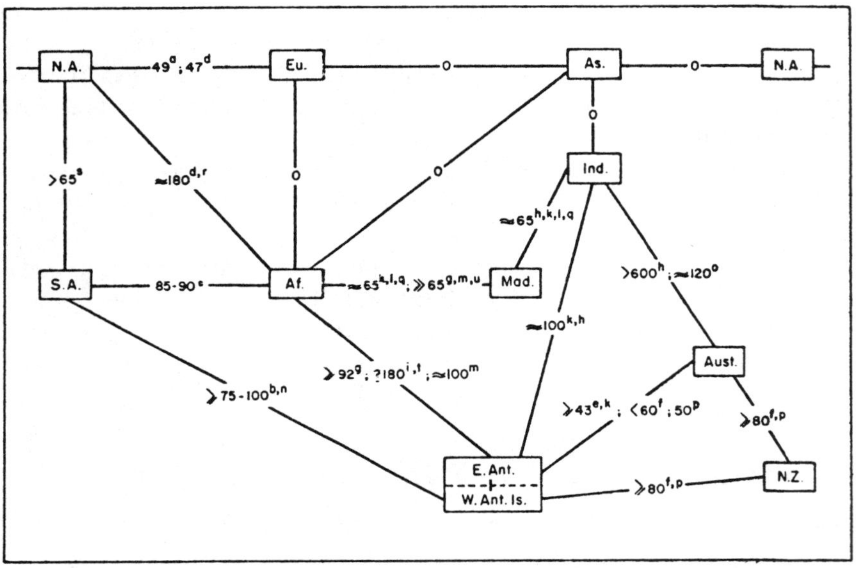

FIGURE 1 Approximate times in m.y. since final separation of major landmasses, exclusive of connections caused by new crust (e.g. Panama) and breaks in surface continuity caused by former epicontinental seas now sundered tectonically. Sea-floor spreading began earlier than final separation in certain cases where landmasses slid apart along R–R transform faults. Biogeographic continuity could have been broken any time from long before to somewhat after the time of onset of drift. a: McKenna, 1972b. b: estimate, could be much more. c: Reyment, this volume, non-palaeontological portion of his evidence (but the rest agrees). d: Pitman & Talwani, 1972. e: Heirtzler et al., 1968; McElhinny & Wellman, 1969. f: Griffiths, 1971. g: Tarling, 1971. h: Crawford, 1971. k: McElhinny, 1970. l: Heirtzler & Burroughs, 1971. m: Le Pichon & Heirtzler, 1968. n: Halpern, 1968. o: Veevers, 1971. p: Griffiths & Varne, 1972. q: Smith & Hallam, 1970. r: Coney, 1971. s: Malfait & Dinkelman, 1972. t: Dietz & Sproll, 1970. u: Green, 1972.

attempted to select the values used here for geophysical or geological reasons, not palaeontological reasons, in order to avoid circularity. Most of the numerical values (in m.y.) accepted here fall within the time span of the Mammalia. It will be important in future analyses, therefore, to inquire whether the biogeographic conditions at those times were those of corridors, filters, or sweepstakes, and whether additional types of dispersal may be envisioned. Preliminary investigations of the Europe/North America connection, which I have called (McKenna, 1971, 1972a, 1972b) the De Geer route after the De Geer Line (in part = Spitsbergen Fracture Zone), indicate that on the basis of fossil mammals as well as geophysics it was indeed a corridor about 50 m.y. ago (Szalay & McKenna, 1971). As can be seen in Table 6, there are a number of complex and changing situations still to be investigated, of which the Australian, Indian, and Madagascar examples are the most attractive. Unfortunately, some recent papers on the origin of the faunas of these areas have been rather naïve (e.g. Martin, 1970; Fooden, 1972) in that little or no account of Simpson's principles has been taken. Increased knowledge of the East Antarctic craton and the West Antarctic archipelago, the Scotia Arc, and various aspects of fragmentation of the Gondwana supercontinent, based on the plate tectonics synthesis, offers hope at present for understanding the physical geography existing at the time of origin of many groups of organisms, both

TABLE 6 Principal Cenozoic corridors, sweepstakes, filters, and arks under the plate tectonic rationale

Corridors	Filters	Sweepstakes	Arks
Europe/China since about 30 m.y.	Isthmus of Panama since about 4 m.y.	Central America/Colombia, whole Cenozoic but increasingly effective until becoming filter	India completes trip northward, docks with Asia in Cenozoic
Europe/North America 49–54 m.y.	Europe/North America > 54 m.y. (climatic)		Australia and New Zealand move northward, the former now beginning to dock with Asia
Each continent within itself except when completely split by epicontinental seas (e.g. Turgai Straits)	Beringia with minor exceptions, whole Cenozoic	Europe/North America briefly, just after 49 m.y.	
?East Antarctica/ Australia, earliest Cenozoic	Turgai Straits, immediately following marine conditions of early Cenozoic	Beringia when flooded	?Arabian peninsula, from Africa to Asia, late Cenozoic (Cooke, 1968)
		Australia/Antarctica in early Cenozoic	
	Various examples within continents, such as late Cenozoic Sahara	Australia/Asia in late Cenozoic	South America moves westward, connects with Central America at about 4 m.y.
	Africa/Eurasia, mid Cenozoic	Africa/Magagascar, whole Cenozoic	
	Africa/Asia, latest Cenozoic	Madagascar/India, early Cenozoic	
		?South America/Africa (only a few forms at most, and possibly none)	
		West Indies/Americas, whole Cenozoic	

plant and animal, even as late as the beginning of the Tertiary, provided that sweepstakes are given adequate consideration. Nevertheless, caution is called for: while it now seems probable that marsupials had a southern origin, it may also be premature to select a particular southern continent, as various recent authors have done (Cox, 1970; Jardine and McKenzie, 1972; Fooden, 1972).

Noah's Arks and Viking Funeral Ships

In his famous paper on mammals and land bridges, Simpson (1940, p. 152) discussed the two-way nature of land bridges and noted that the hypothetical ancient dispersal route between South America and Australia via Antarctica did not pass that test. In a footnote he stated, 'It is conceivable that a bridge might function in one direction by a sort of lock or storm-door action, an otherwise uninhabited region receiving a fauna first from one source, losing that connection, and only then being united with a second continent, so that animals would be transported from the first to the second but not in the other direction. There is, however, no good evidence that such a peculiar sequence of events ever actually happened and it should hardly be postulated except in the absence of any acceptable alternative hypothesis.' Presumably, vertical crustal movements were what Simpson had in mind to produce the physical sequence of land distribution permitting such one-way transfer. Under the plate tectonic synthesis, however, it is possible and even probable that over a period of time much the same one-way effect can occur, generated instead by horizontal, rather than vertical, plate motions.

If the ocean floor moves, carrying away newly created crust from mid-ocean ridges in a sort of bilaterally symmetrical conveyor belt system that is ultimately to sink once more at the oceanic trenches, and if such generating ridges extend beneath continents at certain times and places, then we have a mechanism for occasional continental fragmentation and possible ultimate recombination. Like rocks carried along on moving glaciers, patches of continental crust of almost any size are passively transported. If these are split off from other continental crust, the fauna of the fragment will sample a substantial fraction of the biota of the parent mass, depending mainly on the sizes of the two daughter masses. Then each will move away from the other. As the two masses separate, the nature of the dispersal route changes from a corridor to a stage of filter action, and then finally to sweepstakes conditions. Relative plate motions are generally between 10^{-5} and 10^{-4} km/year, so trips of 1000 to 10,000 km are possible in 100 m.y., which is about half the time mammals have been present on the earth. Inasmuch as other plates would also be moving, motions toward one another would result in oceanic isolation of shorter duration before two continental fragments from distant areas met. When a fragment of one continent came toward another landmass the sequences of sweepstakes, filter, and corridor conditions would be enacted in an order reversed from the events at separation. The two biotas would merge and strongly interact. If its landmass were large enough, a Noah's ark, to use an almost inescapable term, would represent a sort of one-way lock, permitting one-way dispersal of a balanced biota (Fig. 2). This one-way dispersal would involve a considerable time delay, during which the transported biota would evolve, and latitudinal components of motion would conceivably affect available ecospace via the effects of climatic change. The time required for Noah's ark journeys would be of epoch to era magnitude, and,

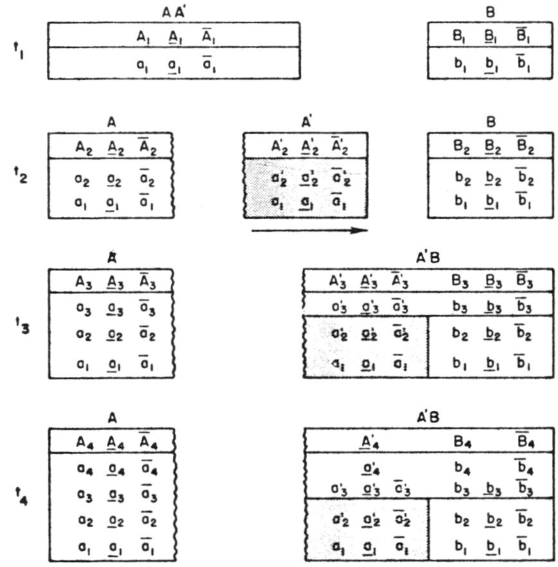

FIGURE 2 Abstract history of two hypothetical landmasses, A and B, and a Noah's ark moving from one to the other, A′. AA′ splits and by time t_3 A′ joins B to form A′B. Waif dispersals in transit are ignored. Extinctions lead to a less numerous fauna on A′B by time t_4. Analysis of the living faunas conducted at time t_4 might erroneously be interpreted to suggest either that at some time before t_4 part of the fauna of 'B' (unrecognized to be A′B) had given rise via a filter or sweepstakes route to the fauna of A, which had then radiated to give rise to A_4, $\underline{A}_4$, and $\bar{A}_4$, or that an ancestor of A'_4 had reached 'B' by sweepstakes action or by an indirect overland filter. Physical evidence of ark motion and collision, coupled with detailed knowledge of the distribution within A′B of fossils (particularly with regard to position on one side or the other of the A′/B suture) would be necessary to decipher the true history. Fossils a_1, $\underline{a}_1$, and $\bar{a}_1$ would have once had a more restricted distribution when living as A_1, $\underline{A}_1$, and $\bar{A}_1$ than would be apparent under the stable continent rationale. Although the fossils a_1, $\underline{a}_1$, and $\bar{a}_1$ occur in the fossil record of 'B', A_1, A_1, and $\bar{A}_1$ never lived there. Their remains were transported as if in a Viking funeral ship (A′), to be beached when A′B ('B') was formed. Still further complications would be generated in case A′B ever separated once more along a line not exactly at the A′/B suture. The obvious analogy here is the history of fender paint in a two-vehicle hit-and-run collision.
Capitals without subscripts stand for landmasses. Capitals with subscripts represent the surviving fauna of the appropriate single or compound landmass at the time indicated by the subscript. Lower case letters indicate fossils, whose occurrence is cumulative. Note that fossils deposited in times t_1 and t_2 ultimately come to be found in discrete parts of A′B, separated by a suture (but perhaps juxtaposed by thrusting or mixed in a tectonic mélange), whereas the fossils of times t_3 and t_4 could occur throughout the landmass.

for this reason, the more long-term the residence of a lineage on any particular landmass the greater the chance that ark boarding would have occurred before relative motion began. The nature of ark boarding, however, would be a matter of interpretation on the basis of Simpson's various palaeobiogeographic principles. The last time of dry-land connection between two moving landmasses is not necessarily the last point of genetic continuity between isolated evolving lineages, nor is the time when separation began necessarily the last time of mutual contact of dry land.

A common misconception about 'continental drift' is that if two landmasses have separated, then just before separation dry land, now rifted, must have connected them. Rifting may theoretically commence in the midst of an epicontinental sea as well as anywhere, however, so that in certain cases isolation may be more ancient than the

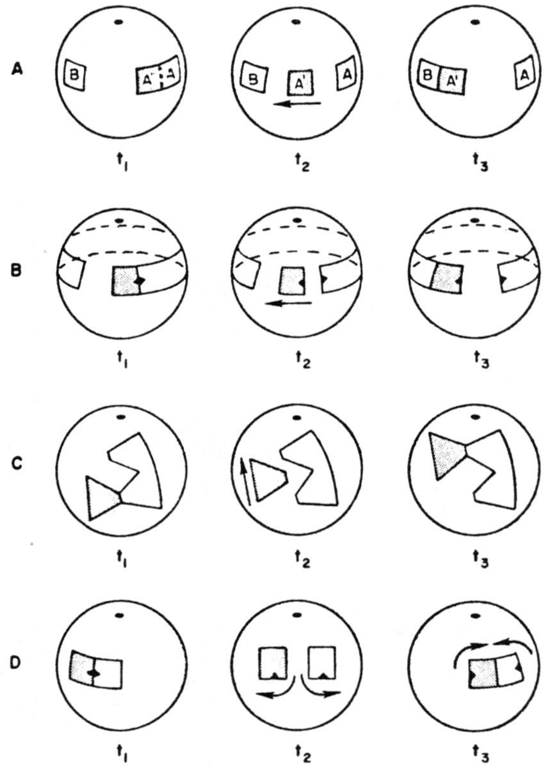

FIGURE 3 Variants of hypothetical Noah's arks. A, simple case; B, somewhat more complicated case in which a fragment of one end of an elongate landmass strikes its other end after a period of isolated drift; C, small scale version of the same (but note that the same end of the ark docks, whereas in B the opposite ends dock); D, small scale exact topological equivalent of B, which, for organism dispersal pattern equivalence, would require complex rotation and translation. Obviously, the time scale and absolute sizes of the landmasses, among other factors, would determine whether these various motions would leave biogeographically detectable results.

time of initiation of drift. For this reason, some of the ark passengers may well have boarded via filter or sweepstakes action while the ark was anchored in the harbor, and other potential passengers will be left behind partly by chance. The presence of *Sphenodon* and certain ratites, but not marsupials, in New Zealand presumably represents a West Antarctic archipelagic sweepstakes dispersal route. On the other hand, the presence of several kinds of monotremes of widely differing morphology in Australia and New Guinea (echidna, platypus, *Ektopodon*) may indicate dry land boarding prior to Australia's Cenozoic northward journey toward a tropical climate, away from Antarctica.

Any landmass splitting away from another and ultimately joining a third would serve to illustrate the Noah's ark principle. Even a separation followed by a reunion at a different contact point would illustrate the point (Fig. 3).

On the basis of geological and geophysical evidence, prime examples of large Noah's arks that have docked during the Cenozoic or are in the process of docking would be Australia and India (Veevers, 1971; Takin, 1972; Crawford, 1971). If one allows two such arks per 100 m.y., one might expect evidence of perhaps a dozen similar situations on the same scale since the beginning of the Palaeozoic. Three arks, now long since docked with North America, have been suggested by J. T. Wilson (1968, Fig. 6). These are composed of certain allochthonous Palaeozoic rocks in Oregon, Florida, and in the area of Newfoundland. On a smaller time scale and involving smaller landmasses and less motion, geological and geophysical evidence suggest that Greenland may have shifted its continental allegiance from North America to Europe a few million years before it began moving westward once more with the North American plate (Pitman & Talwani, 1972). The motion of the Arabian Penninsula (Cooke, 1968) and, on a still smaller scale, the Danakil horst in the Red Sea (Tazieff *et al.*, 1972), certain Mediterranean islands (Hsü, 1971; Alvarez, 1972), and various blocks moved across water gaps throughout the world by transcurrent faulting, might serve as small scale examples.

The grounded Viking funeral ship principle can be seen in Fig. 2 as a consequence of ark docking. The principle is explained in the legend of Fig. 2; it affects fossils, ancient sedimentary environments, and palaeomagnetic measurements alike.

Tilted Latitudes and Filter Bridge History

Irving (1964), McElhinny & Wellman (1969), Creer (1970), and many other authors have provided summaries of 'polar-wander paths' (which might better be known as

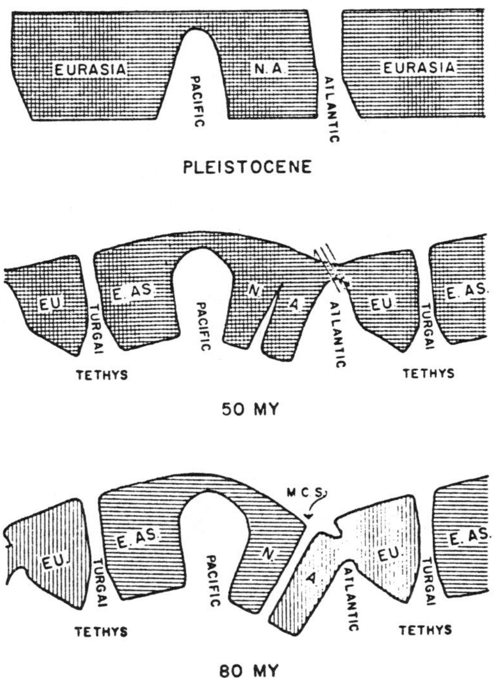

FIGURE 4 Crude 'palaeomercator' sketches of Eurasian and North American landmass distribution (minus India) at three times in the past, incorporating concepts of differential plate motion, 'polar wander' and epicontinental sea redistribution. Beringia was formerly at a higher palaeolatitude. The depicted ancient sinuosity of these landmasses is an artifact of shifted polar coordinates. M.C.S. = North American Mid-Continent Seaway. Lined areas indicate former endemism. Crosshatched area indicates surviving mixed fauna,

'lithosphere-wander paths' instead). To cite only one example having a bearing on mammalian distribution, numerous studies summarized by these authors show that the calculated position of the North American and Eurasian plates (Holarctica) has shifted with regard to the rotational pole and that in the Late Cretaceous the rotational pole (calculated by averaging many virtual geomagnetic pole positions) lay near Beringia. If so, two effects are inescapable. The first is that in Beringia winter darkness would have alternated with summer light to a degree greater than now, and the second is that present latitude belts and land distribution relative to the present coordinate system would make sweeping curves if plotted on maps based upon palaeopolar coordinates. A crudely sketched sequence of 'palaeomercator' projections of Holarctica is shown in Fig. 4. In this figure plate separation in the North Atlantic (Pitman & Talwani, 1972) and epicontinental sea distribution are incorporated. Insofar as palaeolatitude is concerned, in the Late Mesozoic and Early Tertiary the Bering filter bridge would presumably have been a more effective filter for mammals than the North Atlantic De Geer route from Greenland to Europe via Spitsbergen and the Barents Shelf (Szalay & McKenna, 1971; McKenna, 1971, 1972a, 1972b) because of the higher palaeolatitude of the former. During the Tertiary the pole position shifted away from Beringia, compensating to an unknown extent for increased climatic zonation. The same shift re-inforced the Tertiary increase in climatic zonation in the Greenland area, thus reducing chances of sweepstakes dispersal in that area after 49 m.y. when the De Geer route ceased as the Barents Shelf and Spitsbergen cleared northern Greenland and Ellesmere Island by transform fault motion along the complex Spitsbergen fracture zone.

Concluding Remarks

Although Matthew's and Simpson's principles of palaeobiogeography were propounded within the stable continent framework, they are not only compatible with the plate tectonics synthesis but also can be extended to include a type of organism dispersal denied by Simpson (Noah's arks) and a type of transfer of fossils from one landmass to another that Simpson did not consider at all (grounded Viking funeral ships). In addition, latitude belts, a major factor in filtering, have not always been where they are now because of 'polar wandering' of the rotational pole (actually lithosphere motion with regard to the earth's axis).

Organism isolation or dispersal via plate interactions does not necessarily coincide with the time of onset of plate separation or of plate collision, but can occur significantly before or after such events and may not occur at all. Many other factors beside physical geography affect organism dispersal. Palaeobiogeographic analyses utilizing the techniques pioneered by Matthew and by Simpson have much to contribute to a fuller understanding of Natural History, of which Geology and Biology are but component parts.

References

Alvarez, W., 1972. Rotation of the Corsica-Sardinia microplate, *Nature, Phys. Sci.*, **235**, 103–5.

Coney, P. J., 1971. Cordilleran tectonic transitions and motion of the North American plate, *Nature*, **233**, 462–5.

Cooke, H. B. S., 1968. The fossil mammal fauna of Africa, *Quart. Rev. Biol.*, **43**, no. 3, 234–64.
Cox, C. B., 1970. Migrating marsupials and drifting continents, *Nature*, **226**, 767–70.
Crawford, A. R., 1971. Gondwanaland and the growth of India, *J. Geol. Soc. India (Bangalore)*, **12**, no. 3, 205–21.
Creer, K. M., 1970. A review of palaeomagnetism, *Earth-Sci. Rev.*, **6**, 369–466.
Dietz, R. S. and Sproll, W. P., 1970. Fit between Africa and Antarctica: a continental drift reconstruction, *Science*, **167**, 1612–14.
Du Toit, A., 1944. Tertiary mammals and continental drift. A rejoinder to George G. Simpson, *Amer. Jour. Sci.*, **242**, 145–63.
Fooden, J., 1972. Break-up of Pangaea and isolation of relict mammals in Australia, South America, and Madagascar, *Science*, **175**, 894–8.
Green, A. G., 1972. Seafloor spreading in the Mozambique Channel, *Nature, Phys. Sci.*, **236**, no. 63, 19–21, 32.
Griffiths, J. R., 1971. Reconstruction of the south-west Pacific margin of Gondwanaland, *Nature*, **234**, 203–7.
Griffiths, J. R. and Varne, R., 1972. Evolution of the Tasman Sea, Macquarie Ridge, and Alpine Fault, *Nature, Phys. Sci.*, **235**, no. 57, 83–86.
Halpern, M., 1968. Ages of Antarctic and Argentine rocks bearing on continental drift, *Earth Planet. Sci. Lett.*, **5**, 159–67.
Heirtzler, J. R. and Burroughs, R. H., 1971. Madagascar's palaeoposition: new data from the Mozambique Channel, *Science*, **174**, 488–90.
Heirtzler, J. R., Dickson, G. O., Herron, E. M., Pitman, W. C. III, and Le Pichon, X., 1968. Marine magnetic anomalies, geomagnetic field reversals, and motions of the ocean floor and continents, *J. Geophys. Res.*, 73, 2119–36.
Hsü, K. J., 1971. Origin of the Alps and western Mediterranian, *Nature*, **233**, 44–48.
Irving, E., 1964. Paleomagnetism and its Application to Geological and Geophysical Problems. John Wiley and Sons, Inc., New York, xvi + 399 pp.
Jardine N. and McKenzie, D., 1972. Continental drift and the dispersal and evolution of organisms, *Nature*, **235**, 20–24.
Le Pichon, X. and Heirtzler, J. R., 1968. Magnetic anomalies in the Indian Ocean and sea-floor spreading, *J. Geophys. Res.*, **73**, 2101–17.
Malfait, B. T. and Dinkelman, M. G., 1972. Circum-Caribbean tectonic and igneous activity and the evolution of the Caribbean plate, *Geol. Soc. Amer. Bull.*, **83**, no. 2, 251–72.
Martin, P. G., 1970. The Darwin Rise hypothesis of the biogeographical dispersal of marsupials, *Nature*, **225**, 197–8.
Matthew, W. D., 1915. Climate and evolution, *Ann. New York Acad. Sci.*, **24**, 171–318.
Matthew, W. D., 1939. Climate and evolution. Second edition, revised and enlarged. Spec. Publ. New York Acad. Sci., **1**, pp. i–xii, 1–223.
McElhinny, M. W., 1970. Formation of the Indian Ocean, *Nature*, **228**, 977–9.
McElhinny, M. W. and Wellman, P., 1969. Polar wandering and sea-floor spreading in the southern Indian Ocean, *Earth Planet. Sci. Lett.*, **6**, 198–204.
McKenna, M. C., 1971. Fossil mammals and the Eocene demise of the De Geer North Atlantic dispersal route, *Geol. Soc. Amer. Abstr. with Programs*, **3**, no. 7, 644.
McKenna, M. C., 1972a. Was Europe connected directly to North America prior to the middle Eocene? *In:* Advances in Evolutionary Biology, **6**, edited by Hecht, M. (in press).
McKenna, M. C., 1972b. Eocene final separation of the Eurasian and Greenland-North American landmasses, *24th Internatl. Geol. Congr.*, Sect. 7, 275–81.
Peters, J. A., 1971. A new approach in the analysis of biogeographic data, *Smithsonian Contrib. Zool.*, no. 107, 1–28.
Pitman, W. C., III, and Talwani, M., 1972. Sea-floor spreading in the North Atlantic, *Geol. Soc. Amer. Bull.*, **83**, no. 3, 619–46.
Reyment, R. A. (This volume). Palaeontological dating of the opening of the South Atlantic Ocean.
Sigé, M. B., 1971. Les Didelphoidea de Laguna Umayo (formation Vilquechico, Crétacé supérieur, Pérou), et le peuplement marsupial d'Amérique du Sud, *Compt. Rendu Acad. Sci. Paris*, **273**, ser. D, 2479–81.

Simpson, G. G., 1936. Data on the relationships of local and continental mammalian faunas, *J. Paleont.*, **10**, no. 5, 410–14.
Simpson, G. G., 1940. Mammals and land bridges, *J. Washington Acad. Sci.*, **30**, no. 4, 137–63.
Simpson, G. G., 1943. Mammals and the nature of continents, *Amer. J. Sci.*, **241**, 1–31.
Simpson, G. G., 1946. Tertiary land bridges, *Trans. New York Acad. Sci., ser. II*, **8**, no. 8, 255–8.
Simpson, G. G., 1947a. Holarctic mammalian faunas and continental relationships during the Cenozoic, *Bull. Geol. Soc. Amer.*, **58**, 613–88.
Simpson, G. G., 1947b. Evolution, interchange, and resemblance of the North American and Eurasian Cenozoic mammalian faunas, *Evolution*, **1**, no. 3, 218–20.
Simpson, G. G., 1952. Probabilities of dispersal in geologic time. *In:* The problem of land connections across the South Atlantic, with special reference to the Mesozoic, edited by E. Mayr, *Bull. Amer. Mus. Nat. Hist.*, **99**, art. 3, 163–76.
Simpson, G. G., 1953. Evolution and geography. Condon Lectures, Oregon State Syst. Higher Educ., 1–64.
Simpson, G. G., 1960. Notes on the measurement of faunal resemblance, *Amer. J. Sci.*, Bradley Volume, no. 258-A, 300–11.
Smith, A. G. and Hallam, A., 1970. The fit of the southern continents, *Nature*, **225**, 139–44.
Szalay, F. S. and McKenna, M. C., 1971. Beginning of the age of mammals in Asia: the Late Palaeocene Gashato fauna, Mongolia, *Bull. Amer. Mus. Nat. Hist.*, **144**, art. 4, 269–318.
Takin, M., 1972. Iranian geology and continental drift in the Middle East, *Nature*, **235**, 147–50.
Tarling, D. H., 1971. Gondwanaland, palaeomagnetism and continental drift, *Nature*, **229**, 17–21, 71.
Tazieff, H., Varet, J., Barberi, F. and Giglia, G., 1972. Tectonic significance of the Afar (or Danakil) Depression, *Nature*, **235**, 144–7.
Veevers, J. J., 1971. Phanerozoic history of western Australia related to continental drift, *J. Geol. Soc. Australia*, **18**, part 2, 87–96.
Wilson, J. T., 1968. Static or mobile earth: the current scientific revolution, *Proc. Amer. Philos. Soc.*, **112**, no. 5, 309–20.

7

Reprinted from *Soc. Sci. Fennica Commentat. Biol.* **31**:1-8 (1967)

Continental drift and the palaeogeography of reptiles and mammals

Björn Kurtén
Department of Zoology
University of Helsinki/Helsingfors

Introduction

The study of continental drift has recently been revitalized by the research on palaeomagnetism, and there can now be little doubt that drift has in fact occurred (Runcorn 1961, Irving 1964). In the early Mesozoic, the continents of the earth were still gathered in two supercontinents, Laurasia in the north and Gondwanaland in the south, separated by the Tethys sea. Of the two, Gondwanaland was the largest, incorporating the land masses that now form South America, Africa, Antarctica, the Indian peninsula, and Australia; while Laurasia consisted of North America, Europe, and Asia north of the Tethys. Redistribution of the continents commenced in the Mesozoic, and a phase of comparatively rapid drift was enacted in the Cretaceous and early Tertiary. Laurasia split up into North America and Eurasia north of the Tethys. The five southern continents drifted apart, the Indian peninsula being finally added to the land mass of Asia.

There has been little serious study of the effects of these concepts on animal palaeogeography and the evolution of the Tetrapoda. Yet it is obvious that palaeogeography is of basic importance in this respect. There is, for instance, the possibility that land vertebrates arose from quite different stocks of crossopterygians in Laurasia and Gondwanaland. A diphyletic (or polyphyletic) origin of tetrapods has in fact been long advocated by Jarvik (1942).

The present contribution is a comparison between the radiation of the reptiles (late Carboniferous to Cretaceous) and that of the mammals (late

Cretaceous and Cenozoic). These two radiations took place in radically different geographic settings. The reptilian radiation occurred while the land masses of the earth were collected into two enormous blocks, the mammalian while they were split into fragments.

At least thirty-three mammalian orders have been in existence during the 100 million years or so of late Cretaceous and Cenozoic time, so that the relationship between ordinal variety and time is 1:3 for the mammals. The Age of Reptiles lasted more than 200 million years, and in that time about 20 orders were produced, so that the relationship is 1:10.

It may, of course, be maintained that reptilian and mammalian orders are not exactly equivalent. Perhaps it would be more correct to use the reptilian suborder instead, in which case the relationship becomes 1:5 approximately. The mammals still come out superior in the rate of radiation.

Can this be correlated with the greater fragmentation of the continents during the mammalian radiation? What is in fact the relationship between numbers of continents, sizes of continents, and production of tetrapod orders? To answer these questions, we must attempt to determine the geographic origin of the reptilian and mammalian orders.

The reptilian radiation

The probable origin of the reptile orders — in Gondwanaland or Laurasia — is very hard to ascertain in most cases. The fossil record of the earliest reptiles (Carboniferous and early Permian) is so scanty in Gondwanaland that it would be hazardous to draw any conclusions as regards the Cotylosauria, the Pelycosauria, and the Protorosauria. There seems also at present to be little evidence on the probable origin of the Ichthyosauria, Thecodontia, Crocodilia, and Saurischia. As regards the other orders, the following tentative suggestions may be brought forth.

A Gondwanaland origin may perhaps be indicated as regards the following orders:

Chelonia. This is dependent on the interpretation of the Gondwanaland *Eunotosaurus* as an early representative.

Mesosauria. Restricted to Gondwanaland.

Eosuchia. Southern distribution in the Permian, invade Laurasia much later.

Rhynchocephalia. First appearance in Gondwanaland, later in Laurasia.

Ornithischia. First ornithischians in the late Trias of Gondwanaland; the order invades Laurasia in the early Jurassic.

In addition, a Gondwanaland origin may possibly be suggested for the Ictidosauria, though the comparatively poor fossil record of this order makes any conclusions very uncertain. It might perhaps also be thought

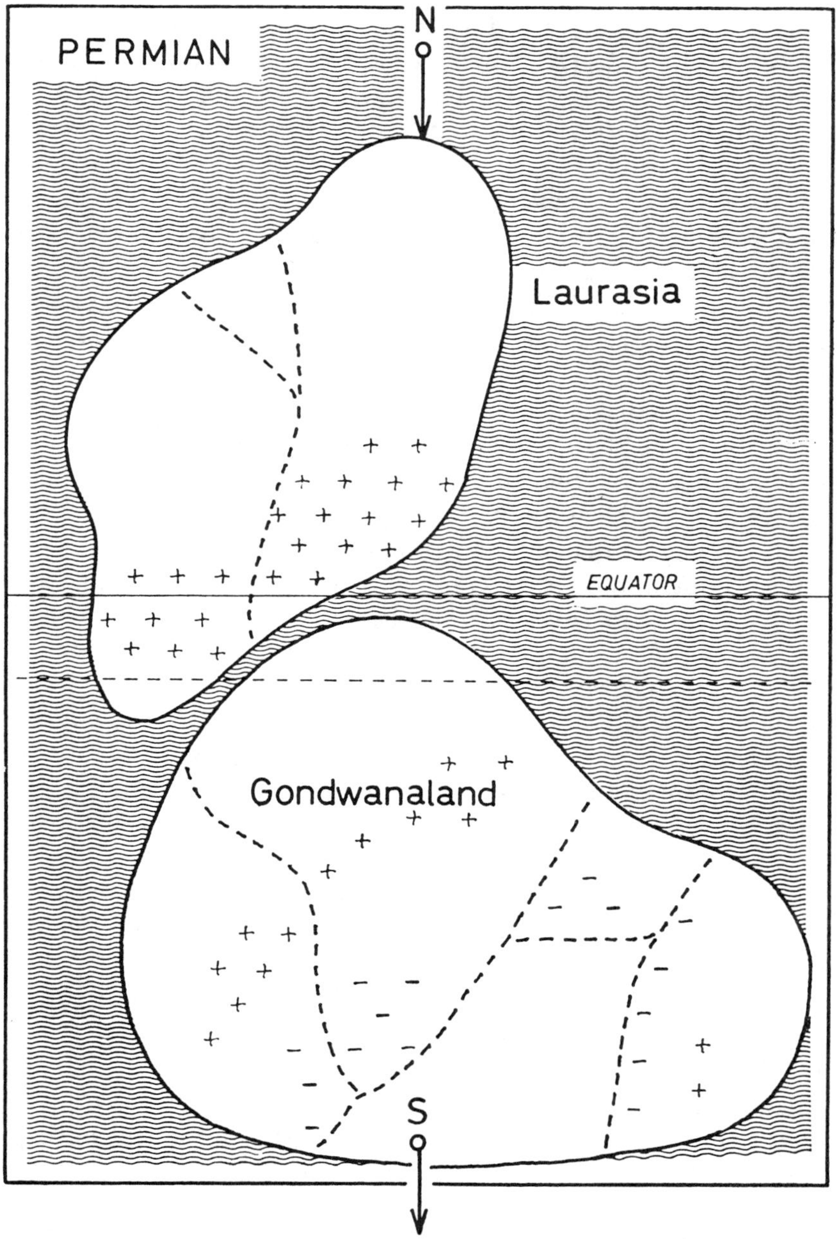

Fig. 1. Diagrammatic representation of Permian palaegeography as indicated by continental drift. Arrows indicate pole movements from early Permian to late Trias; dashed line is late Triassic palaeoequator. Plus signs indicate general distribution of warm-climate indicators (desert sands, evaporties, coral reefs, coal beds); minus signs, cold-climate indicators (tillites, glacial striae). Laurasia consists of North America, Greenland, and Eurasia (Europe in the south, Asia in the north). Gondwanaland is composed of South America, Africa, India, Australia, and Antarctica (with the South Pole; counting clockwise).

that the sudden appearance of full-fledged pterosaurs in the lower Jurassic of Laurasia might reflect an immigration from an unknown evolutionary centre somewhere in the south, but such a reasoning is obviously very tenuous. Still, even if these cases are left, out, there are at least five reptilian orders for which a Gondwanaland origin is indicated.

A Laurasian origin is possible in the following cases:

Sauropterygia. Nothosaurs appear in the early Trias of Laurasia, enter Gondwanaland somewhat later.

Therapsida. Laurasian forms transitional between Pelycosauria and Therapsida may indicate that the origin of this order lay in the northern continent.

Finally, an ergument might be made for a Laurasian origin of the Squamata, based on the Rhaetic gliding kuhneosaurids.

Thus, among those comparatively few reptilian orders which can be assigned to one or the other of the supercontinents, five or six seem to have arisen in Gondwanaland, and two or three in Laurasia.

In the Permian and early Mesozoic, Gondwanaland was not only the largest land mass of its time, but also the one with the most varied climate (fig. 1). Known Permian and Triassic palaeoclimates of Gondwanaland range from arctic (glacial) to tropical in nature, while the smaller Laurasian continent was mainly tropical and temperate. In fact, Gondwanaland might perhaps be compared with Eurasia of the present day, while the role of Laurasia was analogous to that of Africa or South America. Considering all this, it appears only natural that the greatest radiation of tetrapods should have occurred in Gondwanaland.

Intermigration between the two continents obviously was possible, but perhaps intermittent. The land bridge probably lay at the western end of Tethys.

The mammalian radiation

The palaeogeographic situation at the time of the mammalian radiation (late Cretaceous and early Tertiary) was radically different. Both Gondwanaland and Laurasia had split up into subcontinents, and in this phase the fragmentation was maximal. In addition, the great marine transgressions of the late Cretaceous and early Tertiary created large interior seas, dividing some of the remaining continental blocks into islands.

The separate continents were now as follows (fig. 2).

(1) North America had drifted away from Eurasia, but apparently retained contact, intermittently at least, by way of a land bridge across the North Atlantic (1—2). Animal geography indicates that this land bridge was in function up to and including the early Eocene, then foundered.

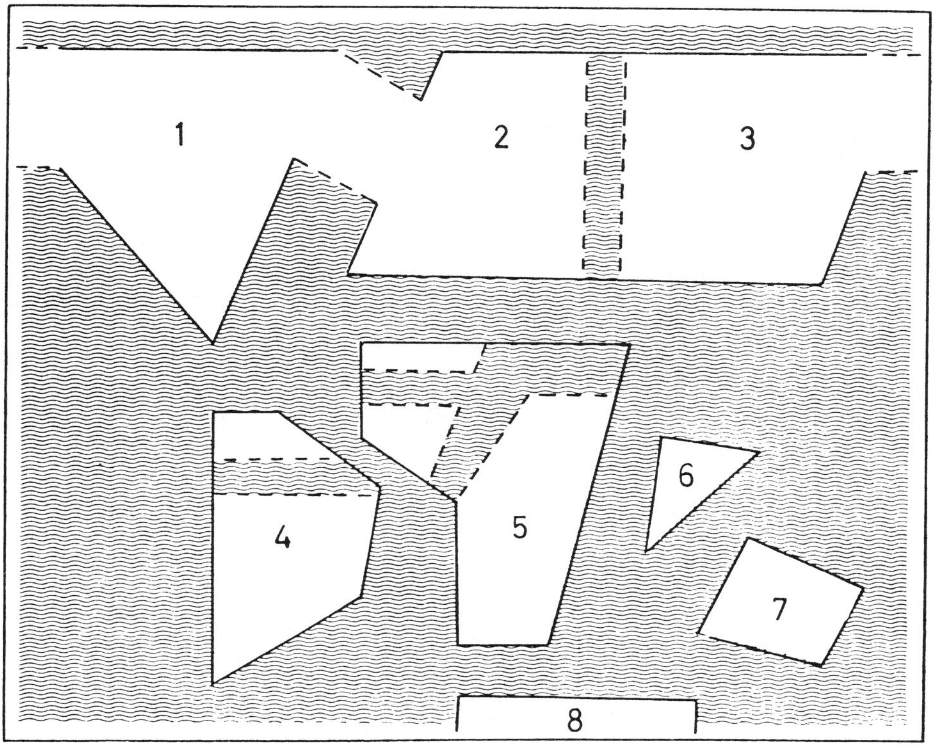

Fig. 2. Palaeography at the Cretaceous-Tertiary transition. Land briges and interior seas in dashed outlines. Areas are 1, North America; 2, Europe; 3, Asia; 4. South America; 5, Africa; 6, Indian peninsula; 7, Australia; 8, Antarctica.

In addition, another intermittent land bridge (the Bering bridge) extended from North America to Asia (1—3).

(2) Although the remaining part of Laurasia, Eurasia, remained a single continental block, it was long divided into a European and an Asiatic part by an interior seaway from Tethys to the Arctic, east of the Ural mountains. This appears to have caused a practically complete migration barrier between the land faunas of Europe and Asia, up to the middle Eocene when the sea began to retreat from this area.

(3) Asia, then, had little or no direct contact with Europe, except by way of a circular route crossing North America.

(4) South America, one of the daughter continents of Gondwanaland, became almost completely isolated at the beginning of the Tertiary. On zoogeographic grounds, however, some Late Cretaceous intermigration with both North America and Australia must be postulated. It probably occurred in the form of »island-hopping» along a so-called sweepstakes route (SIMPSON, 1953): the Gondwanaland remnants probably formed the

stepping stones between South America and Australia. After this, South America remained isolated up to the later Tertiary.

(5) The African continent was an important evolutionary centre. A large part of it was flooded by shelf seas, dividing the continent into islands. It was more or less completely isolated in the Paleogene, apart from some »filter-barrier» migration.

(6) The Indian peninsula forms another fragment of Gondwanaland, which drifted northward and finally came into contact with the Asiatic land mass, piling up the Himalaya in the process. Its zoogeographic role in the early Tertiary is still somewhat obscure. There seems to have been some limited intermigration with Asia.

(7) Australia, though probably in indirect contact across Gondwanaland with South America in the late Cretaceous, has been completely isolated since then.

(8) The role of Antarctica in the evolution of the mammals is little known.

There were thus no less than eight separate nuclei, in various degrees of isolation from each other, to provide areas for mammalian radiation. Let us now consider how the evolution of the mammalian orders was distributed.

Europe, Asia and North America were not completely isolated from each other, and so it is difficult to determine the place of origin of the Holarctic mammalian orders. At present it is only in North America that we have a good fossil record for mammals from the late Cretaceus and the earlier Paleocene. Only in the case of those orders that appear a little later in the fossil record, inferences of this kind are are now possible. Thus it might be suggested, for instance, that the Rodentia arose in North America, and the Lagomorpha in Asia. Otherwise, little can be said of the exact centre of origin for the long array of orders that probably arose in the Holarctic area — the Multituberculata, Insectivora, Deltatheridia, Dermoptera, Chiroptera, Primates, Carnivora, Condylarthra, Perissodactyla, Artiodactyla, Tillodontia, Taeniodontia, Amblypoda, Dinocerata, Pholidota, and possibly the Edentata, though it may also be of South American origin.

South America, an isolated island continent, produced a highly varied fauna with the orders Paucituberculata, Astrapotheria, Pyrotheria, Litopterna, Notoungulata, and perhaps Edentata: the South American origin of the two last-mentioned is less certain. The order Marsupicarnivora is distributed between South America and Australia, and of its orgin nothing more can be said than that it apparently lay in some one of the daughter continents of Gondwanaland.

Africa, in semi-isolation, produced a large crop of orders; their pre-

Eocene history is unfortunately unknown. Here belong the Hyracoidea, Embrithopoda, Proboscidea, and probably also the Sirenia, Desmostyliformes, and Tubulidentata. Among infraordinal taxa of special interest with an African origin may be mentioned the pongid-hominid group, which appears to be represented by *Propliopithecus*.

In the Australian continent only two orders of mammals appear to have evolved, the Peramelina and Diprotodonta. But early Tertiary and Cretaceous Australian faunas are not yet known. Again, there are no orders that on present knowledge can be assigned to the Indian peninsula, or to Antarctica.

The geographic origin of Cetacea is unknown. The chances are that the whales descended from some Holarctic group of land mammals, but an African or Indian origin cannot at present be excluded.

Thus, the three Holarctic continents together produced some 17—18 orders, or on an average, about six orders each; Africa and South America also produced about six orders each; Australia two; India (and Antarctica) none, on present evidence. There is obviously a good correlation between size of area and number of orders produced, just as in the case of Gondwanaland and Laurasia.

From a zoogeographic point of view, the Tertiary was an almost continuous amalgamation process. At the beginning of the period, the eight continents were almost completely isolated from each other; then they gradually came into contact, land bridges were being formed, and the local faunas began to intermigrate. In this process there was a gradual weeding out of ecologically overlapping form, which in some instances led to the complete extinction of entire orders — this is especially marked in the South American fauna. Thus the great number of mammalian orders produced was also strongly influenced by the separation into many disjunct areas; otherwise this remarkable duplication of adaptive types could hardly have been brought about.

Conclusions

Study of the reptilian radiation in the Permian and Mesozoic, and the mammalian in the Cretaceous and Tertiary, in the palaeogeographic settings indicated by continental drift, suggests that the taxonomic variety produced is correlated with the size of the area available, with its environmental variety, and with the number of disjunct areas.

References

DIETZ, R.S., & SPROLL, W. P. 1966. Equal areas of Gondwana and Laurasia (ancient super continents). Nature, vol. 112, 1196–1198.

FALCON, N. L. 1967. Equal areas of Gondwana and Laurasia. Nature, vol. 113, 580.

IRVING, E. 1964. Paleomagnetism and its application to geological and geophysical problems. Wiley, New York.

JARVIK, E. 1942. On the structure of the snout of crossopterygians and lower gnathostomes in general. Zool. Bidr. Uppsala, vol. 21, 235–675.

RUNCORN, S. K. (ed.) 1961. Continental drift. Academic Press, New York & London.

SIMPSON, G. G. 1953. Evolution and geography. Condon Lectures, Oregon State Board of Higher Education, Eugene, Oregon.

Addendum

The outcome of a recent discussion on the areas of Gondwanaland and Laurasia (DIETZ & SPROLL 1966; FALCON 1967) is that the supercontinents were semi-equal in size, Gondwanaland being slightly larger and with a more varied environment.

Part III

COMPARATIVE ANATOMY, CLASSIFICATION, AND PHYLOGENY

Editor's Comments on Papers 8 Through 16

The papers reprinted in Part III are representative of the heart of vertebrate paleontology. They deal with the morphology, classification, and relationships of fossil vertebrates.

Cope has been mentioned previously for his contributions to biostratigraphy. Paper 8 represents one of his many contributions to vertebrate anatomy and genealogy. In it he summarizes his hypotheses as to the relationships among some of the major groups of vertebrates: mammals, reptiles, amphibians ("Batrachia"), bony jawed fishes, sharks, and jawless fishes. Finally, Cope traces the Vertebrata to an ancestral form morphologically similar to the extant amphioxus (Cephalochordata). In this paper appears the first strong argument presented for the derivation of tetrapods (land vertebrates, especially the amphibians) from rhipidistian fishes rather than from dipnoan lung fishes (Schaeffer, 1965). Cope's hypothesis has been extremely important, influencing a whole school of researchers to study the detailed anatomy of the Devonian rhipidistian fish *Eusthenopteron foordi* (see Jarvik, 1980, 1981). However, it has recently been suggested that Cope may have been mistaken, that lung fishes may actually be more closely related to tetrapods (Rosen et al., 1981). This is currently a subject of intense debate (Thomson, 1981).

It is interesting to note that while Cope was concerned with establishing the evolutionary trends and genealogies of organisms (Paper 3, p. 23; Paper 8), he rejected Darwin's theory of natural selection (as did a majority of biologists and paleontologists until well into the twentieth century: see Mayr, 1982). In contrast to Cope, Marsh (Paper 2, p. 490) was an ardent proponent of natural selection. Cope propounded neo-Lamarckian theories in which both the external environment and the use or disuse of certain organs or parts were believed to be the explanation for the gradual evolution of organisms (e.g., Cope, 1889). Cope's "kinetogenesis theory" of evolution greatly influenced the ideas of his admirer Henry Fairfield Osborn (1857-1935), perhaps the leading figure in vertebrate paleontology in the early twentieth century (see Paper 15, pp. 1070, 1074). Although, at least in his later work, Osborn did not accept the principle of acquired characteristics, he also never completely relinquished the idea that environmental conditions can influence the evolution of the "heredity germ" of organisms (see Osborn, 1929, pp. 805-889).

The American John Strong Newberry (1822-1892) originally studied medicine, but then turned his attention to geology and paleontology. He participated in early explorations of the western United States in the 1850s, was a professor of Geology at Columbia University, and was director of the Ohio Geological Survey (1869-1884). His most notable work is on Mesozoic and Cenozoic plants and on fossil, especially Paleozoic, fishes. Paper 9 is his classic summary of the extent of knowledge, in 1873, of the relationships and succession of Paleozoic fishes.

Roy Lee Moodie (1880-1934) is well known for his work on paleopathology, paleoichnology (fossil footprints and other trace fossils), and fossil amphibians. Paper 10 is a selection from his work on amphibians in which, instead of addressing the more popular subject of the relationships of amphibians to other tetrapods, he attempts to trace the ancestry of a certain group of modern amphibians, the salamanders (Caudata).

Samuel Wendell Williston (1852-1918) began his career in vertebrate paleontology as a collector for O. C. Marsh. He received the M.D. in anatomy and also made many contributions to entomology. In his work on vertebrates, Williston concentrated primarily on the Reptilia. Paper 11 is a summation of many of his ideas concerning reptilian phylogeny and classification. The basic higher-level relationships laid out in this paper were accepted by most workers for the next sixty years, and even today the outlines of Williston's classification are widely used.

Marsh's (mentioned several times previously, see Paper 2) most enduring scientific contributions were his numerous descriptions and reconstructions of early Tertiary mammals, Cretaceous birds, and Mesozoic reptiles. Paper 12 is an excerpt from Marsh's monograph on the dinosaurs of North America in which he synthesized a lifetime of work on these animals. In this excerpt Marsh presents detailed anatomical comparisons among several different types of dinosaurs, offers some of his famous skeletal restorations of well-known European dinosaurs, and briefly discusses the affinities of the Dinosauriiformes.

Paper 13 shifts to a discussion of the origin of the living "descendants" of the dinosaurs, the birds. Its author, William King Gregory (1876-1970), one of the outstanding anatomists of his time, was particularly interested in the origins and functional anatomy of major groups of vertebrates (see also Paper 18). In Paper 13 Gregory reviews and examines a number of functional explanations for the origin of flight in birds. Over the last 100 years, there have been two basic schools of thought regarding the origin of flight in birds (Lewin, 1983; Padian, 1982). According to one, the arboreal theory hypothesizes that birds are derived from a tree-dwelling animal, through a gliding stage. The cursorial theory, supported by the other, suggests that birds evolved from a bipedal, running animal whose forelimbs were free to develop into wings. In Paper 13, Gregory argues persuasively for the arboreal theory. During the 1970s, however, the cursorial theory again came into vogue, largely through the work of John H. Ostrom of Yale University (Ostrom, 1974, 1976, 1979). Quite recently a detailed evolutionary/functional model has been proposed for the evolution of flight (and thus of birds) in a ground-running, bipedal dinosauriiform (Caple, Balda, and Willis, 1983), which corroborates Ostrom's views. According to this latest model, all intermediate stages, from a totally cursorial animal to the fully developed bird with powered flight, are fully functional. Unlike previous models, many of which are reviewed in Paper 13, no leap of faith (e.g., how a wing designed for gliding could be transformed into one for powered flight) is necessary to account for intermediate forms.

In Paper 14, the origin of another major group of vertebrates, the Mammalia, is discussed by Robert Broom. Broom (1866–1951) was born in Scotland and studied medicine at Glasgow University. He went to Australia (1892) and then to South Africa (1897), where he lived for the rest of his life. He was curator of Paleontology at the Transvaal Museum (1934–1951). Broom is particularly well known for his work on mammallike reptiles and for his studies of early man, in which he concentrated on the excavations of the South African australopithecine sites of Sterkfontein, Kromdraai, and Swartkrans (cf. Paper 16, the first description of *Australopithecus*).

Paper 14 is the summary of a lecture that Broom presented to the Section of Biology of the New York Academy of Sciences in 1913. Broom first reviews previous work on the origin of mammals and then discusses his own researches. Paper 14 is of interest not only for the substance of Broom's lecture, but also as a glimpse of a typical scientific meeting in the early twentieth century attended by leading figures in vertebrate paleontology (W. K. Gregory, W. D. Matthew, R. C. Osburn, H. F. Osborn). As a side note, the middle Permian, which Broom dated as 18 million years old (Paper 14, p. 303; cf. Walcott, 1893; Osborn, 1916), is now generally dated as approximately 260–270 million years old (Odin, 1982).

Henry Fairfield Osborn (1857–1935) was one of the most prolific, outspoken, egotistical (see Osborn, 1930), and well-known vertebrate paleontologists who ever lived. Born in Fairfield, Connecticut, he attended Princeton University and while a student there organized his first fossil-collecting expeditions to Colorado and Wyoming. He was later to lead numerous other expeditions (especially notable were the Central Asiatic Expeditions of the American Museum of Natural History, New York City, in the 1920s). Most of Osborn's long career was spent at the American Museum, where he was first curator of the Department of Vertebrate Paleontology and then president of the museum. He also served as president of the American Association for the Advancement of Science. An extremely able museum administrator, Osborn was innovative in assembling mounted skeletons and life reconstructions of fossil animals for public exhibition. He was offered the position of secretary (i.e., head) of the Smithsonian Institution in Washington, D.C., which he turned down because he felt that accepting would interfere with his research.

Outstanding among Osborn's numerous scientific contributions (he published more than 800 papers, books, and monographs) was his development and elaboration of Cope's work on the Tritubercular Theory of mammalian dental evolution (Paper 15). Osborn's tooth-cusp nomenclature, based on his theory of the origin and elaboration of mammalian molars from simple conical crowns (see Paper 15), is now almost universally used, although the underlying theory has since been greatly modified (see Hershkovitz, 1977; Bown and Kraus, 1979).

With Paper 16 we explore the origin of man. Professor Raymond Arthur Dart was born in Brisbane, Australia in 1893 and studied medicine at the University of Sydney. He was briefly at the University of London and in 1923 took the position of professor of Anatomy at Witwatersrand University, South Africa, where he still resides. In 1925 Dart described *Australopithecus* (Paper 16), the genus most closely related to our own (i.e., *Homo*). The immediate reaction of both the scientific and lay communities was either to be openly hostile or to simply ignore Dart's work on the "missing link." However, subsequent finds, many made by Broom (mentioned above), Louis B. S. Leakey (e.g., Leakey, 1959; Leakey, Tobias, and Napier, 1964) and his family, and most recently by Donald C. Johanson (Johanson and White, 1979) have vindicated Dart's original views that *Australopithecus* is extremely near, and perhaps on, the direct ancestry of man (see Ciochon and Corruccini, 1983, for a number of recent articles on this and related subjects). As an historical note it should be pointed out that, in agreement with most scientists of the time, Dart considered Piltdown man (Dawson and Woodward, 1913), *Eoanthropus dawsoni,* to be an authentic find (Paper 16). Subsequently, Piltdown man was exposed as a gigantic hoax (Weiner, 1955). A modern human skull and an orangutan jaw had been planted together in an English Pleistocene gravel deposit.

REFERENCES

Bown, T. M., and M. J. Kraus, 1979, Origin of the Tribosphenic Molar and Metatherian and Eutherian Dental Formulae, in *Mesozoic Mammals: The First Two-thirds of Mammalian History,* J. A. Lillegraven, Z. Kielan-Jaworowska, and W. A. Clemens, eds., University of California Press, Berkeley, pp. 172–181.

Caple, G., R. P. Balda, and W. R. Willis, 1983, The Physics of Leaping Animals and the Evolution of Preflight, *Am. Naturalist* **121:**455–476.

Ciochon, R. L., and R. S. Corruccini, eds., 1983, *New Interpretations of Ape and Human Ancestry,* Plenum Press, New York and London, 912p.

Cope, E. D., 1889, The Mechanical Causes of the Development of the Hard Parts of the Mammalia, *Jour. Morph.* **3:**137–277.

Dawson, C., and A. S. Woodward, 1913, On the Discovery of a Palaeolithic Human Skull and Mandible in a Flint-bearing Gravel Overlying the

Wealden (Hastings Beds) at Piltdown, Fletching (Sussex), *Geol. Soc. London Quart. Jour.* **69:**117-144.
Hershkovitz, P., 1977, *Living New World Monkeys (Platyrrhini), with an Introduction to Primates,* vol. 1, The University of Chicago Press, Chicago and London, 1132p.
Jarvik, E., 1980, 1981, *Basic Structure and Evolution of Vertebrates,* Academic Press, London and New York, vol. 1 (1980), 592p.; vol. 2 (1981), 350p.
Johanson, D. C., and T. D. White, 1979, A Systematic Assessment of Early African Hominids, *Science* **203:**321-330.
Leakey, L. S. B., 1959, A New Fossil Skull from Olduvai, *Nature* **184:**491-493.
Leakey, L. S. B., P. V. Tobias, and J. R. Napier, 1964, A New Species of the Genus *Homo* from Olduvai Gorge, *Nature* **202:**7-9.
Lewin, R., 1983, How Did Vertebrates Take to the Air? *Science* **221:**38-39.
Mayr, E., 1982, *The Growth of Biological Thought,* Harvard University Press, Cambridge, Mass., and London, 986p.
Odin, G. S., ed., 1982, *Numerical Dating in Stratigraphy,* 2 vols., John Wiley & Sons, Chichester, England, 1068p.
Osborn, H. F., 1916, The Origin and Evolution of Life upon the Earth, *Sci. Monthly* **3:**5-22.
Osborn, H. F., 1929, *The Titanotheres of Ancient Wyoming, Dakota, and Nebraska,* U.S. Geological Survey Monograph 55, Government Printing Office, Washington, D.C., 953p.
Osborn, H. F., 1930, *Fifty-two Years of Research, Observation and Publication: 1877-1929,* Charles Scribner's Sons, New York, 172p.
Ostrom, J. H., 1974, The Origin of Birds, *Ann. Rev. Earth Planetary Sci.* **3:**55-77.
Ostrom, J. H., 1976, Some Hypothetical Anatomical Stages in the Evolution of Avian Flight, *Smithsonian Contr. Paleobiology* **27:**1-21.
Ostrom, J. H., 1979, Bird Flight: How Did It All Begin? *Am. Scientist* **67:**46-56.
Padian, K., 1982, Macroevolution and the Origin of Major Adaptations: Vertebrate Flight as a Paradigm for the Analysis of Patterns, *North American Paleontol. Conv., Third, Proc.* **2:**387-392.
Rosen, D. E., P. L. Forey, B. G. Gardiner, and C. Patterson, 1981, Lungfishes, Tetrapods, Paleontology, and Plesiomorphy, *Am. Mus. Nat. History Bull.* **167:**163-275.
Schaeffer, B., 1965, The Evolution of Concepts Related to the Origin of the Amphibians, *Systematic Zoology* **14:**115-118.
Thomson, K. S., 1981, A Radical Look at Fish-Tetrapod Relationships, *Paleobiology* **7:**153-156.
Walcott, C. D., 1893, Geologic Time, as Indicated by the Sedimentary Rocks of North America, *Jour. Geology* **1:**639-676.
Weiner, J. S., 1955, *The Piltdown Forgery,* Oxford University Press, London, 226p.

8

Reprinted from *Am. Philos. Soc. Proc.* **30:**278–281 (1892)

ON THE PHYLOGENY OF THE VERTEBRATA

E. D. Cope

(Read before the American Philosophical Society, October 7, 1892.)

I have traced the origin* of the Mammalia to the Theromorous reptiles of the Permian epoch, for the following reasons. The latter include the Pelycosauria, Cotylosauria, Procolophonina and perhaps other orders. In both classes there is only one postorbital arch of the skull, and this is the zygomatic. In both (excepting Prototheria and Procolophonina †)

* Proceeds. Amer. Philos. Soc., 1884, p. 43.
† Seeley, Philos. Trans. Royal Soc., 1889, 269.

the coracoid element is of reduced size, and is coössified with the scapula. In both (except Cotylosauria) the capitular articulation of the ribs is intercentral. In both, the humerus has distal condyles and epicondyles, and there is an entepicondylar foramen in the Pelycosauria as in the lower Mammalia. The posterior foot is constructed in the Pelycosauria almost exactly like that of the Prototheria. The single occipital condyle of the reptiles is not found in the Mammalia, but in some of the Lacertilia (Uroplates, Gecco) there are two condyles, the median (basioccipital) portion of the single condyle being rudimental. The Pelycosauria could not, however, have given origin to the Prototheria, since in that subclass of mammals there is a well-developed coracoid. But in the Procolophonina this element is developed as in the Prototheria. Moreover, the Pelycosauria and the Procolophonina have the interclavicle, which is an element of membranous origin, while in the Prototheria we have the corresponding cartilage bone, the episternum. This element is present in the Permian order of the Cotylosauria, which is nearly related to the Pelycosauria. This order has, however, single-headed ribs, springing from the diapophyses, which is not usual in the Mammalia. But in some Cotylosauria the diapophyses are short, and in the Monotremata the postcervical ribs are single-headed, so this character may not prove an insurmountable one. It is evident that the Mammalia were derived from some type probably referable to a Permian reptilian order of the Theromorous series, although to which one is not yet known.

The Reptilia have been supposed by Hæckel to have taken their origin from the Batrachia. I have indicated that it is probable that the Batrachian order, which stands in this relation to the Reptilia, is the Embolomeri of the Permian epoch. This conclusion rests on the following considerations. The Reptilian order of the Cotylosauria approaches the Batrachia of the subclass Stegocephali in the overroofing of the posterior regions of the skull ; in the presence of vomerine teeth, and in the absence of obturator foramen of the pelvis. In some Cotylosauria (Diadectidæ) the stegocephalian intercalary bone of the skull is well developed. But in the Cotylosauria, the vertebral column consists mainly of centra, while in the Stegocephali it consists entirely or partly of intercentra. But in the Embolomeri the centra are well developed, and are larger than the intercentra anterior to the pelvis. Hence this is the only order of Stegocephali from which the Reptilia could have been derived.

Hæckel derived the Batrachia from the Dipnoi (Dipneusta), and I followed him in this belief, being strengthened in it by Huxley's ascription of an autostylic suspensorium of the mandible* to both divisions. This phylogeny is questioned by Pollard † and by Kingsley ‡ who would see the ancestry of the Batrachia in the Crossopterygian fishes on embryological grounds derived from a study of Polypterus. In support of their

* Proceedings Zoölogical Society of London, 1876, p. 59.

† Anatomischer Anzeiger, vi, p. 338, 1891.

‡ American Naturalist, 1892, p. 679. Kingsley would also derive the Dipnoi from Crossopterygia.

view I would cite the absence of the maxillary arch in the Dipnoi, and its full development in the Stegocephali, which are the ancestral Batrachia. The large development of the dorsal and anal fins in the Dipnoi is not favorable to the Hæckelian view; nor do the paired fins approach as nearly to the limbs of Batrachia as do those of some other fishes. It has been shown by Huxley that the suspensorium of the Batrachia is hyostylic in its earliest stages, and that it becomes autostylic at a later period of development. The Batrachia may then have originated from a hyostylic Teleostomous fish; *i. e.*, one with complete maxillary arch. Among Teleostomata we naturally look for forms with limbs which approach nearest the Batrachian type, and in which median fins are feeble or wanting. Such are the Rhipidopterygia, which include the families of Holoptychiidæ, Tristichopteridæ, Osteolepididæ, Cœlacanthidæ and perhaps some others. These families, except the last, abounded in the waters of the Devonian period, at the time when the ancestors of the Batrachia also

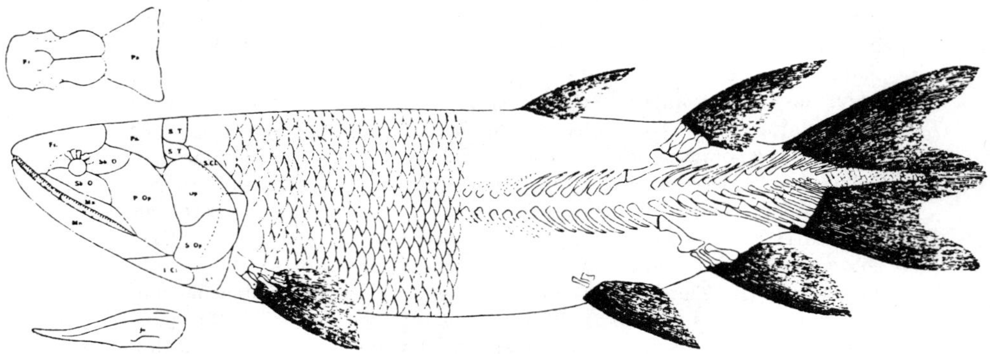

Fig. 1. *Eusthenopteron foordii* Whiteaves; ⅓ natural size. Devonian of New Brunswick. From Whiteaves.

existed. All of them agree in possessing the median fins of greatly reduced proportions, and the mesodermal or internal elements of the paired fins more like the limbs of the Batrachia than are those of any known fishes. The constitution of the superior cranial wall is a good deal like that of the stegocephalous Batrachia. The characters of the fins can be learned from the accompanying figure of the *Eusthenopteron foordii* Whiteaves, one of the Tristichopteridæ. The pectoral fin well-nigh realizes Gegenbaur's theory of the derivation of the Chiropterygium from the Archipterygium.

The question of the ancestry of the Batrachia cannot be considered to be yet settled.

The ancestral type of fishes is probably the Ichthyotomous order of the subclass of sharks (Elasmobranchii).* They are hyostylic, and have cranial

*Cope, Proceedings Amer. Philos. Soc., 1884, p. 585.

segmentation, the basioccipital element being conspicuous. The fins are all primitive, and those of all other types of fishes might have been derived from them. Opposed to this estimate of their relation to other vertebrates is the fact that they have not been yet found prior to the Carboniferous period. But our knowledge of the fishes of the Devonian is yet very imperfect. The types ancestral to the Pisces must have existed in the Silurian, and forms which may well have fulfilled this function have been discovered there. I refer to the Agnatha, which have been traced to the summit of the Devonian. The Silurian Agnatha are the Pterasprididæ, which display the lowest type, and the Cephalaspididæ, and these were succeeded by the Pterichthyidæ in the Devonian. There is a wide gap between these forms and any of the fishes, and nothing can be affirmed plausibly with regard to the phylogeny. There are superficial resemblances between the dorsal and ventral dermal scuta of the Pterichthyidæ and the Arthrodirous Dipnoi, but there is no considerable affinity between those divisions.

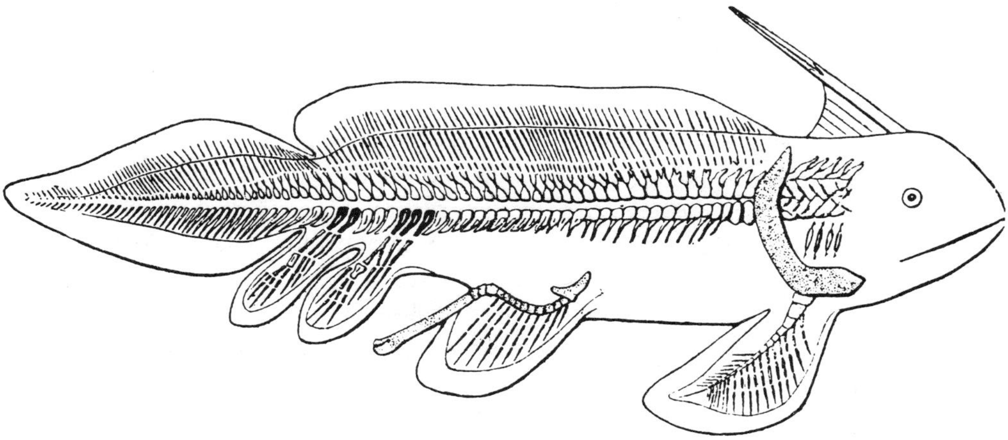

Fig. 2. *Xenacanthus dechenii*, one of the Ichthyotomi; restored by Dr. H. E. Sauvage. From the Coal Measures of France.

The extinct Agnatha agree with the existing lampreys in the absence of lower jaw and pectoral (scapular) arch, and both must be traced, in accordance with Hæckel's phylogeny, to the Acrania, which is now represented by the amphioxus (genus Branchiostoma). This order is easily the ancestor of existing Vertebrata, and shows points of resemblance to both Tunicata and worms. It has been suspected by Dohrn to have undergone degeneration, which may have been the case, since this phenomenon is so abundantly exhibited by both Tunicata and worms. It is not difficult to believe with Kowelewsky, that the Acrania were derived from the Tunicata. Semper has suspected, on the other hand, that the ancestors of the Vertebrata are to be found in the Annelide worms.

9

Reprinted from *Geol. Survey Ohio Rept.* **1**:247-258 (1873)

THE

CLASSIFICATION AND GEOLOGICAL DISTRIBUTION

OF OUR

FOSSIL FISHES.

J. S. Newberry

So little is generally known in regard to American fossil fishes, that I have thought the notes which I now give upon some of them would be more interesting and intelligible if those into whose hands they will fall could have a more comprehensive view of this branch of palæontology than they afford. I shall therefore preface the descriptions which follow with a few words on the geological distribution of our Palæozoic fishes, and on the relations which they sustain to fossil forms found in other countries, and to living fishes. This seems the more necessary, as no summary of what is known of our fossil fishes has ever been given, and the literature of the subject is so scattered through scientific journals and the proceedings of learned societies, as to be practically inaccessible to most of those who will be readers of this report.

I. THE ZOOLOGICAL RELATIONS OF OUR FOSSIL FISHES.

To the common observer, the class of Fishes seems to be well defined and quite distinct from all the other groups of vertebrate animals; but the comparative anatomist finds in certain unusual and aberrant forms peculiarities of structure which link the Fishes to the Invertebrates below and Amphibians above, in such a way as to render it difficult, if not impossible, to draw the lines sharply between these great groups. *Amphioxus*, the lowest of the Fishes, has a structure much simpler, and organs and senses more rudimentary, than some of the two higher of the invertebrate classes, Mollusks and Crustaceans; while in *Lepidosiren* we find an organization that is as much amphibian as piscine. In the judgment of some anatomists, the *Dipnoi*—which include *Lepidosiren* of South America, *Protopterus* of Africa, and *Ceratodus* of Australia—should be set off in a distinct class by themselves, or be united with the *Ganocephala* (the Carboniferous salamanders), as holding an intermediate position between the true Fishes and Amphibians. Nearly all fishes, however, possess in common a structure which may be very readily defined. They have a fusiform body, adapted to rapid motion through a resisting medium, and they are aquatic in habit and respiration. From their mode of life—floating freely as they do in a fluid having approximately the same specific gravity as their own—they have no occasion for limbs, such as serve to support the body in antagonism to gravitation, and act as locomotive organs in the higher vertebrates. Their movements are therefore effected only by fins, into which the anterior and posterior extremities and tail are developed. As the highest efficiency of these locomotive organs is incompatible with their use as organs of prehension, they are never employed for seizing their prey, but this is accomplished by an unusual development of the mouth and its appendages. Not being required to sustain the weight of the body, the framework is much less rigid than in terrestrial animals, and the bones are softer and more elastic. This last feature in their organization has prevented the perfect preservation of many fishes in the fossil state, and has greatly increased the difficulty of the study of some groups, especially the sharks, of which the cartilaginous skeletons have generally disappeared, and the teeth, fin-rays and dermal tubercles, disconnected, and perhaps widely scattered, are all that remain.

Various schemes of classification of fishes have been proposed by different zoologists; the changes and additions made to the earlier systems having, for the most part, followed closer study and a better knowledge of their anatomy. The first scientific system of classification proposed was that of Cuvier, who divided the class of fishes into three great orders—the *Chondropterygii* (cartilaginous fishes), the *Acanthopterygii* (fishes with spined fins), and *Malacopterygii* (fishes with soft fin-rays). Agassiz, who followed Cuvier in time, divided fishes into four orders, according to the character of their scales, viz: *Cycloidii* (fishes with rounded, entire scales, as the salmon), *Ctenoidii* (fishes with serrated scales, as the perch), *Ganoidii* (fishes with brilliant, glistening, enameled scales, such as the gar-pike), and *Placoidii* (fishes with dermal tubercles or plates, such as the sharks and rays). Johannes Muller divides fishes into five orders, viz: *Leptocardia* (amphioxus), *Cyclostomata* (lampreys), *Teleostei* (bony fishes), *Ganoïdei* (ganoid fishes), and *Selachii* (sharks and rays). Professor Richard Owen proposed to divide the vertebrates into two classes—*Hæmatocrya*, cold-blooded, and *Hæmatotherma*, warm-blooded animals—the class *Hæmatocrya* including Fishes, Amphibians and Reptiles. In Owen's classification, fishes may be said to comprise three of the five sub-classes of *Hæmatocrya*, viz: *Dermopteri*, *Teleostomi* and *Plagiostomi*—the first including Agassiz's *Ganoidii* and Muller's *Leptocardia;* the *Teleostomi* including the *Cycloidii* and *Ctenoidii* of Agassiz; the *Plagiostomi*, the sharks, rays and chimæras. Of these sub-classes of fishes, Owen makes ten orders, viz: I. *Cirrostomi* (lancelet); II. *Cyclostomi* (lamprey and hag); III. *Malacopteri* (soft finned fishes, herring, salmon, etc.); IV. *Acanthini* (cod, etc.); V. *Acanthopteri* (perch, etc.); VI. *Plectognathi* (file-fish, trunk-fish, etc.); VII. *Lophobranchii* (sea-horse); VIII. *Ganoidei* (ganoids); IX. *Holocephali* (chimæra); X. *Plagiostomi* (sharks and rays): his order XI., *Protopteri*, being considered transitional. Other systems have been framed by Hæckel, Gunther, Lutken, Cope, Gill, etc.,* but perhaps the most satisfactory, inasmuch as it includes a discussion of fossil forms, is that of Prof. Huxley, proposed within the past year. Huxley divides the class PISCES into two sub-classes, of which the first includes only *Amphioxus*, and is equivalent to "*Leptocardia*" of Hæckel. In this subdivision "the notochord extends to the anterior end of the body. There are no skull, auditory or renal organs, such as exist in the higher vertebrata. The heart is a simple tube, and the liver is saccular." In the second sub-class of Huxley (=*Pachycardia*, Hæck.), "the notochord extends behind the pituitary fossa. A skull, auditory and renal organs are developed. The heart is divided into auricular and ventricular chambers. The liver has the ordinary structure." The orders into which Huxley divides the class PISCES are as follows:

1. PHARYNGOBRANCHII (*Amphioxus*).
2. MARSIPOBRANCHII (Lamprey and Hag).
3. ELASMOBRANCHII (Sharks, Rays and Chimæras).
4. GANOIDEI (Ganoids).
5. TELEOSTEI (Bony Fishes).
6. DIPNOI (*Lepidosiren*, etc.).

When we add fossil forms to the living groups of fishes, and trace the history of this class from its first appearance in far-off geological ages up to the present time, some very interesting facts are brought out; to a few of which I will very briefly allude.

In the first place, we learn from the study of fossil fishes that in all probability no fishes swam in the most ancient Silurian sea; that they began to appear in the sea of the Upper Silurian age, and in the Devonian existed in large numbers, attained a magnitude equal to that of any

* Professors Gill and Cope are agreed in separating *Amphioxus*, *Myxine* and *Petromyzon* from fishes, and in forming three classes out of the old class of fishes, viz: PISCES (which includes the sub-classes *Teleostei*, *Ganoidei* and *Elasmobranchii*), MARSIPOBRANCHII and LEPTOCARDII. (Smithsonian Miscell. Coll., No. 247; Annals and Mag. Nat. Hist., Vol. IX., 1872.)

fishes of the present day, and possessed, in some instances, a structure scarcely less highly organized than that of the highest group of modern fishes.

In tracing the history of fishes through the successive geological ages, we also find that in later times they have been greatly diversified by divergence from some of the earlier and simpler forms; and that while the number of individuals may not be greater now than in the Jurassic or Carboniferous periods, the number of genera and species is certainly much greater now than formerly. This multiplicity of form has apparently been produced by differentiation; that is, by the exaggeration of certain characters possessed by a remote ancestor, in different groups of his descendants, until nearly all traces of relationship and common parentage are lost. It will be readily understood from these remarks that fossil fishes, in many instances, present "synthetic types"; that is, a single genus and species will be found to combine characters which are distributed among different families, and perhaps orders, of living fishes; hence, before a truly comprehensive and intelligent view can be taken of the class of fishes, it is essential that the fossil forms, which are so often connecting links, and which fill many of the gaps that exist in our present fauna, should be carefully co-ordinated with living species. The study of fossil fishes becomes, therefore, intensely interesting and highly instructive to the zoologist. It is to be regretted, however, that the rarity of the remains of fishes and the imperfection of their preservation make it impossible that they should solve all questions with regard to the relationship of living fishes, or of the life-history of the class; but every year new light is thrown upon the structure and relationship of fishes by the discovery of some new or unusually well preserved fossil. I have, therefore, ventured to hope that the remains of the remarkable fishes which are for the first time made known in this report, will be objects of more than mere idle curiosity, and that they will contribute something to the better understanding of the organization and genesis of all the great groups of fishes.

As I have before said, the dismembered condition and incomplete preservation of many fossil fishes, and the entire absence of their soft parts, make it exceedingly difficult to determine in all cases accurately their relationship to each other and to living fishes. Enough, however, has been learned of their structure to permit us to group many of them with a good degree of confidence; and quite a number of families and many more genera and species have been established to receive them.

One of the most striking results of the study of fossil fishes is the discovery that in the earlier geological ages the order which includes much the largest part of our present ichthyic fauna (the Teleosts) had no existence, and that it was only introduced in the Cretaceous period. We also learn that Placoderms and Elasmobranchs constituted the earliest groups of fishes: that during the Devonian, Carboniferous and Jurassic ages the Ganoids attained great development, and that they subsequently dwindled away—giving place to the Teleosts—until now they are represented by but seven genera on the face of the globe.

Of existing Ganoids, by far the larger number are found in North America. Five of the seven genera occur on this continent, and four of the five nowhere else. The living Ganoids are distributed among the genera *Lepidosteus* (Gar-pikes), *Amia* (Dog-fish of the Lakes), *Accipenser* (Sturgeons), *Scaphirhynchus* (Shovel-fish), and *Spatularia* (Paddle-fish)—all of which are North American—and *Polypterus* and *Calamichthys*, found in Africa.

The persistence of the ancient fishes in North America is doubtless due to the fact that certain parts of this continent are the oldest known portions of the earth's surface. The Canadian highlands have, so far as we know, never been submerged since the beginning of the Silurian, and nearly all of the interval between the Mississippi and Atlantic has existed as dry land since the close of the Carboniferous age. Hence, though exposed to extreme alternations of climate—as in the Tertiary and Glacial periods—parts of our continent have apparently always remained as asylums where some representatives of its ancient tribes of fishes have found a safe retreat, and where the chain of descent has been unbroken from remote geological ages.

A general view of the zoological relations and geological distribution of fossil fishes may be gathered from the following table, and the notes which follow it:

Table showing Classification and Geological Distribution of Fossil Fishes.

	Palæozoic.				Mesozoic.			Cenozoic.	
	L. Sil.	U. Sil.	Dev.	Carb.	Trias.	Juras.	Cret.	Tert.	Rec't.
Dipnoi									*
Teleostei							*	*	*
Placodermi		*	*						
Ganoidei									
a. Amiadæ									*
b. Lepidosteidæ			*	*	*	*	*	*	*
c. Crossopterygidæ			*	*	*	*			*
d. Chondrosteidæ			* ?			*		*	*
e. Cephalaspidæ		*	*						
f. Pycnodontidæ				*	*	*	*	*	
Acanthodidæ			*	*					
Elasmobranchii									
a. Squalli		*	*	*	*	*	*	*	*
b. Rajæ						*	*	*	*
c. Chimæroidi			*			*	*	*	*
Marsipobranchii									*
Pharyngobranchii									*

So far as is at present known, no fishes belonging to the orders *Pharyngobranchii* or *Marsipobranchii* have been found in the fossil state. This is probably due to the soft and perishable nature of their skeletons and integuments. It is evident that no traces of *Amphioxus* would remain if buried in sediments at the bottom of the water it inhabits; and the same is true of the Hag (*Myxine*), though if Lampreys existed in former times it is somewhat singular that no traces of their horny teeth should have been discovered.

The *Elasmobranchii* seem to be represented in the earliest traces of fishes known, and they have continued to exist in large numbers, and apparently with no very radical change of structure, from the Devonian and Carboniferous ages to the present day. From the fact that the skeleton in the *Elasmobranchii* is generally cartilaginous, it has almost always disappeared in the process of fossilization; and the more indestructible organs—the teeth, spines, and dermal tubercles—have alone resisted decay. These are also usually disconnected and often widely separated. Hence it is exceedingly difficult to say with certainty to what order of Elasmobranchs many of these *disjecta membra* belong—whether the fishes that wore them were Sharks, Rays or Chimæroids. The teeth will, however, frequently permit us to decide this question, and it would seem that the most ancient Elasmobranchs were mainly sharks. Of these we find great numbers of spines, teeth and dermal ossicles in the sediments of the Carboniferous sea, and many, though not as many, in those of the Devonian and Upper Silurian. The teeth of these ancient sharks present all the varieties of crushing, cutting and piercing organs; and it seems that there was even a greater variety among the sharks of the ancient seas than in those of the present day. It will also be remarked, that these old-time sharks were much more generally provided with defensive spines; some of which attained dimensions entirely unparalleled in our present fauna. Another characteristic of this ancient group of cartilaginous fishes is, that their superficial bony organs were generally highly ornamented. The spines to which I have referred are often most elegantly sculptured, and the dermal tubercles are embellished in various ways to a degree rarely observed in living species. Agassiz has shown that a large part of the ancient sharks are zoologically allied to *Cestracion*—the Port Jackson shark—so that this isolated genus, of which there are now but two or three species, may be considered, like the Ganoids and Chimæroids, as a relic of a fauna once immensely developed, but now on the eve of total extinction.

The Chimæroids are now represented by the two genera *Chimæra* and *Callorhynchus*—two species only of the former and one of the latter being known. Numerous teeth of extinct Chimæroids have been found in the Tertiary, Cretaceous and Jurassic rocks of the Old World, and it was long ago shown by Sir Philip Egerton, who described these fossil forms under the names of *Edaphodon*, *Passalodon*, etc., that the Chimæroids were numerous and attained large size in the Mesozoic seas. But up to the date of the publication of this report, no unmistakable remains of Chimæroids have been announced as discovered in Palæozoic rocks. It will be seen, however, that in the fossils which are figured and described in the present volume as different species of *Rhynchodus*, we have satisfactory proof of the existence in the Devonian seas of Chimæroids, apparently closely resembling in structure the best known of the more recent *Holocephali*. As has been remarked elsewhere, the Conodonts obtained by Pander from the Lower Silurian rocks of Russia, were described by him as the teeth of Elasmobranchs, and as affording evidence of the existence of sharks or rays at a much earlier period in the world's history than had been before suspected. Since the character of these organs is still under discussion, they can not be accepted as conclusive evidence of the truth of Prof. Pander's generalization.

The *Ganoidei*. The order of Ganoids,as at present constituted, includes among its living and fossil genera a multitude of forms, some of which seem to have very little in common. The most striking characteristics of the living Ganoids are the possession of bony or horny scales covered with enamel, the multiple valves of the *bulbus arteriosus*, the non-decussating optic nerves, and the abdominal ventral fins. It has also been ascertained that they generally have unequally-lobed or heterocercal tails, in which the vertebral column is prolonged into the superior lobe. The latter character is, however, by no means universal; although, so far as known, the caudal fins of Ganoids differ in structure from those of Teleosts. Whatever differences they may exhibit among themselves, they may in all cases be said to be *vertebrated;* that is, penetrated to a greater or less degree by the extension of the vertebral column; but in many of the Ganoids of ancient times (e. g. *Cœlacanthini*) the tail was equally divided by the prolongation of the spine—a phase of structure to which McCoy has applied the term *diphycercal.*

A large part of the Ganoids, both living and extinct, are provided with a coat of mail in a series of rhomboidal, bony and enameled scales, of which a typical example may be seen in the Gar-pike (*Lepidosteus*). But to this prevalent character we have a striking exception in the "Dog-fish" of the lakes (*Amia calva*), which is an undoubted Ganoid, but is provided with thin, circular, imbricated scales, very similar to those of most osseous fishes. On recurring to the past history of the order, we find both of these varieties so fully represented in fossil forms that we are able to arrange them in two great groups—the Cycliferous and Rhombiferous Ganoids. If it contained only these two forms, the order of *Ganoidei* would be well defined; but we must make it hold among living fishes the strange group composed of the Sturgeon, the Shovel-fish and the Paddle-fish, in which the body is either naked or covered with large plates, and is never, properly speaking, scaled. To these fishes the name Chondrostean Ganoids is given, from the preponderance of cartilage in their structure. The co-ordination of the extinct forms with those I have enumerated is a matter of no little difficulty, since they present marked peculiarities in their exoskeletons, and their internal anatomy is necessarily unknown. By Prof. Huxley the order of *Ganoidei*, when made to include both fossil and living forms, is divided into the following sub-orders:

1. The *Amiadæ*, having a single representative in the rivers and lakes of North America (*Amia calva*), and no member of the group is certainly known to exist in a fossil state. The characteristics of this sub-order, as given by Huxley, are cycloid scales, preoperculum, single median jugular plate, branchiostegal rays, non-lobate paired fins, and heterocercal tail.

2. The *Lepidosteidæ* include fishes which have rhomboidal, enameled scales, a preoperculum, branchiostegal rays, non-lobate paired fins, and heterocercal tails. Fishes of this group occur in the rivers and lakes of North America. Several species of the genus *Lepidosteus* also occur in Tertiary rocks. In the Cretaceous, Jurassic and Triassic strata, the *Lepidosteidæ* are represented by numerous genera—*Dapedius*, *Lepidotus*, *Æchmodus*, etc.—and in the Palæozoic formations by *Palæoniscus*, *Amblypterus* and *Eurylepis* in the Carboniferous, and perhaps *Cheirolepis* in the Devonian.

3. The *Crossopterygidæ*. The most striking character in this group is found in the paired fins, which are lobate; that is, having their solid, central portions covered with scales, and appearing like prolongations of the abdominal walls. The scales may be cycloid or rhomboid. The dorsal fins are either two in number, or, if single, very long, or composed of many subdivisions. There are no branchiostegal rays, the jugular plates are two principal, with sometimes several supplementary ones. The tail may be heterocercal or traversed centrally by the vertebral column (diphycercal). The only members of the sub-order now living are *Polypterus* of the Nile, and *Calamichthys*, which inhabits the rivers of Senegal. One of the most remarkable groups of this sub-order, found in the fossil state, is that which has been designated as the family of *Cœlacanthini*. The fishes of this group are found in the Chalk, where

they are represented by *Macropoma*, in the Jurassic, represented by *Undina*, and in the Permian and Carboniferous, where the genus *Cœlacanthus* occurs, of which three species are described in the present volume.

All of these last-mentioned fishes are characterized by hollow fin-rays —whence their name—and by paired, elliptical, jugular plates, two dorsals which are sustained by palmated, inter-spinous bones, and by diphycercal tails, through which the vertebral column extends, and bears at its extremity a minute, supplementary caudal fin. In the Devonian and Carboniferous formations a large number of genera represent the *Crossopterygidæ*, viz.: *Osteolepis*, *Diplopterus*, *Glyptolæmus*, *Megalichthys*, *Holoptychius*, *Rhizodus*, *Dipterus*, *Phaneropleuron*, and probably the American genus *Onychodus*. In some of the fishes of this sub-order the bodies of the vertebræ are ossified, and it is the opinion of Prof. Huxley that the passage from the Fishes to the Amphibians was through this group.

4. The *Chondrosteidæ*. In the fishes of this sub-order the body is naked, or more generally protected by bony plates, which cover more or less of the surface. Neither the pectoral nor the ventral fins are lobate. The branchiostegal rays are few or absent; the tail is heterocercal. The teeth are small or absent. In this group are the sturgeons (*Accipenser*) which inhabit the rivers of all the northern hemisphere, and *Spatularia* and *Scaphirhynchus*, found only in North America. These constitute all the living members of the sub-order. In the Jurassic rocks the sturgeons are represented by *Chondrosteus*, and, as I think, in the Devonian, by *Macropetalichthys*.

5. The *Cephalaspidæ*. These form a remarkable group of fishes which occur only in the Devonian and Upper Silurian rocks, and include, perhaps, the oldest well-defined fishes yet known. The tpye of this sub-order is *Cephalaspis*, which has been made well known to the students of geology through the descriptions given by Agassiz and Hugh Miller. In *Cephalaspis* the head and anterior portions of the body were covered by a broad, shovel-shaped, bony shield. The posterior portion was scaled. The other members of the group are *Pteraspis*, *Auchenaspis*, *Menaspis*, *Scaphaspis*, etc. They are all small, and their relations to living fishes are quite uncertain. Prof. Huxley has pointed out the resemblance which they bear to the Chondrosteans; comparing *Cephalaspis* with *Scaphirhynchus*, and *Pteraspis* with *Spatularia*. It is a somewhat remarkable fact that, with the exception of one well-marked species of *Cephalaspis* (*C. Dawsoni*), found by Dr. Dawson in the Devonian rocks of Gaspe, none of the *Cephalaspidæ* have been discovered in America.*

* An exhaustive monograph of the *Cephalaspidæ* has been published by Messrs. Lancaster & Powrie in the volumes of the Palæontographical Society for 1870–71.

6. The *Placodermi*. This is the name given by Pander to the group of fossil fishes called *Placoganoids* by Owen, which includes *Coccosteus*, *Pterichthys*, *Asterolepis* and *Heterostius* of Pander, and also the gigantic fishes, for the first time fully described in this report, and found as yet only in the Devonian rocks of Ohio—*Dinichthys* and *Aspidichthys*. In the Placoderms, the head and anterior portions of the body were protected by bony plates, which were for the most part united by sutures, and had the external surface studded with points or bosses of enamel. The posterior portion of the body was either naked, or covered with angular, enameled scales. The vertebral column was generally cartilaginous; but a *Coccosteus*, with true bony vertebræ, is reported to have been recently discovered.*

The remains of these singular fishes were first found in Russia (1813) by Prof. Asmuss; and by him and by Prof. Eichwald several genera were described under the names of *Asterolepis*, *Bothriolepis*, *Homostius* and *Heterostius*. The fragmentary condition of the specimens upon which these descriptions were based created considerable confusion of nomenclature. *Asterolepis* of Eichwald is now known as *Pterichthys*, and *Homostius* has been superseded by *Asterolepis*; the last two names having been firmly fixed to the icthyolites which now bear them by the graphic descriptions of Hugh Miller. Similar fossil fishes have since been found in considerable numbers in Scotland by Miller, Peach and others, in Bohemia by Barrande, and in this country by the writer. No well-marked specimens of *Pterichthys*, *Coccosteus* or *Asterolepis* have yet been met with in America, but I have a few detached plates, obtained from the Corniferous limestone at Delaware, which probably belong to *Coccosteus;* and the great fishes of the Huron shale, described in this report under the names of *Dinichthys* and *Aspidichthys*, exhibit such resemblances in structure to the Placoderms of the Old World, that we are compelled to place them in the same sub-order.

The Placoderms are eminently characteristic of the Devonian age; and the larger members of the group were not only the most highly organized, but were, from their size and armament, the most formidable of then existing animals.

The affinities of the *Placodermi* are still matters of doubt and discussion among zoologists. The cartilaginous condition of the vertebral column is the only feature in their structure that indicates a low or embryonic organization; and this is a character of somewhat doubtful significance. The vertebræ are also reported to have been sometimes ossified, as in the

* Murchison's Siluria, p. 478.

case referred to above. The amount of true bony tissue that composed their cephalic, dorsal and ventral shields, and their massive jaws, was far in excess of that secreted by most of our so-called bony fishes. Prof. Huxley, who has discussed this question with characteristic sagacity, compares the bony armor of the Placoderms with that of the plated Siluroids, and is disposed to consider them as connecting links between the physostome Teleosts—the highest of the order—and the typical Ganoids.

7. The *Acanthodidæ.* These were small fusiform fishes, of which the remains are not rare in the Palæozoic strata of the Old World, but have never yet been met with in America. Their geological range is from the base of the Devonian to the Permian. A large number of species have been described by Agassiz, Sir Philip Egerton, Mr. Powrie, Profs. Kner, Roemer and others, in the genera *Acanthodes*, *Diplacanthus*, *Climatius*, *Parexus*, etc.

The Acanthodians derive their name from the numerous and relatively large spines which are set at the anterior margins of the fins, and elsewhere on the body. These spines are implanted in the integuments like those of sharks; and like them also are often ornamented and some times set with rows of denticles. The surface of the body in the Acanthodians was covered with closely-set, shagreen-like scales; the spine was "notochordal," and the cranium mostly cartilaginous.

By Prof. Agassiz the Acanthodians were regarded as Ganoids; but Mr. Powrie considers them as Elasmobranchs, while Prof. Huxley is disposed to regard them as annectant forms between the Elasmobranchs and Ganoids.

8. The *Pycnodontidæ* are also fishes of somewhat questionable relations, though generally considered as Ganoids. In form and in some points of structure they resemble the Plectognath-Teleosts—*Balistes*, *Ostracion*, etc.—but the spinal column was cartilaginous, and the dentition consisted of a series of bony and enameled bosses set in a kind of pavement, and adapted to crushing mollusks and crustaceans. The Pycnodonts range from the base of the Carboniferous series to the Tertiary, but are now all extinct. Few Pycnodont fishes have as yet been found in the rocks of America. I have identified *Platysomus* among the fossils collected from the Coal Measures of Illinois, and have seen the teeth of undescribed Pycnodonts collected from the Lower Carboniferous of Arizona by Mr. G. K. Gilbert, from the Cretaceous of Brazil by Prof. Hartt, and from the Green Sand of New Jersey by Prof. Marsh.

ERRATUM

Page 256, line 15 from the bottom of the page should read "The type of . . . ."

10

Reprinted from *Am. Naturalist* **42:**361–373 (1908)

THE ANCESTRY OF THE CAUDATE AMPHIBIA

DR. ROY L. MOODIE

UNIVERSITY OF KANSAS

THE phylogeny of the group of vertebrates commonly called amphibians or batrachians has been, and still is to some extent, among the most obscure of the problems connected with the descent of animals. So far as I am aware there have been but few statements as to the possible ancestry of the Amphibia and no attempt has been made to set forth in detail the series of structures through which the animals have passed from the beginning of their line to the present. During the course of an extensive investigation on the extinct Amphibia of North America, the writer has reached some interesting conclusions in regard to the ancestry of at least one group of the modern Amphibia and these conclusions are here set forth in detail. Some of the facts offered in support of these conclusions have been given in other connections.

So far as at present known, the first trace of vertebrate life on earth is that of fishes in Ordovician rocks in Colorado and in two places in Wyoming, *i. e.,* the Black Hills and the Big Horn Mountains. In the Devonian, if the impressions from the Catskill are properly interpreted, the fishes had given rise to a quadrupedal type of animals which are usually known as the Amphibia. Only impressions of footprints in the sandstone are known, but these are quite instructive. There are no traces of the bones

FIG. 1. Restoration of *Micrerpeton caudatum* Moodie from the Coal Measures of Illinois. Length of form, 49 mm. Spider is a composite restoration and is based on actual specimens and on the results of Roemer, Scudder, Beecher and Melander. ×2⅓.

of these quadrupeds known in rocks of Devonian or of Mississippian age, although Lohest (1) some years ago called attention to some remains which he thought were amphibian from the Devonian rocks of France. Thevenin has recently cast doubt upon this interpretation by Lohest (2) and the figures as given by Lohest would seem not to be amphibian, but fish remains.

In rocks of the lower part of the Pennsylvanian occur, in North America, the first evidences of the bones of quadrupeds. The Amphibia show, even thus early, that their line had divided into three and possibly four distinct groups which are usually known as the Branchiosauria, the Microsauria, the Aistopoda and the Stereospondyli. The presence of the last group is indicated by the two vertebræ from the Carboniferous of Nova Scotia named by Professor Marsh *Eosaurus canadensis* and also by certain fragments of large ribs and large skulls from the coal beds of Ohio. The Temnospondyli did not make their appearance until the latter part of the Pennsylvanian, when their remains are found in Carboniferous rocks of Kansas, Illinois and Pennsylvania.

The Stereospondyli and the Temnospondyli are specialized side branches of the amphibian or reptilian stem and will not concern us further here. The Aistopoda are regarded by the writer at present as being specialized microsaurians. This would not favor the view of Wiedersheim as to the descent of the Cœcilians. This point needs further investigation and will be discussed elsewhere. The Microsauria Dr. Gadow has placed in the subclass Proreptiliæ and, in the opinion of the writer, his classification represents the correct facts, but further discussion will be postponed. The group to which the reader's attention is here invited is the one usually known as the Branchiosauria, representatives of which are found only in Western Europe and in North America, where a single specimen is known from Illinois. The group is a very small one both as to the size of the individual members and as to the number of species.

The Branchiosauria are distinguished from all of the other Amphibia-like vertebrates by the presence of short, heavy, straight ribs. The Microsauria always have long, thin, curved ribs. The Aistopoda usually are destitute of ribs or when present they are curved and slight. In conversation recently with Dr. Gadow he made the interesting suggestion that the characters presented by the ribs might be used as a basis on which to separate the group heretofore known as Stegocephala into two main divisions. One of the divisions would be the true Amphibia with the Branchiosauria as the representative order among the early forms, in which case the term Branchiosauria would be a misnomer, and the other Stegocephala would be included under some such name as the Pro-

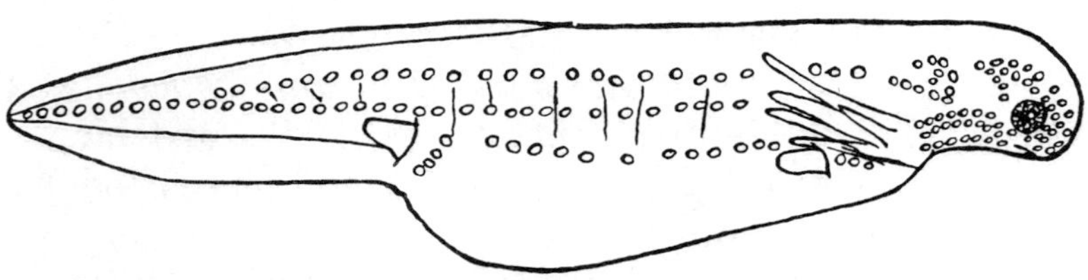

FIG. 2. Outline of a Larva of *Necturus*, showing the arrangement of the lateral line systems. After Platt.

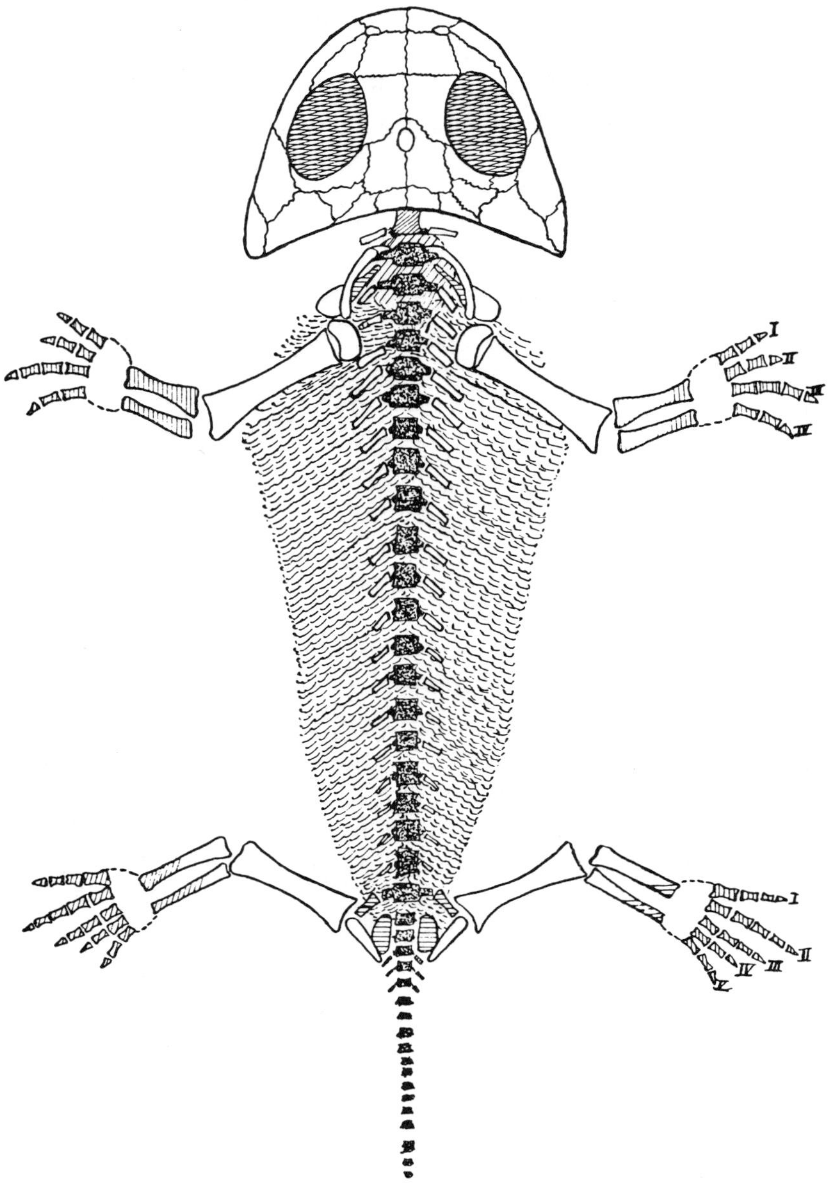

FIG. 3. Restoration of the Skeleton of *Micrerpeton caudatum* Moodie. The feet are conjectural and are based on the studies of Credner. The broad ventral armature is well developed in this form. × 3½.

reptilia or would be distributed among the reptiles proper. The former group is characterized by the short straight ribs borne on stout transverse processes and the latter group is characterized by the long curved ribs borne intercentrally. There seem to be some difficulties in this classification, but it is hoped that they may be cleared up by future investigation.

The Branchiosauria, as has been several times suggested by various authors, represent the ancestral forms of at least the tailed division of the Amphibia. A suggestion as to the ancestry of the tailless forms will be given elsewhere. The conclusion that the Branchiosauria are the direct ancestral forms to the modern Caudata is based on several characters. These characters are: the structure of the skull, the structure and form of the vertebræ and ribs, the number of digits, the arrangement of the phalangeal elements, the characters of the pectoral and pelvic girdles, the character of the lateral line system, the structure and form of the long bones and finally the shape of the body; all of which will be discussed below.

It has been suggested on embryological grounds that the Amphibia are a degenerate group (3) and this is borne out, in so far as the Caudata are concerned, by a study of the cranium of the early and recent forms as well as by other structures. The cranium of the Branchiosauria is almost identical in structure with that of the Microsauria and it exhibits a completely roofed-over skull with only five openings, namely, those for the orbits, the nostrils and the pineal opening.

The elements forming the roof of the skull are quite constant in the Branchiosauria and also in the Microsauria and they differ only in position and relations. Practically the same elements form the skull roof in the two groups, but the forms differ in other important respects. In the skull of the Branchiosauria we find in the median line a row of elements in pairs, which extend the entire length of the skull. These are (beginning anteriorly): the premaxillæ, the nasals, the frontals, the parietals and the supraoccipitals. On the side of these elements lie others which vary in their position to some extent; the median elements are fixed and not so likely to

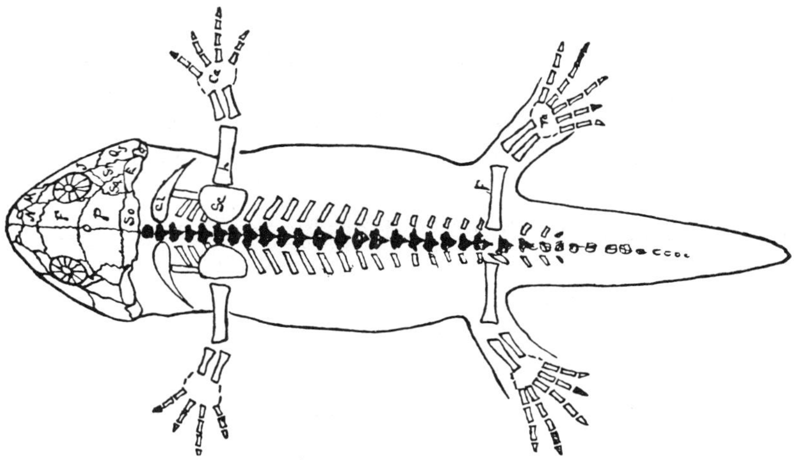

FIG. 4. Restoration with Outline of the *Branchiosaurus fayoli* Thevenin based on the figures given by Thevenin. The last phalanx in digit I of the hand is conjectural and the last two on digit II and the last one on digit IV of the foot are conjectural. From the Upper Carboniferous of France. ×2.

vary. Anterior to the orbit there are the prefrontal, the lachrymal, the maxilla, and lateral to the orbit lies the jugal. Posterior to the orbit lie the postorbital, the postfrontal, the squamosal, the supratemporal, the opiotic, and on the postero-lateral border of the skull lie the quadrato-jugal and quadrate.

In the skull of the modern Caudata we find an arrangement of the elements which is quite similar to that described for the Branchiosauria. In the skull of Megalobatrachus (Fig. 10), for instance, we find the following

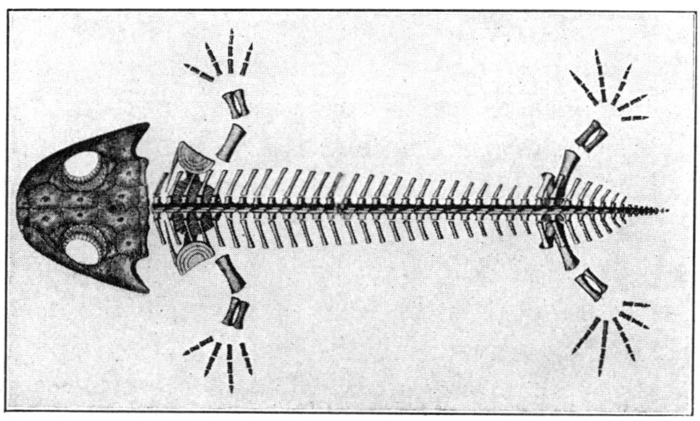

FIG. 5. Restoration of *Branchiosaurus amblystomus* Cred. from the Permian of Saxony. After Credner.

bones paired in the median line: the premaxillæ, the nasals, the frontals, the parietals, and on the posterior end of the skull occur the exoccipitals. In this series it is clear that the supraoccipital elements have disappeared. Lateral to the median paired row there are the maxilla, the prefrontal, the squamosal, and in allied forms the jugal, quadratojugal and quadrate. The epiotic is wanting in all skulls of modern amphibians except some coccidians and the elements are not so firmly united as in the skull of the Branchiosauria. The skull of the Caudata exhibits a weaker, more degenerate condition, than is found among the early forms. The lower surface of the skull is practically the same in the two groups, especially in the possession of a large parasphenoid. Teeth on the palate bones are lacking, for the most part, in the modern forms, which is another evidence of degeneracy.

If the structure of the vertebræ is now examined it will be seen that in this particular there is close affinity between the modern Caudata and the Paleozoic Branchiosauria. The vertebræ of the modern Caudata are amphicœlous and opisthocœlous, with the notochord rarely persistent and with the transverse processes springing off boldly from the body of the centrum. Similar conditions are observed in the Branchiosauria where the vertebræ, so far as they are known, are amphicœlous and the transverse process is unusually strong with the notochord probably persistent. The transverse processes give support to the short, heavy, stout ribs which lay in the flesh and made acute angles posteriorly with the vertebral column. The ribs in the Branchiosauria and in the later Caudata are indistinguishable so far as form is concerned. The point of difference is that in the modern forms the ribs are usually weaker posteriorly and are, for the most part, confined to the presacral region of the body. This is another instance in which the Caudata show their degenerate structure. On the basis of the ribs alone the Branchiosauria may well be separated from the Microsauria and from all other of the so-called Stegocephala, and would be considered as the true Amphibia,

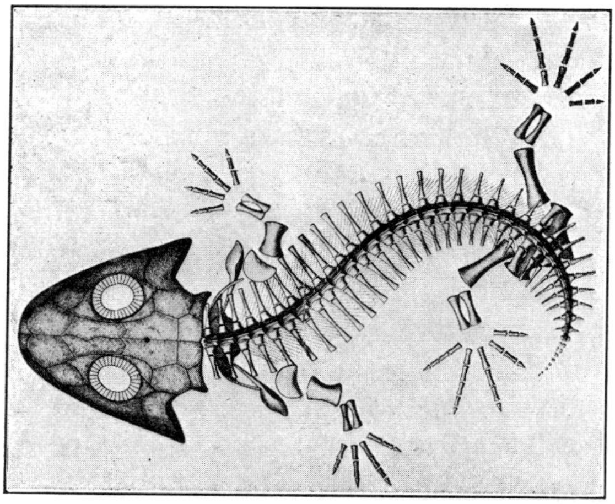

FIG. 6. Restoration of *Melanerpeton* from the Permian of Saxony. After Credner.

while all other forms would be excluded from this class and would have to be placed among the Reptilia, where they probably belong; whether they are to be considered as a subclass of the Reptilia or not, does not concern us here.

In the forelimb there are never more than four digits either in the Paleozoic Branchiosauria nor in the later Caudata. So far as I am aware, there is no paleontological evidence to show that there ever were more than four digits in the hand of the caudate Amphibia and in this discussion the first digit will be considered as No. I., the second digit as No. II. and so on. Fritsch, it is true, has figured five digits in the hand of *Branchiosaurus salamandroides* Fr. (Fig. 7), from the Permian of Bohemia, (4) but in giving us a glimpse into the material on which his restoration is based there are only four digits preserved in the hand, although Fritsch says the first had been lost. I doubt if there ever were more than four, and the digits in Fritsch's specimen were all preserved. In all of the Permian forms from Saxony Credner has figured four digits in the hand (Figs. 5–6).

The specimens recently described by Thevenin from the Upper Carboniferous of France (5) show only four digits in the hand. In Fig. 4 a restoration of the *Branchiosaurus fayoli* Thevenin has been attempted. The reconstruction is based on the figures as given by Thevenin. Only one phalanx is restored and that is the distal one of the first digit. All of the rest were preserved. In the Wealden of Belgium the *Hylæobatrachus croyii* Dollo (Fig. 8) exhibits but four digits in the hand. In *Andrias scheuchzeri* Tschudi (Fig. 9) from the Miocene of Switzerland, there are but four digits in the hand, and in all of the modern caudate forms, so far as I am aware, there are but four digits and the phalangeal formula for all, both in Paleozoic and in later forms, is universally 3-3-4-3.

In the foot the Branchiosauria show the same agreement with the modern Caudata. There are always five digits and they usually have the phalangeal formula 4-5-4-3-3. Neither the carpus nor the tarsus is fully ossified in the Branchiosauria or in the later Caudata, although the carpus and tarsus are partly osseous in the Amblystomatidæ.

The pectoral girdle of the Branchiosauria consists usually of four elements, three paired and one unpaired. These are: the single interclavicle, the clavicles, the coracoids (cleithra?) and the scapulæ. Among the modern

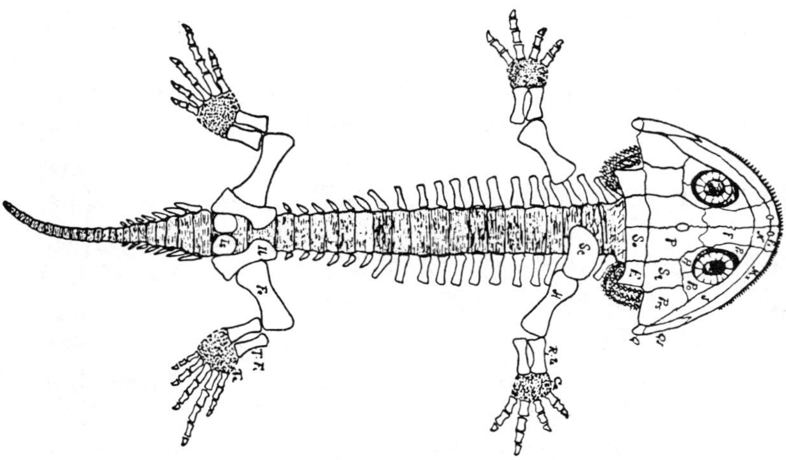

FIG. 7. Restoration of *Branchiosaurus salamandroides* Fr. from the Permian of Bohemia. Modified after Fritsch.

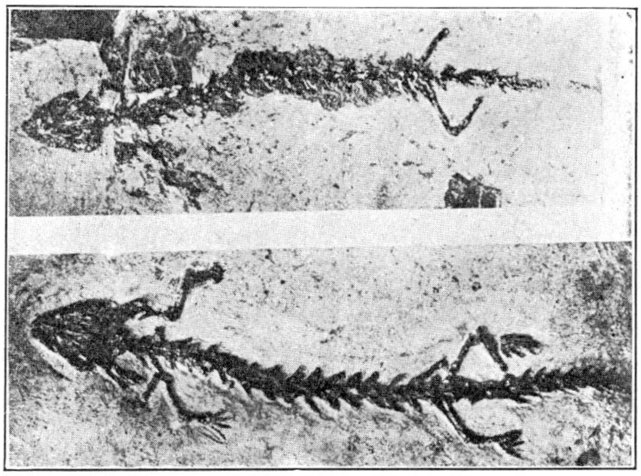

FIG. 8. *Hylæobatrachus croyii* Dollo from the Wealden of Bernissart. After Dollo.

caudate forms as well as among the Miocene forms all of the elements with the exception of the scapulæ have become cartilaginous, so that there remains in the modern caudate amphibians only a bony scapula. This seems to be the case in the Wealden Hylæobatrachus, but the single specimen is not well enough preserved to furnish definite information on this point. The pelvic girdle has, apparently, undergone no change in the evolution of the Caudata. The pubis is never ossified in either the Branchiosauria or in the latter Caudata and the ischium is usually plate-like, while the ilium is more or less rounded. The sacral rib, of which there is never more than one pair, is usually well developed.

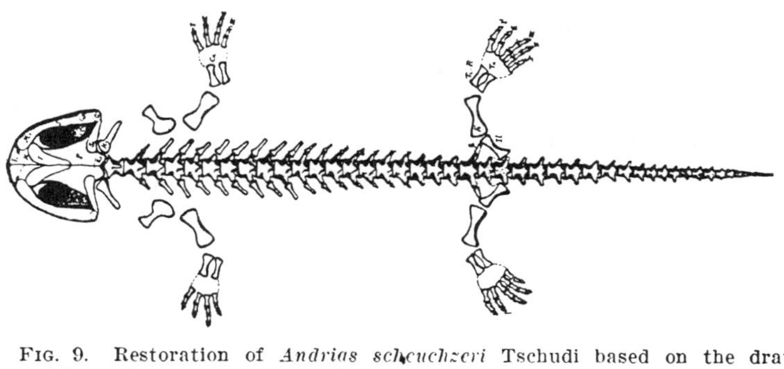

FIG. 9. Restoration of *Andrias scheuchzeri* Tschudi based on the drawings published by von Meyer in "Fauna der Vorwelt." × ¼.

In an essay now in press on "The Lateral Line Organs in Extinct Amphibia" the writer has called especial attention to the character of the lateral line organs as they are preserved in the single specimen of a branchiosaurian, Micrerpeton, from the Coal Measures of Illinois. A restoration of this form has been attempted in Fig. 1. The animal, as preserved, measures only 49 millimeters. It is apparently an adult, as there are no evidences of branchiæ and the limb bones are well formed. The spider shown in the restoration is a composite and is

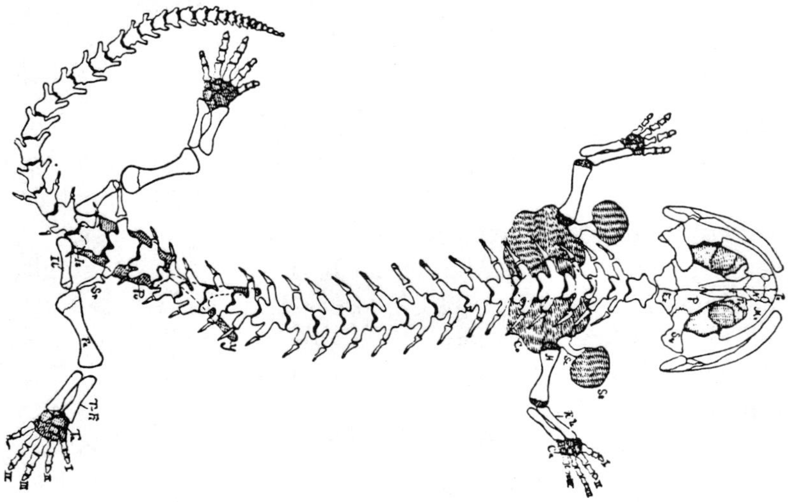

FIG. 10. The Skeleton of *Megalobatrachus japonicus* ——. After Osawa.

based partly on actual specimens partly on the results of Roemer, Beecher, Scudder and Melander. The body of the spider which was used as a model was about a half an inch in length. The particular characters in the amphibian which deserve mention now are the two lines representing the lateral line organs. There is seen a median lateral line which starts at the tip of the tail and runs forward. The other begins somewhat anteriorly and springs boldly from the median lateral line at a distance of a few millimeters from the tip of the tail. These lines represent rows of pigmented scales which show on the specimen as dark lines. The sense organs were undoubtedly located beneath these specialized scales, much as the lateral line organs are protected in some of the fishes. In Necturus (Fig. 2) the arrangement of the lateral line organs is almost identical with what has just been described for Micrerpeton. I have suggested elsewhere that this similarity in the lateral line organs of these two forms may be indicative of relationship between the groups to which these forms belong. There is no reason why the type of lateral line should not have been preserved in the amphibians, since we know that in some of the fishes certain types of peculiar lateral line structure have persisted for nearly as great a length of time.

In the structure of the limb bones and ribs the Branchiosauria are much like the modern Amphibia in that the bones are formed almost entirely of perichondrium. There are never any bony epiphyses in any of the Amphibia. In some of the toads Parsons has seen the cartilaginous epiphyses calcified, but they are never osseous. The endochondrium is not at all or but slightly developed in the Amphibia, and in the fossilized bones this condition causes a concavity at the ends. In the Microsauria the endochondrium is more fully formed, although there are some in which the ends of the limb bones are nearly flat.

The form of the body in the Branchiosauria is strikingly salamandrine, as may be seen by referring to figures 1 and 4. In Micrerpeton the tail is quite long and almost equals the length of the presacral region. In the *Branchiosaurus fayoli* Thevenin (Fig. 4), the tail is shorter but the form is much the same. The body of the Branchiosaurus (Fig. 4) is more stout than that of the Micrerpeton (Fig. 3) and the ribs are longer.

In summing up all of the characters as presented by the Caudata and the Branchiosauria it seems most probable that the Caudata are but degenerate Branchiosaurians and the changes which have taken place in the skeleton are mostly brought about by the loss of certain parts.

Summary

1. The Stegocephala will probably have to be divided into two groups, one of which will be considered as Amphibia and the members of the other group will have to be placed among the Reptilia but can not all be placed under any single head since they represent divergent types and must be placed in with the reptilian groups to which they have probably given origin.

2. The Branchiosauria are the ancestral forms of at least the caudate Amphibia. The Branchiosauria are first known in the Pennsylvanian of North America and they present characters which separate them clearly from all other groups of the so-called Stegocephala.

3. The characters which the Branchiosauria have in common with the Caudata are: short, straight ribs; stout transverse processes springing from the body of the vertebra; practically the same number of presacral vertebræ; the same skull structure; the degenerate character of the pectoral girdle and the close correspondence of the pelvic girdle; the same number of the digits and the same phalangeal formula in the Branchiosauria and Caudata; the similar characters presented by the lateral line system; the structure of the long bones and the shape of the body.

4. There can be but little doubt that the Caudata are the direct descendants of the Branchiosauria, of which they are but degenerate forms.

Bibliography

1888. Lohest, M. *Annales de la Société Geol. de Belgique,* Tome XV, p. cxx.

1906. Thevenin, A. *Annales de Paleontologie,* Tome I, Fascicule III, p. 13.

1907. Hatta. On Gastrulation in Petromyzon. *Journ. of the College of Science,* Tokyo.

1889. Fritsch, Anton. Fauna der Gaskohle. Bd. II.

1906. Thevenin, A. *Annales de Paleontologie,* Tome I, Fascicule III, p. 9, Plate I.

11

Reprinted from *Jour. Geology* **25:**411-421 (1917)

THE PHYLOGENY AND CLASSIFICATION OF REPTILES

S. W. WILLISTON
University of Chicago

Not many years ago it was the fashion to construct phylogenetic trees, often of wonderful design, for almost every group of animal and vegetable life. Because of the failure of so many of them, the practice has somewhat fallen into desuetude in recent years, and it is only hesitatingly that I have ventured, for the first time, to express in tabular form my own views of the phylogeny and classification of the reptiles. I do so the more readily, however, as I have no startling novelties to offer.

Phylogenetic schemes are always useful when constructed with proper discrimination and with due regard to the known and the unknown. They often furnish some residue of permanent knowledge, some real contribution to taxonomy and the doctrine of evolution; or, if not, by their failure they limit the field of legitimate speculation.

Especially has our greatly increased knowledge of the early land vertebrates, both at home and abroad, rendered it possible, I believe, to approximate more closely the real origin of many forms of vertebrate life than it was a dozen years ago. It was not long ago that we were seeking the beginning, or at least the early stages, of all reptile life in the order Rhynchocephalia. Because we found, or thought that we found, in *Sphenodon*, or *Hatteria*, as the genus was long called, the most generalized or primitive characters among living reptiles, it was not unnaturally assumed that its immediate ancestral stock was the most primitive reptilian type of the past. And when Credner discovered a quarter of a century ago, far back in the Permian rocks, another very primitive reptile, it was also assumed, too readily, that it was of the same stock. Upon that error, and it was an error, was built an elaborate edifice with the Rhynchocephalia as its cornerstone, until, as so

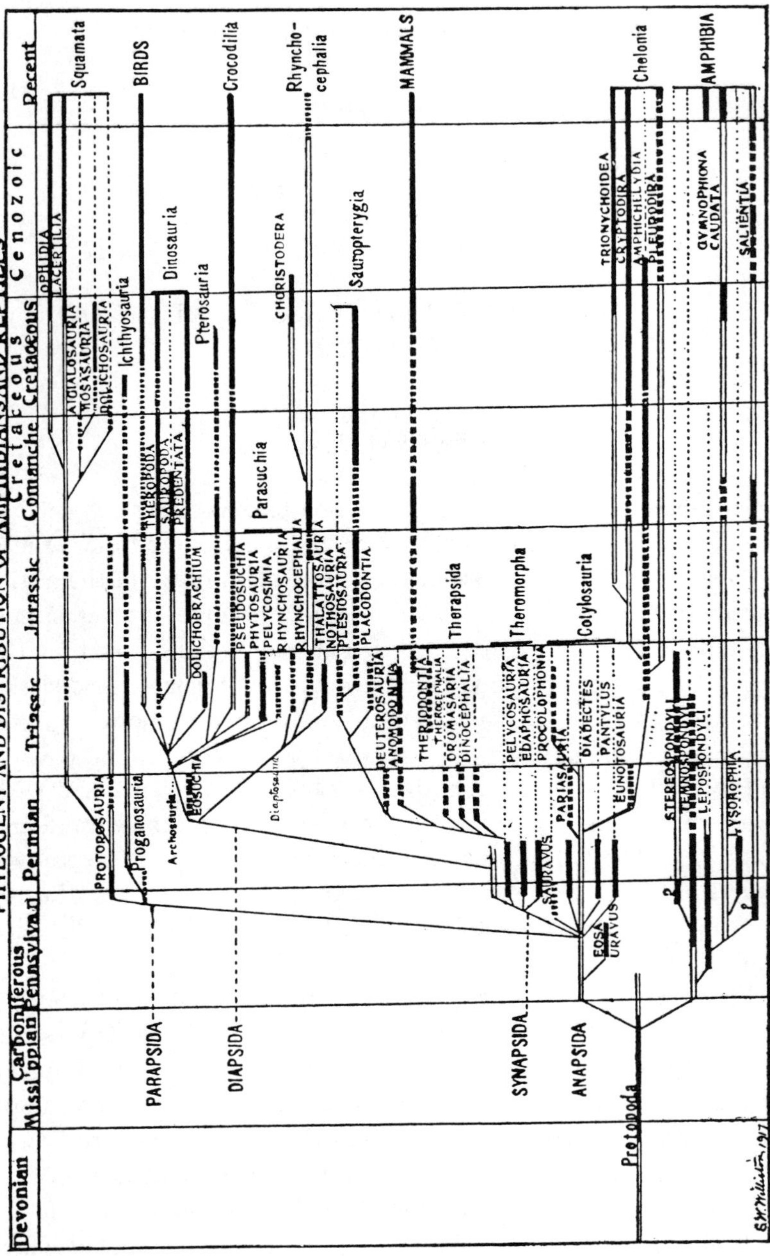
PHYLOGENY AND DISTRIBUTION OF AMPHIBIANS AND REPTILES
Devonian
Carboniferous
Missi'pplan
Pennsylvan'
Permian
Triassic
Jurassic
Cretaceous
Comanche
Cretaceous
Cenozoic
Recent
PARAPSIDA
DIAPSIDA
SYNAPSIDA
ANAPSIDA
Protopoda
Squamata
BIRDS
Dinosauria
Pterosauria
Crocodilia
Parasuchia
Rhyncho-cephalia
Sauropterygia
MAMMALS
Therapsida
Theromorpha
Cotylosauria
Chelonia
AMPHIBIA
North America
Elsewhere
Unknown

FIG. 1.

often in such cases, it was found that the foundation was insecure, and the edifice toppled. Everything was made to fit the uncertain base. Baur made the Proganosauria an integral part; Baur and Osborn found in the ichthyosaurs a mere wing; Broom and Osborn added a true cotylosaur; and Baur and Case built in the pelycosaurs; *Protorosaurus* and *Pleurosaurus* were mere chinks; and everybody (except Cope) united with them the lizards and snakes. I do not mention these names in any invidious spirit; they are all of men justly famous for their work in paleontology.

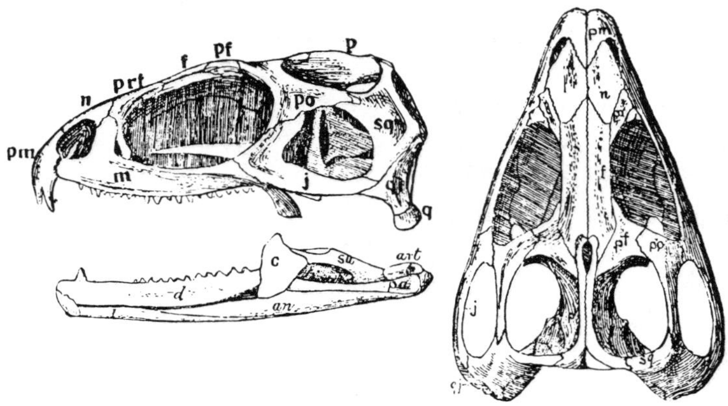

FIG. 2.—*Sphenodon*, skull from side and above. Diapsida. Recent

We all have the same proclivity, to find or to think that we find those things for which we are seeking. As Baur himself has said, they show "wie leicht man sich taüschen lassen kann, wenn man durch eine allgemeine gültige Anschauungsweise beeinflusst wird." A growing skepticism of the Rhynchocephalian affinities of *Paleohatteria* disclosed little support for the far-reaching conclusions based upon it; and, one by one, other assumptions have fallen by the wayside. A careful examination of the type specimens of *Paleohatteria* assured me that the genus was really a member of the Theromorpha. Watson holds the same opinion; and Huene has urged its relationship to the Pelycosauria.

It was Cope who, years ago, first suggested that in the temporal region of the skull the surest criteria for the classification of the Reptilia are to be found. Woodward carried the suggestion further, and showed their availability, but it was Osborn and McGregor who first applied them definitely. They assumed too much, as we have seen, but the credit is due to Osborn, more than to anyone else, for the foundation of a true reptilian phylogeny, and to him we owe especially a better knowledge of the double-arched reptiles. He has called them the **Diapsida**, and there is no better name for them. After the elimination of the forms which we are sure do not belong with them, we are all now, I think, in accord as to their phyletic unity. It is only in details that further research (and there is much yet to be done) will be of value. The separation of the great group called the dinosaurs, first proposed by Seeley and warmly espoused of late by von Huene, into two co-ordinate divisions, the Saurischia and Ornithischia, has much to commend it. Valid arguments for the phyletic unity of the pterosaurs, crocodiles, and pseudosuchians have been offered by Huene, and their close relations with the phytosaurs and other "thecodonts" is probable. All of these have certain annectant characters, which, to me at least, are impressive. Altogether they constitute the phyletic group of the Archosauria, a term we owe to Cope.

The Diaptosauria, the other component group of the Diapsida, is much smaller than when Osborn named it; and of the few forms that are left, one, the Thalattosauria of Merriam, is discordant and uncertain. It has been offered as a connecting link between the true rhynchocephalians and the squamate reptiles; I would rather shift it bodily to the Parapsida. The Diaptosauria are the more generalized diapsids; the distinction between them and the Theromorpha is not great. The earliest assured member of the Diapsida has been carried back, I believe, no farther than late Permian, in *Youngina*, to which Broom has given the inappropriate group name of Eosuchia, inappropriate because *Eosuchus* Dollo is a true crocodile.

The origin of the Diapsida, thanks chiefly to Baur and Case, seems clear. These authors thought that the Pelycosauria were really a part of the Rhynchocephalia, and for years they were classed among them in our textbooks. It was an error, but the error has shown, definitely I think, how the Diapsida arose, by the simple separation in reptiles with a lower temporal vacuity of the orbito-squamosal arch, leaving an upper temporal opening. The bones here are usually loosely connected, so loosely indeed in *Dimetrodon* that it is only within the past few years that we have become sure that the separation was not permanent; a permanent separation, one that would admit a knife blade even, and the deed is done. Whether or not *Ophiacodon* is a real example of this beginning, and I believe that it is, from such forms as *Mycterosaurus* the step to the diapsid type is trivial. If I am correct, and I am confident that I am, we then have the origin of the double-arched reptiles in early or middle Permian times from theromorph reptiles with a single, typically lower, temporal opening, possibly from forms not unlike *Paleohatteria*.

This group or subclass, which, with due modifications of the original concept, may properly bear the name **Synapsida** given to it by Osborn, includes scores of well-known genera of the orders Theromorpha, Therapsida, and doubtless also the Sauropterygia. It is the group that gave origin to the mammals, and has long since been extinct. In its simplest and most primitive types, together with all primitive characters of the skeleton, it has a single temporal opening bounded below by the jugal, above by the postorbital and squamosal. This opening, I believe, arose by the separation of the squamosal and jugal, and not by a definite perforation of any bone. And this opening is the sole character by which the group is ultimately distinguished from the Cotylosauria, its ancestral stock.

The evolution of the theromorph type through the dinocephalian, therocephalian, and theriodont to the mammalian seems assured by the African discoveries. In this evolution the structure of the temporal region has undergone changes of which we do not yet feel sure. The theriodont and anomodont reptiles, like the plesiosaurian, may have the temporal opening extending from the jugal to the parietal, apparently homologous with the combined openings of the diapsid forms, or, as Watson has suggested, with neither the synapsid nor diapsid; but, tracing the development as a whole, I should sooner believe that they all arose from the primitive type like that of *Dimetrodon* or *Mycterosaurus*, that is, from a typically lower opening bounded above by the postorbito-squamosal arcade. In none of these forms does the quadratojugal enter into the opening; in some it has entirely disappeared; and the former, I believe, was the primitive condition. The opening, I believe, first appeared below and behind the orbit, at the apex of the squamosal. In the Cotylosauria the squamosal is a large bone extending far down on the side of the skull and back of the quadrate. In *Dimetrodon* and *Sphenacodon*, indeed, the quadratojugal is almost confined to the posterior side of the quadrate. Its tendency was to disappear in this type of skull, and only in some Diapsida did it become a part of the lower arch.

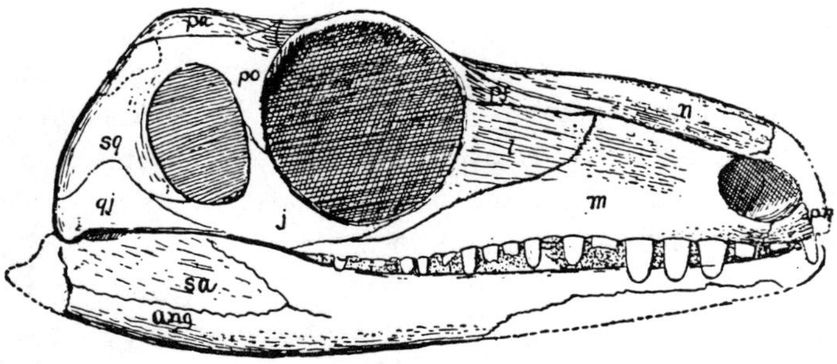

FIG. 3.—*Mycterosaurus*, skull, natural size. Synapsida, Permocarboniferous

It seems now evident that the temporal openings arose in yet another way: by the primitive separation of the postorbito-squamosal arcade from the parietal in the stegocrotaphous skull; and it seems very probable that this type of skull arose very early in geological history, as early as the lower opening, and before the separation of the upper arch in the diapsid skull. In these forms, or in most of them at least, an additional temporal bone was retained long after it was lost in other groups. And this is one of the reasons why I believe that the Ichthyosauria and the Squamata arose from a common or allied stem, direct from the Cotylosauria. For this phylum I propose the name **Parapsida.** As we have seen, Baur and others, because of the many primitive characters of the Ichthyosauria, believed that the order came from the original double-arched stem, that the lower temporal opening had been secondarily closed. Cope, in 1896, asserted that the ichthyosaurs had an independent origin from the Cotylosauria, and so indicated in a phylogenetic diagram. Broom adopted this view in 1901. In 1904 I quoted Cope's views with approval. More recently von Huene has reached the same conclusion, finding in the Proganosauria or Mesosauria, as here accepted, either the ancestral stock or one closely allied to it. Baur, it is true, in 1887 considered the Proganosauria as the ancestral stock of the ichthyosaurs, but Baur's Proganosauria later included *Paleohatteria*, believed to be a double-arched reptile. Merriam accepted the Cotylosaurian ancestry, and Sollas very recently has voiced his approval.[1] Indeed, the opinions now seem to be unanimous and further discussion is superfluous.

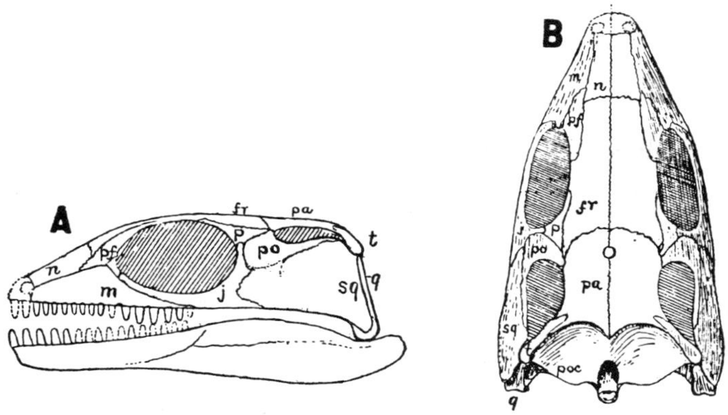

FIG. 4.—*Araeoscelis*, skull, from side (*A*), and above (*B*), natural size. Parapsida Permocarboniferous.

No order of reptiles has been the subject of more dispute than the Squamata. The apparent absence of the quadratojugal, and the presence of an additional bone in the upper temporal region, together with the freely movable quadrate, have been explained in various ways. At one time the arch articulating with the proximal end of the quadrate was considered the lower one, and the two bones the squamosal and quadratojugal. Again, the

[1] Sollas makes a rather curious error in saying that I was prepared to accept the view, with certain reservations, of the direct descent of the ichthyosaurs from the Stegocephalia. What I said was that the two bones in the temporal region of the ichthyosaurs point to a direct origin from the stegocrotaphous reptiles.

lower arch with the quadratojugal has been supposed to be lost, or replaced by a ligament, and the upper bones called by various names: supratemporal, squamosal, prosquamosal, etc. This latter view is the one now generally accepted, and, according to it, the lizards have arisen from a primitively diapsid type which, by the loss of the lower arch and the acquirement of streptostyly, has become secondarily single-arched. Twelve years ago I ventured the opinion that the two temporal bones of the squamate skull are the tabular and squamosal, the former a bone unknown or unrecognized in other reptiles since Triassic times. This view has nowhere obtained approval except by Broom. For years past von Huene, Broom, and I have repeatedly urged that the Lacertilia are a more primitive type of reptiles than the Rhynchocephalia, and, years ago, I ventured the prediction that the order would eventually be discovered in the Permian. The discovery of *Araeoscelis* in the Permocarboniferous of Texas seems to fulfil that prediction. *Araeoscelis* has a single temporal opening bounded quite as in the lizards, but with a fixed quadrate, the broad temporal region below unperforated. The cervical ribs are single-headed and attached to the centra. I am convinced that the *Araeoscelis* type of skull, by the simple emargination of the lower border of the squamosal and the consequent streptostyly, gave origin to the Lacertilia. Watson has also shown that *Pleurosaurus*, a Jurassic genus which has long been located among the Rhynchocephalia, likewise has a single, upper temporal vacuity, with a fixed quadrate, but the squamosal narrower. He believes that the genus was ancestrally related to the Squamata, though he differs from me in the interpretation of the bones of the temporal region, adopting the original Baur view of the presence of squamosal and quadratojugal. This group I have called the Protorosauria, believing that Seeley was correct in his original interpretation of *Protorosaurus*.

Whatever may be the interpretation, one thing is evident: even earlier than the origin of the upper vacuity in the Diapsida a simple upper vacuity, as in the lizards, had developed, but with the temporal region imperforate below, and this type persisted in *Pleurosaurus* to late Jurassic. If the Squamata did not originate in this way, then *Araeoscelis* and *Pleurosaurus* and probably

Protorosaurus must represent an entirely new order of reptiles, which very properly may be associated with the Ichthyosauria in the division I call the Parapsida. It might be urged that the Diapsida originated from such a type by the development of a lower vacuity after the upper one had been evolved. The arguments against this view are too many and too potent; I need not repeat them.

Admitting three chief groups of reptiles arising in late Pennsylvanian or early Permian times, we have yet another, one which by general consent is ancestral to all later amniota, the Cotylosauria, and their direct descendants the Chelonia.

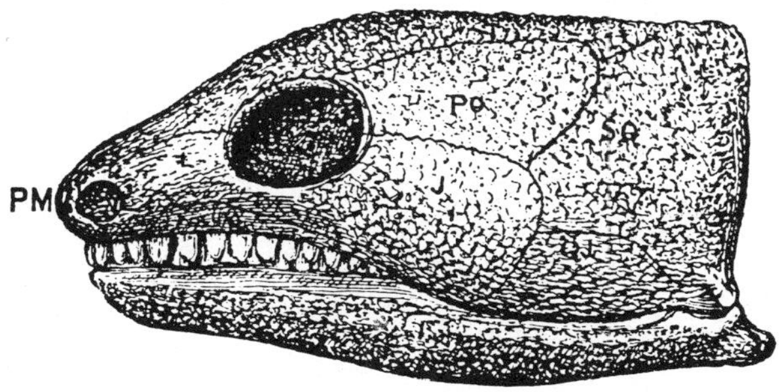

FIG. 5.—*Pantylus*, skull, from side, three-fourths natural size. Anapsida Permocarboniferous.

In this group the temporal region of the skull is wholly imperforate, and for the most part completely roofed over, a group which very properly may be called the **Anapsida.** It was Baur who first asserted that the turtles could not have originated from reptiles with a perforated temporal region. Cope derived the order directly from the Cotylosauria, through the Chelydosauria, an order based upon a misapprehension. I approved and emphasized Baur's views in 1904. Case and Hay both hold the same opinion, and only recently Watson has forcibly and convincingly presented the claims of *Eunotosaurus*, from the Permian, as a real connecting link between the two orders. And Broom and others are of the same opinion. This unanimity of opinion renders further discussion superfluous; Watson has presented the arguments.

We have, then, at least four main divisions or subclasses of the class Reptilia, all beginning in Paleozoic times, and all represented by their direct descendants today; the birds, crocodiles, and tuatara of the Diapsida; the mammals of the Synapsida; the lizards and snakes of the Parapsida; and the turtles of the Anapsida. They suggest the following linear arrangement of the known groups, the doubtful or poorly known ones, perhaps entitled to ordinal rank, printed in italics:

Anapsida	Diapsida
Cotylosauria	Rhynchocephalia
Chelonia	*Rhynchosauria*
Synapsida	*Thalattosauria*
Theromorpha	*Choristodera*
Therapsida	Phytosauria
Sauropterygia	*Pseudosuchia*
Placodontia	Crocodilia
Parapsida	Pterosauria
Ichthyosauria	Dinosauria
Squamata	"*Eosuchia*"
Protorosauria (*Aracoscelidia, Acrosauria*)	

I am aware that other general phylogenetic schemes of the Reptilia have been proposed, especially by Boulenger and Goodrich, but long years of study of the reptiles has convinced me that, while all may have features worthy of consideration, the chief reliance must be placed upon the skull structure, especially that of the cranial and temporal regions. As von Huene and others have urged, these parts are the most conservative, and least liable to homoplastic duplication. Next to the skull, the ribs are conservative. In all the Archosauria the double-headed dorsal ribs are attached to the diapophyses. In the Diaptosauria (except Thalattosauria), the Synapsida (except the Sauropterygia), and Anapsida the rib tubercle articulates with the arch, the capitulum with the intercentral space, while in the Parapsida, so far as known, the ribs are attached more or less exclusively to the centra. Other characters originally proposed as distinctive between the Diapsida and the Synapsida in the wider sense have been proven to be invalid. We know that the primitive foot structure is nearly that of the lizards and *Sphenodon* of today, that the reduction of the phalanges in the theriodonts and turtles is purely homoplastic. The supposed relationship between the turtles and the plesiosaurs is also purely homoplastic; nor am I aware of any constant character in the girdles, limbs, or ventral ribs. Goodrich, in a recent paper on the phylogeny of reptiles, relies greatly upon the structure of the feet, especially of the fifth metatarsal. I cannot accept his contentions, some reasons for which I have published elsewhere.

Just when the animals we call reptiles arose in geological history we do not know; certainly it was in early Pennsylvanian times, probably in Mississippian. That they arose from what we call the Amphibia, forms with temnospondylous vertebrae, is certain, though there is probably not much more reason for calling the ancestral stock Amphibia than Reptilia. I prefer to call it, provisionally, Protopoda. It was ancestral to both, and both classes have advanced since their divergence, the Amphibia some, the Reptilia much. Could we find, as some time we hope that we may, in mid-Mississippian or late Devonian times, a skeleton of one of those ancestral creatures, we should perhaps not call it by the name of any known order; it would be the old question over again of the differences between animals and plants. At present we know the Protopoda only by their footprints. As it is, we are dealing chiefly with archaic forms, even by the close of Pennsylvanian times, forms which have retained in various degrees their primitive characters while adding or losing others in different ways. Just as the most primitive mammals now living have become highly specialized in the loss of teeth, so too the amphibians, as we know them in Paleozoic times, were more or less specialized or degenerated. The only known distinctive characters between the two classes, as represented by their known skeletons in Permocarboniferous times, are found in the atlas and feet, and doubtless some time we shall find these differences bridged over. For a long time we relied upon the open palate of the Amphibia, but Watson has deprived us of that support, and now we are compelled to group the characters at length in any differential diagnosis of the two classes. However, with nearly every character in common to the Amphibians and reptiles, we are never in doubt as to which class a given form belongs if we know it well enough, for all the common characters have never been found in the same specimen, and doubtless never will be in any specimen from rocks later than the Mississippian.

ERRATUM

Page 420, column 1 of the table, Williston has stamped "Proganosaurs" between "Squamata" and "*Protorosauria*."

12

Reprinted from pages 227-237 and Plates LXXVI, LXXVII, LXXVIII, LXXIX, LXXX, LXXXI, LXXXII, LXXXIII, LXXXIV, and LXXXV of *U.S. Geol. Survey 16th Ann. Rept.*, pp. 133-414 (1896)

THE DINOSAURS OF NORTH AMERICA

O. C. Marsh

[*Editor's Note:* In the original, material precedes and follows this excerpt.]

PART IV.

CONCLUSION.

The brief review of North American Dinosaurs given in the preceding pages, in connection with the accompanying illustrations, will make the reader acquainted with the more important type specimens of this interesting group of reptiles, as now known from this continent. To discover and bring together these remains, representing several hundred individuals, from widely separated localities and various geological horizons, has been a long and laborious undertaking, attended with much hardship, and often with danger, but not without the pleasure that exploration in new fields brings to its votaries. These researches, especially in the West, have been continued by the writer more than a score of years, and have led him across the Rocky Mountains a still greater number of times. The field work thus prosecuted has been of great service in the subsequent study of the specimens secured, especially in determining the natural position in life of each animal investigated.

In comparing the type specimens of these various animals, one with another, as they were found and as they appeared when removed from the vesture of their entombment, many questions have suggested themselves that can not be answered in the present limited paper. Resemblances and differences are striking, both in structure and form, in these ancient reptiles, but the true meaning of such features is a difficult problem to solve. On the interpretation of characters thus exhibited in these animals depend both the laws of their classification and theories of their origin.

COMPARISON OF CHARACTERS.

In the concluding part of the present paper a number of plates (LXXVI–LXXXI) have been given with a view to illustrate especially the corresponding parts of various animals of different orders, showing the wide divergence in some points of nearly allied forms, and the approach in particular features of types clearly distinct.

In Pl. LXXVI four skulls of as many typical genera of herbivorous dinosaurs (Triceratops, Claosaurus, Camptosaurus, and Diplodocus) are represented, with a cast of the brain cavity of each in position. All are so drawn that they can be readily compared, thus exhibiting in a striking manner both the diminutive size of the brain in each in proportion to that of the skull, and also the form of the brain cavity, when seen from above. In the next plate (LXXVII) the brain casts alone of several dinosaurs, as seen from the side, are exhibited, and with them for comparison the corresponding cast of a young alligator. The special features of the dinosaur brain are well shown in these two plates.

Pl. LXXVIII will make clear the wide divergence of forms of teeth in four different families of predentate dinosaurs. The typical genera,

Camptosaurus, Claosaurus, Stegosaurus, and Triceratops, each have teeth of a distinct type, yet it seems possible to trace the gradation of one to the other through different intermediate forms. In Triceratops the teeth have two distinct roots, a feature unknown in any other reptiles, living or extinct, but characteristic of mammals.

The series of pubic bones of herbivorous dinosaurs shown on Pl. LXXIX is especially instructive, as they indicate how the anterior and posterior elements of the pubis may vary in the Predentata, and thus afford good characters for classification. The same is true, but in a less degree, of the ischia represented on Pl. LXXX, which all pertain to one group of dinosaurs. The comparison may even be carried much further, as in the two other orders (the Theropoda and Sauropoda) some families have ischia of the type here represented, as shown on Pls. XXVIII and XXXV.

The pelves represented on Pl. LXXXI, pertaining to the three predentate genera, Camptosaurus, Triceratops, and Stegosaurus, will supplement the facts presented on the two preceding plates. The series might be much further extended, and prove equally instructive. This will be done by the writer in the monographs now in preparation, as in these the whole subject of dinosaurian reptiles will receive careful consideration.

RESTORATIONS OF EUROPEAN DINOSAURS.

The remaining restorations of dinosaurs in this paper are four in number, and represent some of the best-known European forms, types of the genera Compsognathus, Scelidosaurus, Hypsilophodon, and Iguanodon. These outline restorations have been prepared by the writer mainly for comparison with the corresponding American forms, but in part to insure, so far as the present opportunity will allow, a more comprehensive review of the whole group. The specimens restored are all of great interest in themselves, and of special importance when compared with their nearest American allies.

COMPSOGNATHUS.

PLATE LXXXII.

The first restoration, that of *Compsognathus longipes* Wagner, 1861, shown one-fourth natural size on Pl. LXXXII, is believed to represent fairly well the general form and natural position, when alive, of this diminutive carnivorous dinosaur that lived during the Jurassic period. The basis for this restoration is (1) a careful study of the type specimen itself, made by the writer in Munich in 1881; (2) an accurate cast of this specimen, sent to him by Professor von Zittel; and (3) a careful drawing of the original, made by Krapf in 1887. The original description and figure of Wagner (Bavarian Academy of Sciences, 1861), and those of later authors, have also been used for some of the details.

No restoration of the skeleton of this unique dinosaur has hitherto been attempted.[1]

Compsognathus has been studied by so many anatomists of repute since its discovery that any attempt to restore the skeleton to a natural position will be scrutinized from various points of view. Interest in this unique specimen led the writer long ago to examine it with care, and he has since made a minute study of it, as related elsewhere, not merely to ascertain its anatomy, but also to learn, if possible, what its relations are to another diminutive form, Hallopus, from a lower horizon in America, which has been asserted to be a near ally. Both are carnivorous dinosaurs, probably, but certainly on quite different lines of descent.

The only previous attempt to restore this remarkable dinosaur was by Huxley when in America in 1876. He made a rapid sketch from the Wagner figure, and this was enlarged for his New York lecture. This sketch represents the animal sitting down, a position which such dinosaurs occasionally assumed, as shown by the footprints in the Connecticut Valley, which Huxley examined in place at several localities with great interest.

In the present restoration of Compsognathus (Pl. LXXXII) the writer has tried to represent the animal as walking in a characteristic lifelike position.

SCELIDOSAURUS.

PLATE LXXXIII.

The second of these restorations is that of *Scelidosaurus Harrisonii* of Owen, shown one-eighteenth natural size on Pl. LXXXIII. This reptile was an herbivorous dinosaur of moderate size, related to Stegosaurus, and was its predecessor from a lower geological horizon in England. This restoration is essentially based upon the original description and figures of Owen (Palæontographical Society, 1861). These have been supplemented by the writer's own notes and sketches, made during examinations of the type specimen now in the British Museum.

Scelidosaurus is a near relative, as it were, of one of the American forms, Stegosaurus, now represented by so many specimens that the skull, skeleton, and dermal armor are known with much certainty. The English form usually called Omosaurus is still more nearly allied to Stegosaurus, perhaps identical with it.[2]

A restoration of the skeleton of Scelidosaurus by Dr. Henry Woodward will be found in the British Museum Guide to Geology and Palæontology, 1890, p. 19. The missing parts are restored from Iguanodon, and the animal is represented as bipedal, as in that genus.

---

[1] The remains of the embryo within the skeleton of Compsognathus, first detected by the writer in 1881 while examining the type specimen, is not represented in the present restoration. This unique fossil affords the only conclusive evidence that dinosaurs were viviparous.

[2] The generic name Omosaurus was preoccupied by Leidy in 1856.

In the present outline restoration of Scelidosaurus the writer has endeavored merely to place on record his idea of the form and position of the skeleton when the animal was alive, based on the remains he has himself examined. In case of doubt, as, for example, in regard to the front of the skull, which is wanting in the type specimen, a dotted outline is used, based on the nearest allied form. Of the dermal armor, only the row of plates best known is indicated. The position chosen in this figure (Pl. LXXXIII) is one that would be assumed by the animal in walking on all four feet, and this is believed to have been its natural mode of progression.

HYPSILOPHODON.

Plate LXXXIV.

The third of these restorations, that of *Hypsilophodon Foxii* Huxley, 1870, given in outline one-eighth natural size on Pl. LXXXIV, has been made with much care, partly from the type specimen, and in part from other material mostly now in the British Museum. The figures and description by the late Dr. Hulke[1] were of special value, although the conclusions of the writer as to the natural position of the animal when alive do not coincide with those of his honored friend, who did so much to make this genus of dinosaurs, and others, known to science. The restoration by Dr. Hulke represented the animal as quadrupedal.

In the case of Hypsilophodon a number of specimens are available instead of only one. This makes the problem of its restoration a simpler matter than in Scelidosaurus. Moreover, there is in America a closely allied form, Laosaurus, of which several species are known. A study of the genus Laosaurus, and the restoration of one species given on Pl. LVII, will clear up several points long in doubt.

Huxley and Hulke both shed much light on this interesting genus, Hypsilophodon; indeed, on many of the Dinosauria. The mystery of the dinosaurian pelvis, which baffled Cuvier, Mantell, and Owen, was mainly solved by them, the ilium and ischium by Huxley, and the pubis by Hulke. The more perfect American specimens have demonstrated the correctness of nearly all their conclusions.

IGUANODON.

Plate LXXXV.

The fourth restoration here given, that of *Iguanodon Bernissartensis* Boulenger, 1881, one-fortieth natural size, has been made in outline for comparison with American forms. It is based mainly on photographs of the well-known Belgian specimens, the originals of which the writer has studied with considerable care during several visits to Brussels. The descriptions and figures of Dollo[2] have also been used in the preparation of this restoration. A few changes only have been introduced in

[1] Philosophical Transactions, 1882.

[2] Bulletin Royal Museum of Belgium, 1882–88.

the accompanying plate, based mainly upon a study of the original specimens.

Besides the four genera here represented, no other European dinosaurs at present known are sufficiently well preserved to admit of accurate restorations of the skeleton. This is true, moreover, of the dinosaurian remains from other parts of the world outside of North America.

AFFINITIES OF DINOSAURS.

The extinct reptiles known as dinosaurs were for a long time regarded as a peculiar order, having, indeed, certain relations to birds, but without being closely allied to any of the groups of known reptiles. Megalosaurus and Iguanodon, the first dinosaurian genera described, were justly considered as representing two distinct families, one including the carnivores, and the other the herbivorous forms.

With the discovery and investigation of Cardiodon (Cetiosaurus) and its allies in Europe, and especially of the gigantic forms with similar characters in America, it became evident that these reptiles could not be placed in the same families with Megalosaurus or Iguanodon, but constituted a well-marked group by themselves. It was this new order, the Sauropoda, as the writer has named them, that first showed definite

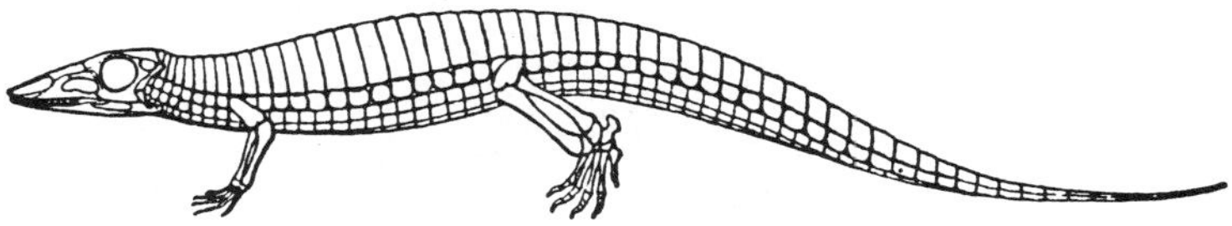

FIG. 56.—Restoration of *Aëtosaurus ferratus* Fraas; with dermal armor of the limbs removed. One-eighth natural size.

characters allying them with other known groups of reptiles. In 1878 he pointed out that the Sauropoda were the least specialized of the dinosaurs, and gave a list of characters in which they showed such an approach to the Mesozoic crocodiles as to suggest a common ancestry at no very remote period.[1]

AFFINITIES WITH AËTOSAURIA.

Again, in 1884, the writer called attention to the same point, and also to the relationship of dinosaurs with the Aëtosauria, as he has named them, a group of small reptiles from the Triassic of Germany showing strong affinities with crocodilians.[2] A restoration of one of these small animals is shown in fig. 56. In the same communication he compared with dinosaurs another allied group, the Hallopoda, which he described from the lower Jurassic of America, but had not then fully investigated. Subsequent researches proved the latter group to be of the first importance in estimating the affinities of dinosaurs, and in figs. 59 and 60 are restorations of the fore and hind limbs of the type species (*Hallopus victor*).

[1] American Journal of Science, Vol. XVI, p. 412, November, 1878.
[2] Report British Association, Montreal Meeting, 1884, p. 765.

AFFINITIES WITH BELODONTIA.

Another group of extinct reptiles, which may be termed the Belodontia, were considered in the same paper as allies of the Dinosauria. They are known from the Trias of Europe and America, and the type genus, Belodon, has been investigated by many anatomists, who all appear to have regarded it as a crocodilian, an opinion that in the light of our present knowledge may fairly be questioned.

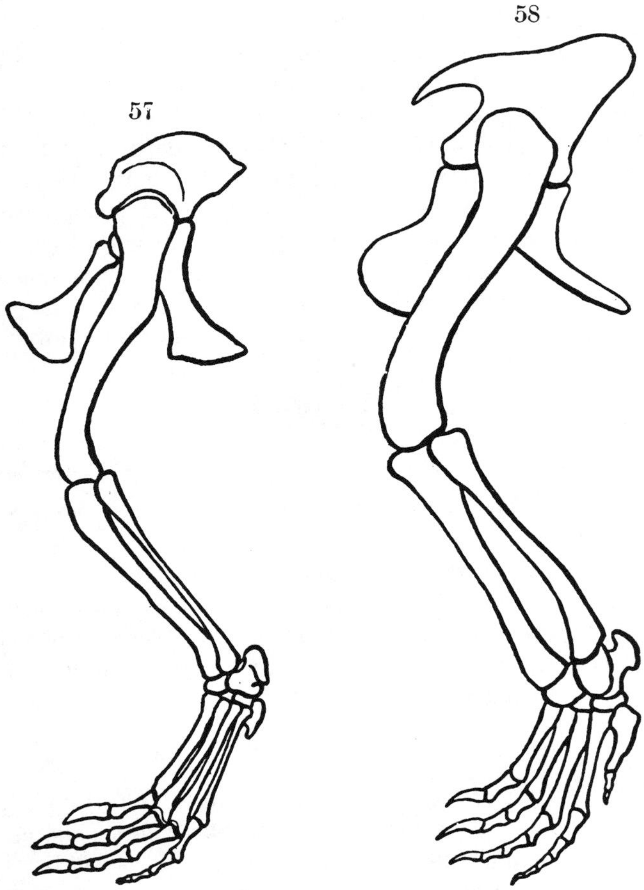

FIG. 57.—Diagram of left hind limb of *Alligator mississippiensis* Gray; seen from the left; in position for comparison with dinosaurs. One-fourth natural size.
FIG. 58.—Diagram of left hind limb of *Aëtosaurus ferratus*; in same position. One-half natural size.

AFFINITIES WITH CROCODILIA.

The relations of these various groups to the true crocodiles on the one hand and to dinosaurs on the other is much too broad a subject to be introduced here, but attention may at least be called to some points of resemblance between the dinosaurs and these supposed crocodilian forms that seem to indicate genetic affinities.

If some of the characteristic parts of the skeletons of these groups are compared, e. g., of the true Crocodilia as existing to-day, the Belodontia, the Aëtosauria, and the Hallopoda, and all with the corresponding portions of the more typical dinosaurs, the result may indicate in some measure the relationship between them. Taking first the pelvis and hind limb, as being especially characteristic. it will be seen in

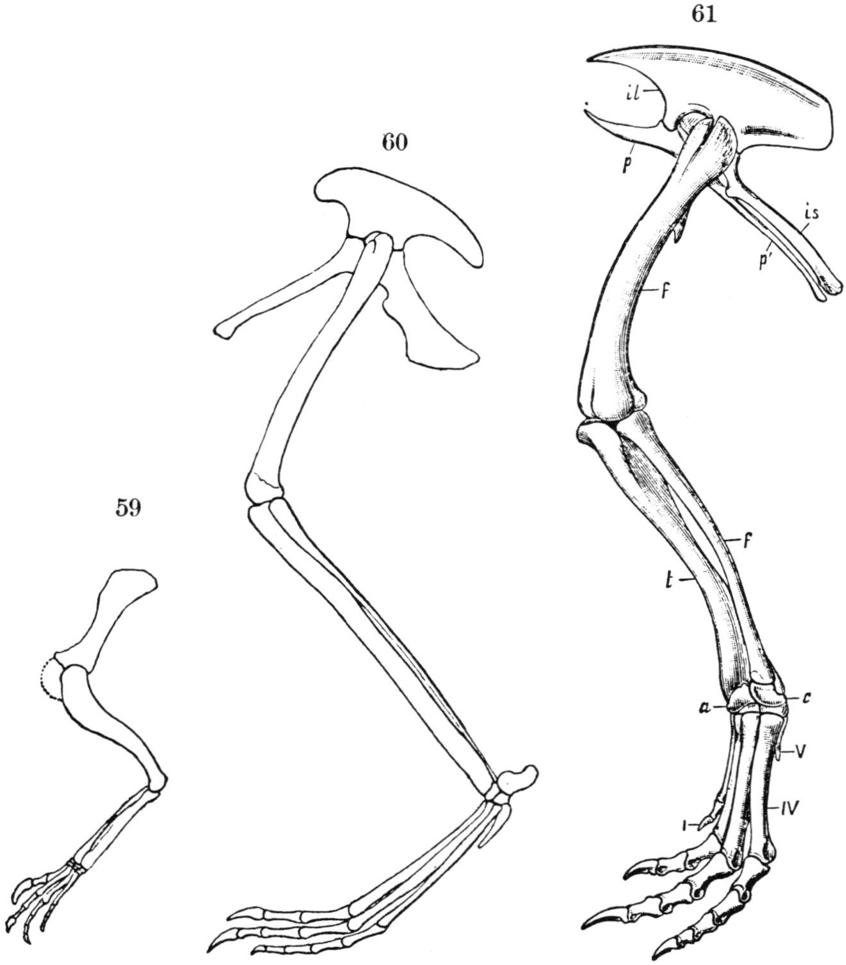

FIG. 59.—Diagram of left fore limb of *Hallopus victor* Marsh; seen from the left.
FIG. 60.—Diagram of left hind limb of same individual. Both figures are one-half natural size.
FIG. 61.—Left hind leg of *Laosaurus consors* Marsh; outside view. One-sixth natural size.
a, astragalus; *c*, calcaneum; *f*, femur; *f′*, fibula; *il*, ilium· *is*, ischium; *p*, pubis; *p′*, postpubis; *t*, tibia; *I*, *IV*, *V*, first, fourth, and fifth digits.

the existing alligator, as represented in fig. 57, that the pubic bone is excluded from the acetabulum, articulating with the ischium only, and not at all with the ilium. The calcaneum, moreover, has a posterior extension. In Aëtosaurus, as shown in fig. 58, the pubic bone forms part of the acetabulum, as in dinosaurs and birds, and this is a noteworthy difference from all the existing crocodiles. The hind foot, however, is of the crocodilian type, with the calcaneum showing a postérior projection.

In Belodon, only the pelvis of which is here represented (fig. 62), the pubis contributes a very important part to the formation of the acetabulum, and to the entire pelvic arch. The latter differs from the pelvis of a typical dinosaur mainly in the absence of an open acetabulum, but a moderate enlargement of the fontanelle at the junction of the three pelvic elements would practically remove this difference. A more erect position of the limb, leading to a more distinct head on the femur, might possibly bring about such a result. The feet and limbs of Belodon are crocodilian in type.

Bearing these facts in mind, the diagram representing the restored fore and hind limbs of the diminutive Hallopus (figs. 59–60) shows first

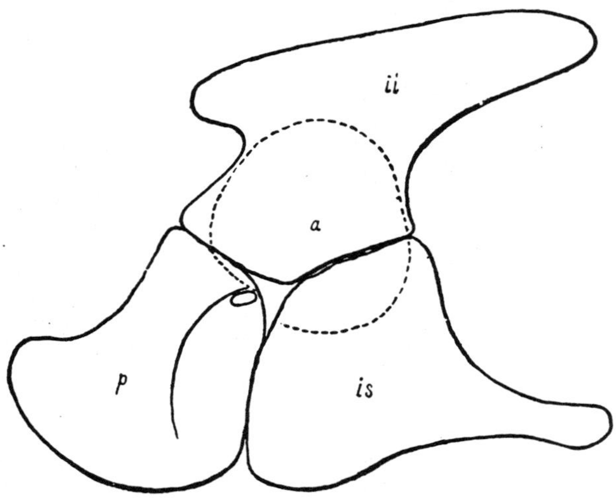

FIG. 62.—Diagram of pelvis of *Belodon Kapffi* von Meyer; seen from the left. One-fourth natural size.

a, acetabular surface within dotted line; *il*, ilium; *is*, ischium; *p*, pubis.

of all the true dinosaurian pelvis, with the pubic bone taking part in the open acetabulum, and forming an important and distinctive element of the pelvic arch. The delicate posterior limb and foot, evidently adapted mainly for leaping, as the generic name suggests, are quite unique among the Reptilia, but the tarsus, especially the calcaneum, recalls strongly the same region in the orders already passed in review.

Just what this posterior extension of the calcaneum signifies in this case it is difficult to decide from the evidence now known. It may be merely an adaptive character, as Hallopus appears in nearly every other respect to be a true carnivorous dinosaur. It may, however, be an inheritance from a crocodilian ancestry, preserved by a peculiar mode of life. Whatever its origin may have been, it was certainly, during the life of the animal, an essential part of the remarkable leaping foot to which it belonged, and in which it has since kept its position undisturbed. The presence of such an element in the foot of this diminutive dinosaur certainly suggests that the group Hallopoda, which the writer has here considered a suborder, stands somewhat apart from the typical Theropoda, but not far enough away to be excluded from the subclass Dinosauria, as defined in the present paper.

The genus Plateosaurus (Zanclodon), which is from essentially the

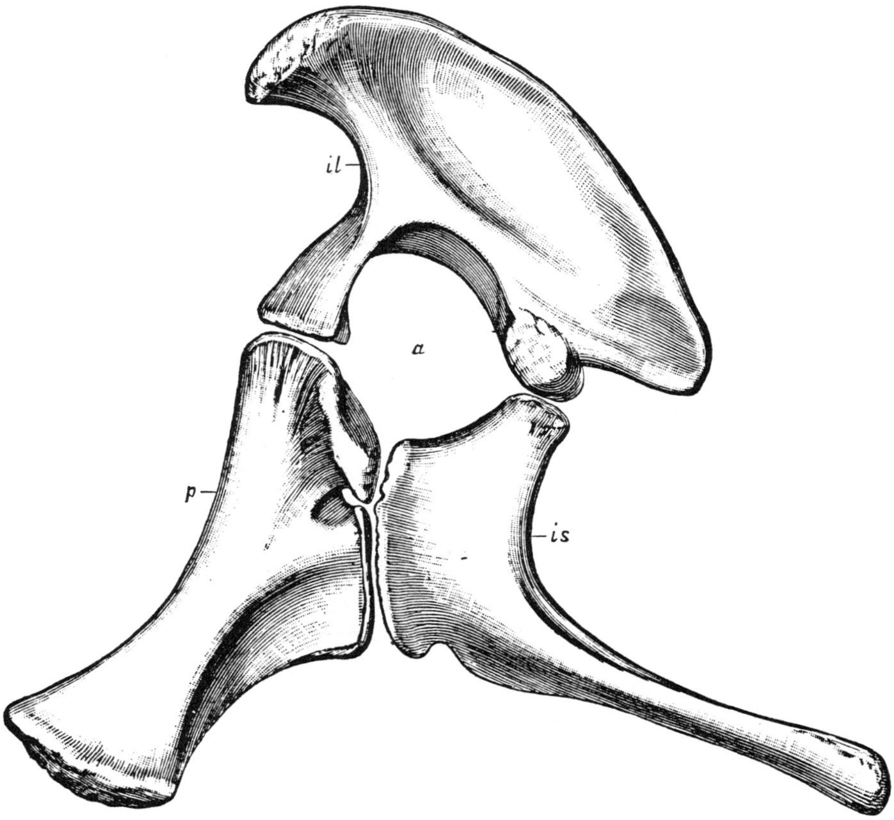

FIG. 63.—Pelvis of *Morosaurus lentus* Marsh; seen from the left. One-eighth natural size. *a*, acetabular opening; other letters as in fig. 62.

same geological horizon in Germany as Aëtosaurus and Belodon, is one of the oldest true dinosaurs known, and a typical member of the order Theropoda. In the pelvic arch of this reptile the ilium and ischium are in type quite characteristic of the group to which it belongs, but the pubic elements are unique. They consist of a pair of broad, thin plates united together so as to form an apron-like shield in front, quite unlike anything known in other dinosaurs. The wide pubic bones of Belodon, and the corresponding plates in some of the Sauropoda (Morosaurus, fig. 63), indicate that this feature of the reptilian pelvis may have been derived from some common ancestor of a generalized primitive type. The known transformations of this same pelvic element in one other order of dinosaurs (the Predentata) show that the modifications here suggested are well within the limits of probability. The hind limb of one genus of this order is shown in fig. 61 (p. 233).

64

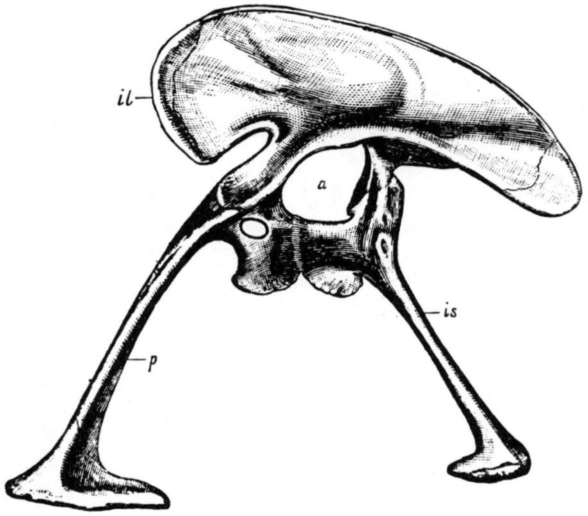

65

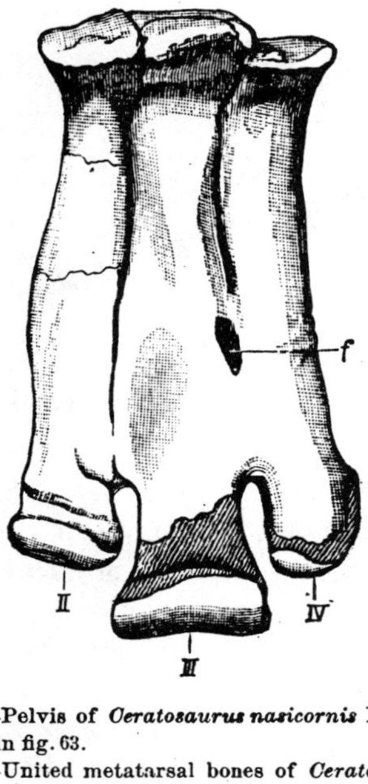

66

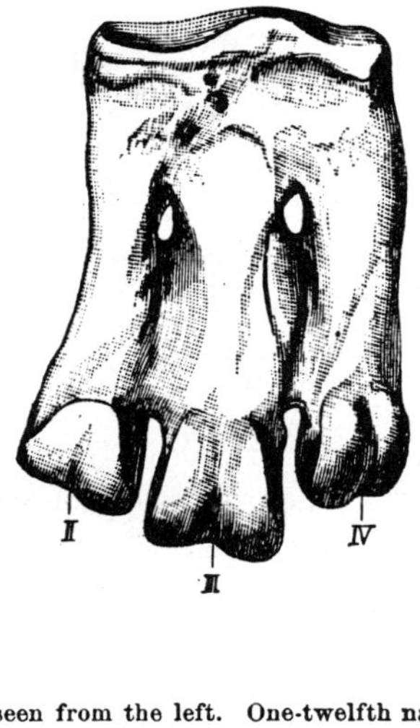

FIG. 64.—Pelvis of *Ceratosaurus nasicornis* Marsh; seen from the left. One-twelfth natural size. Letters as in fig. 63.

FIG. 65.—United metatarsal bones of *Ceratosaurus nasicornis;* left foot; front view. One-fourth natural size.

FIG. 66.—United metatarsal bones of great penguin (*Aptenodytes Pennantii* G. R. Gr.); left foot: front view. Natural size.

The skulls of Aëtosaurus and Belodon both show features characteristic of some of the dinosaurs, especially of the Sauropoda, but these features need not be discussed here.

AFFINITIES WITH BIRDS.

The relation of dinosaurs to birds, a subject of importance, must also be postponed for another occasion. One point, however, may be mentioned in this connection. The pelvic bones of all known birds, living and extinct, except the genus Archæopteryx, are coossified, while in all the known dinosaurs they are separate, excepting Ceratosaurus (fig. 64) and Ornithomimus. Again, all known adult birds, living and extinct, with possibly the single exception of Archæopteryx, have the metatarsal bones firmly united (fig. 66), while all the Dinosauria, except Ceratosaurus (fig. 65), have these bones separate. The exception in each case brings the two classes near together at this point, and their close affinity is thus rendered more than probable.

These few facts will throw some light on the affinities of the reptiles known as the Dinosauria. The problem is certainly one of much difficulty, and the writer hopes soon to discuss it more fully elsewhere.

PART V.

CLASSIFICATION OF DINOSAURIA.

In the present review of the dinosaurs the writer has confined himself mainly to the type specimens which he has described, but has included with them other important remains where these were available for investigation. The extensive collections in the museum of Yale University contain so many of the important type specimens now known from America that they alone furnish an admirable basis for classification, and it was mainly upon these that he first established the present system, which has since been found to hold equally good for the dinosaurs discovered elsewhere. In the further study of these reptiles it was also necessary to examine both the European forms and those from other parts of the world, and he has now studied nearly every known specimen of importance. These investigations have enabled him to make this classification more complete, and to bring it down to the present time.

Many attempts have been made to classify the dinosaurs, the first being that of Hermann von Meyer in 1830. The name Dinosauria, proposed for the group by Owen in 1839, has been generally accepted, although not without opposition. Hæckel, Cope, and Huxley followed, the last in 1869 proposing the name Ornithoscelida for the order, and giving an admirable synopsis of what was then known of these strange reptiles and their affinities. Since then, Hulke, Seeley, Lydekker, Gaudry, Dollo, Baur, and others have added much to our knowledge of these interesting animals.

PLATE LXXVI.

DINOSAURIAN SKULLS; SHOWING SIZE OF BRAIN.

396

PLATE LXXVII.

BRAIN CASTS OF DINOSAURS.

398

PLATE LXXVIII.

TEETH OF PREDENTATE DINOSAURS.

400

PLATE LXXIX.

PUBIC BONES OF PREDENTATE DINOSAURS.

402

PLATE LXXX.

ISCHIA OF PREDENTATE DINOSAURS.

404

PLATE LXXXI.

PELVES OF PREDENTATE DINOSAURS.

PLATE LXXXII.

EUROPEAN DINOSAURS.—THEROPODA.

COMPSOGNATHIDÆ.

PLATE LXXXIII.

EUROPEAN DINOSAURS.—PREDENTATA.

SCELIDOSAURIDÆ.

PLATE LXXXIV.

EUROPEAN DINOSAURS.—PREDENTATA.

HYPSILOPHODONTIDÆ.

PLATE LXXXV

EUROPEAN DINOSAURS.—PREDENTATA.

IGUANODONTIDÆ.

1

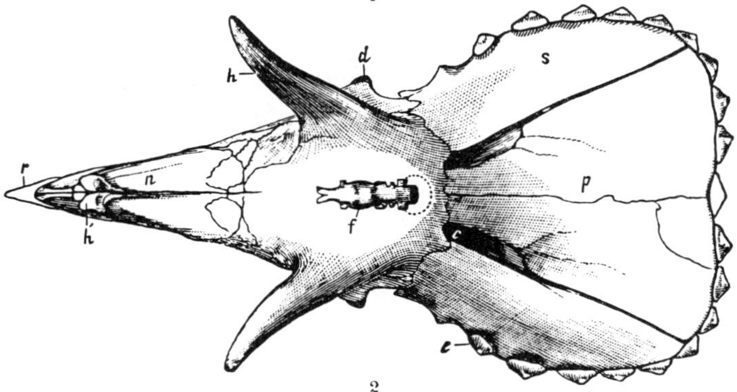

2

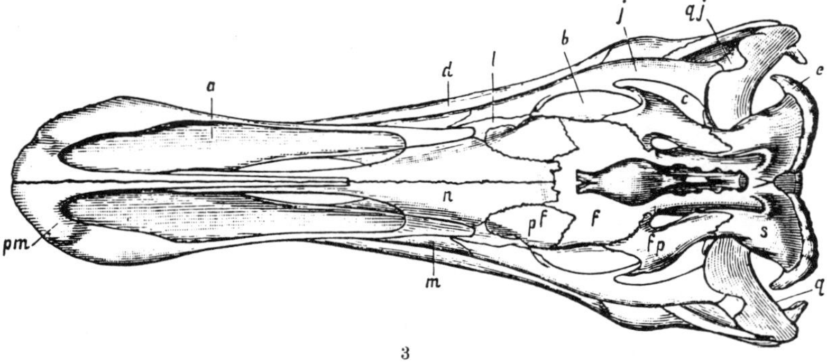

3

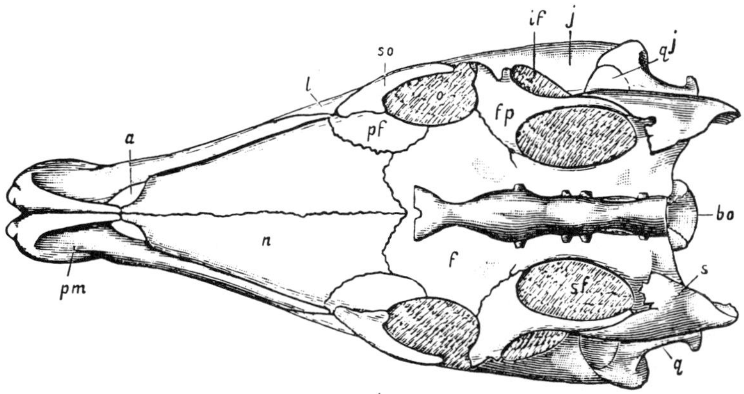

4

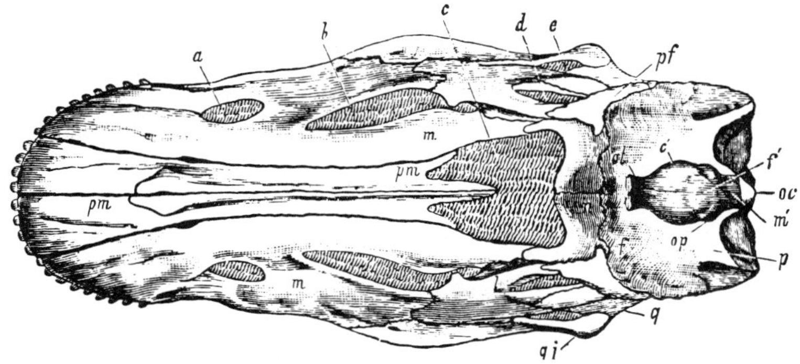

SKULLS OF DINOSAURS SHOWING SIZE OF BRAIN. CAMPTOSAURUS, CLAOSAURUS, DIPLODOCUS, AND STERRHOLOPHUS.

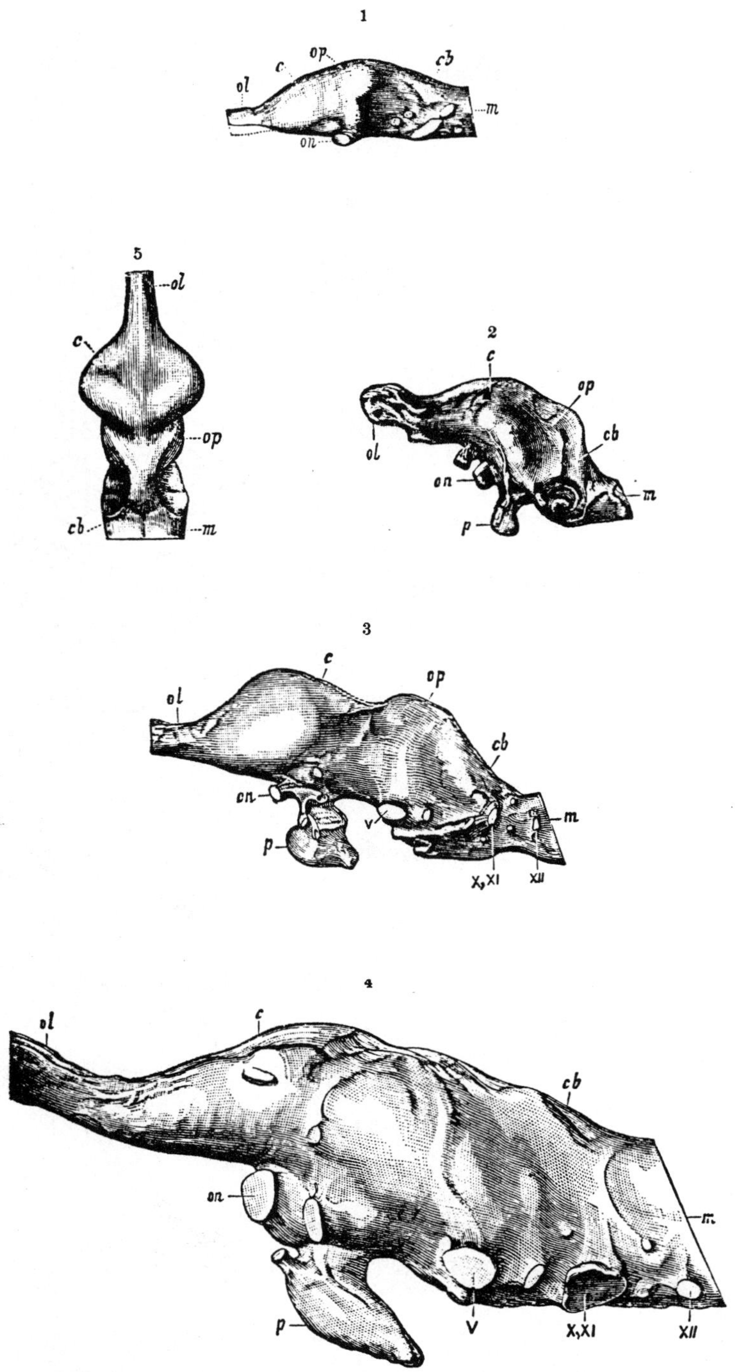

BRAIN CASTS OF DINOSAURS. CERATOSAURUS, CLAOSAURUS, STEGOSAURUS, TRICERATOPS, AND RECENT ALLIGATOR.

Plate LXXVIII

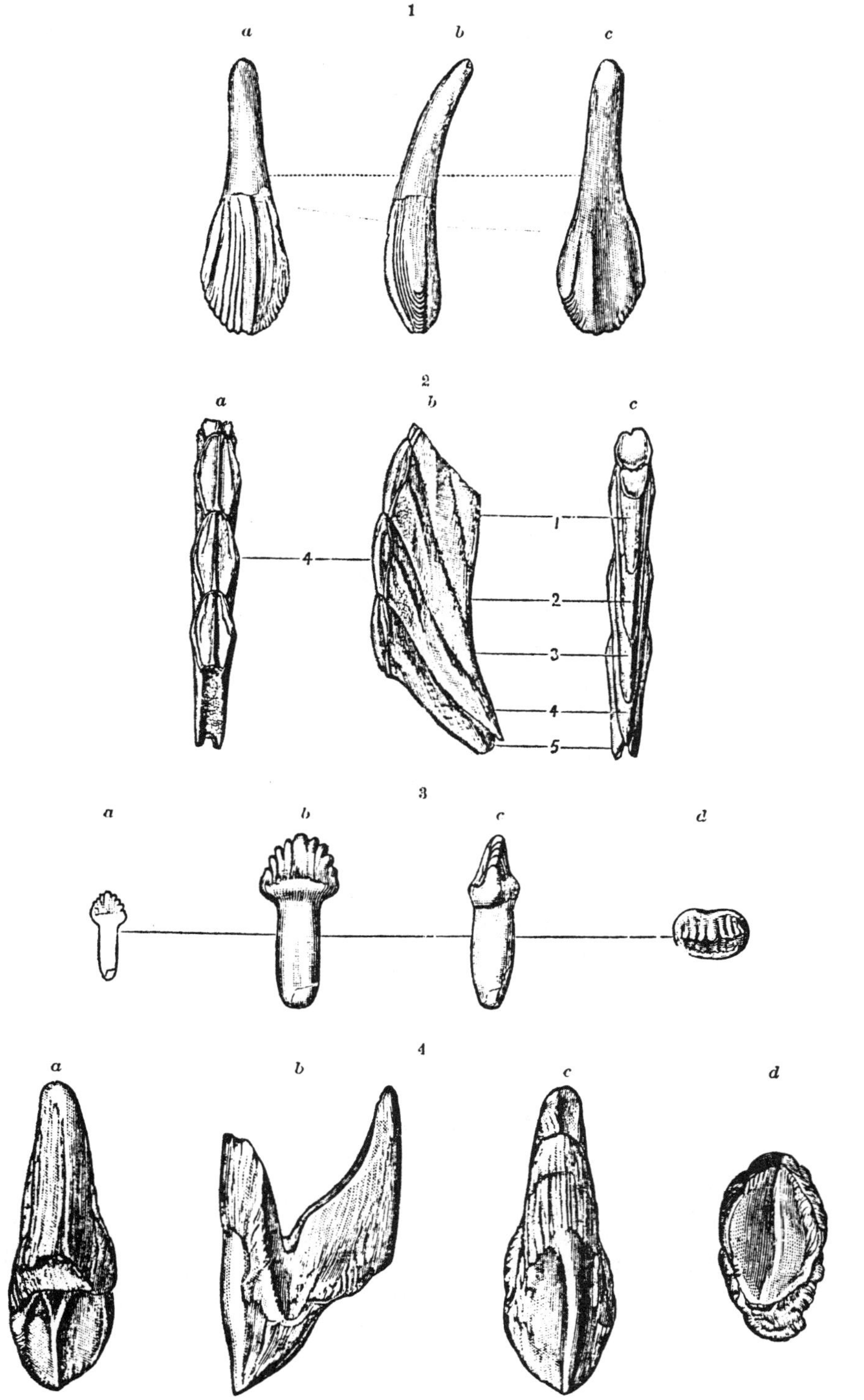

TEETH OF PREDENTATE DINOSAURS. CAMPTOSAURUS, CLAOSAURUS, STEGOSAURUS, AND TRICERATOPS.

Plate LXXIX

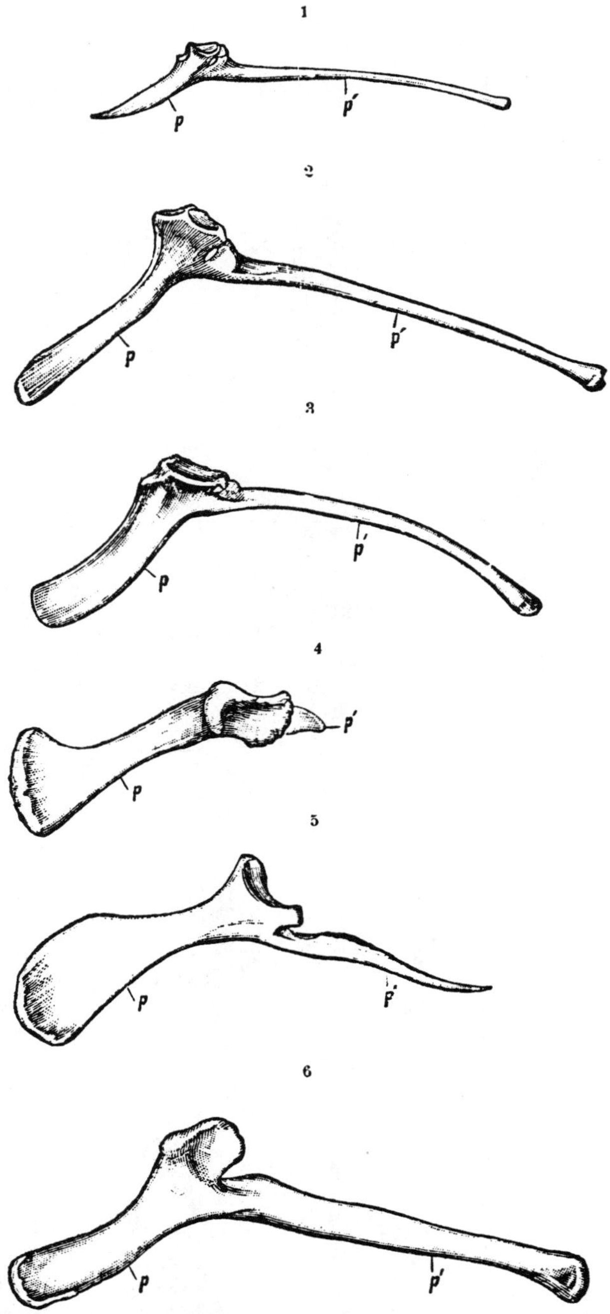

PUBES OF PREDENTATE DINOSAURS. CAMPTOSAURUS, CLAOSAURUS, DRYOSAURUS, LAOSAURUS, STEGOSAURUS, AND TRICERATOPS.

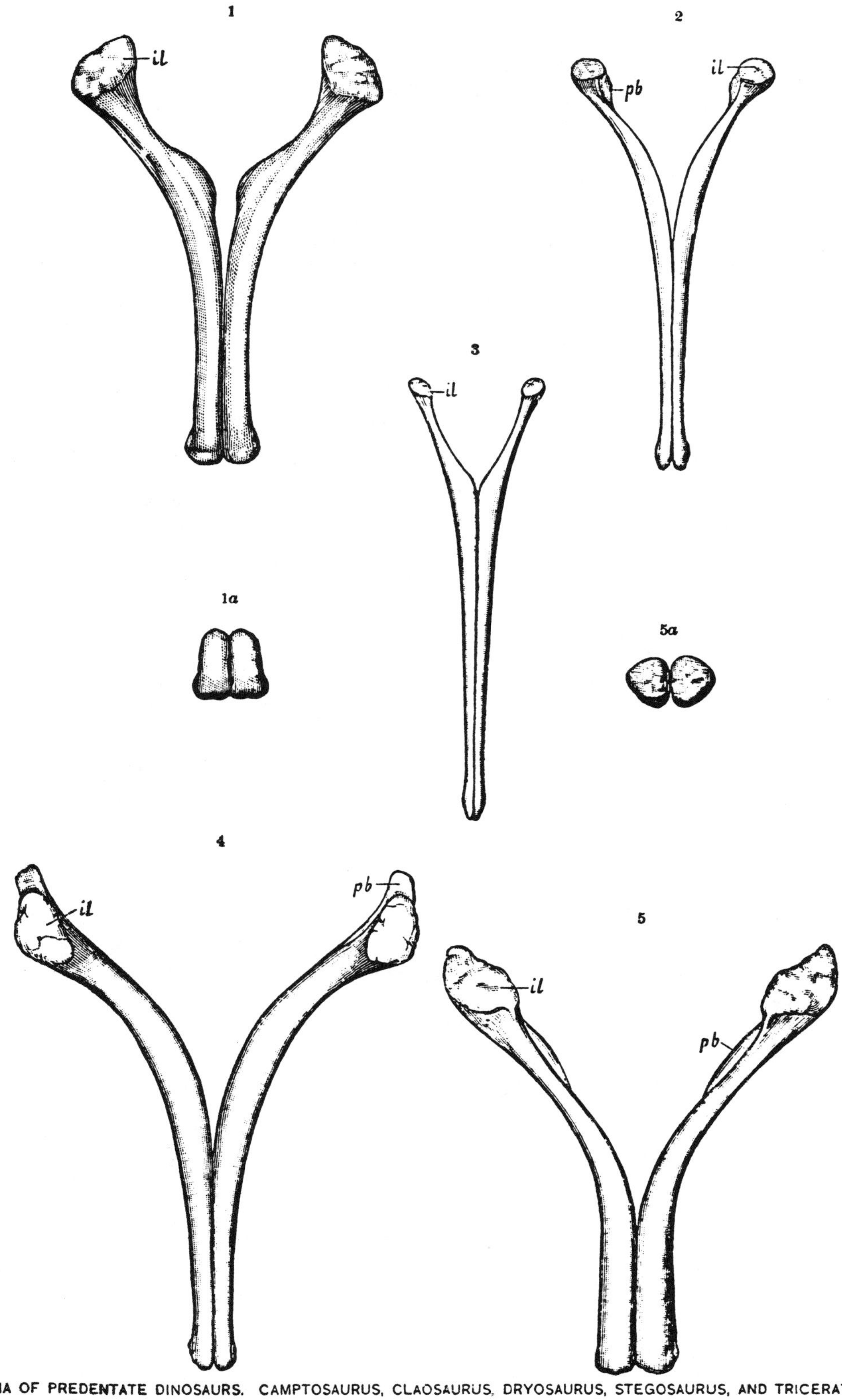

ISCHIA OF PREDENTATE DINOSAURS. CAMPTOSAURUS, CLAOSAURUS, DRYOSAURUS, STEGOSAURUS, AND TRICERATOPS

Plate LXXXI

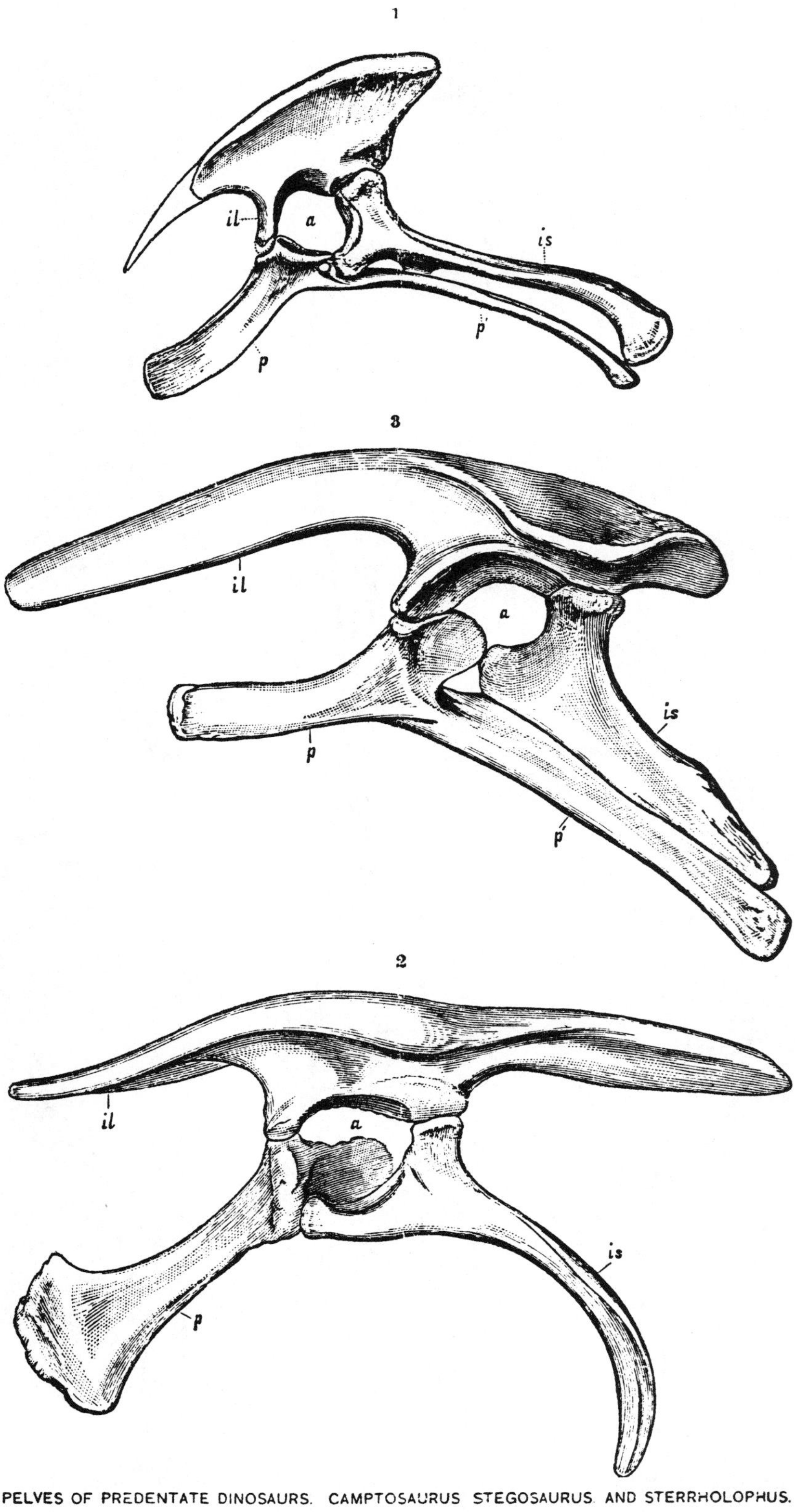

PELVES OF PREDENTATE DINOSAURS. CAMPTOSAURUS STEGOSAURUS. AND STERRHOLOPHUS.

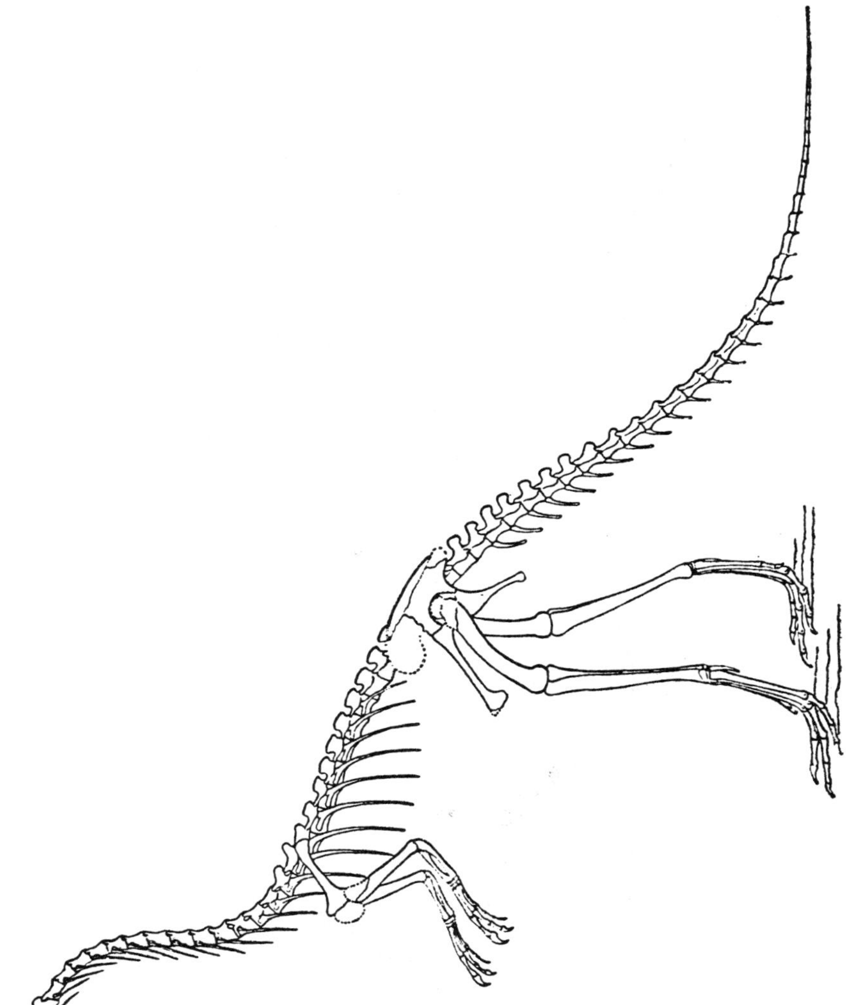

RESTORATION OF COMPSOGNATHUS LONGIPES Wagner.

One-fourth natural size. Jurassic, Bavaria.

Plate LXXXII

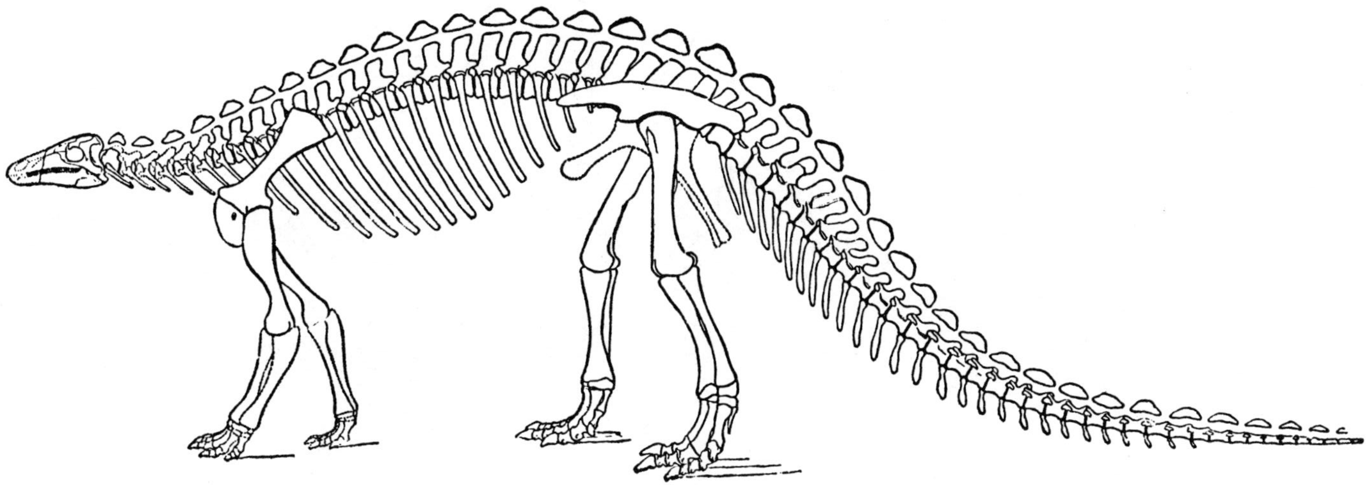

Plate LXXXIII

RESTORATION OF SCELIDOSAURUS HARRISONII Owen.
One-eighteenth natural size. Jurassic, England.

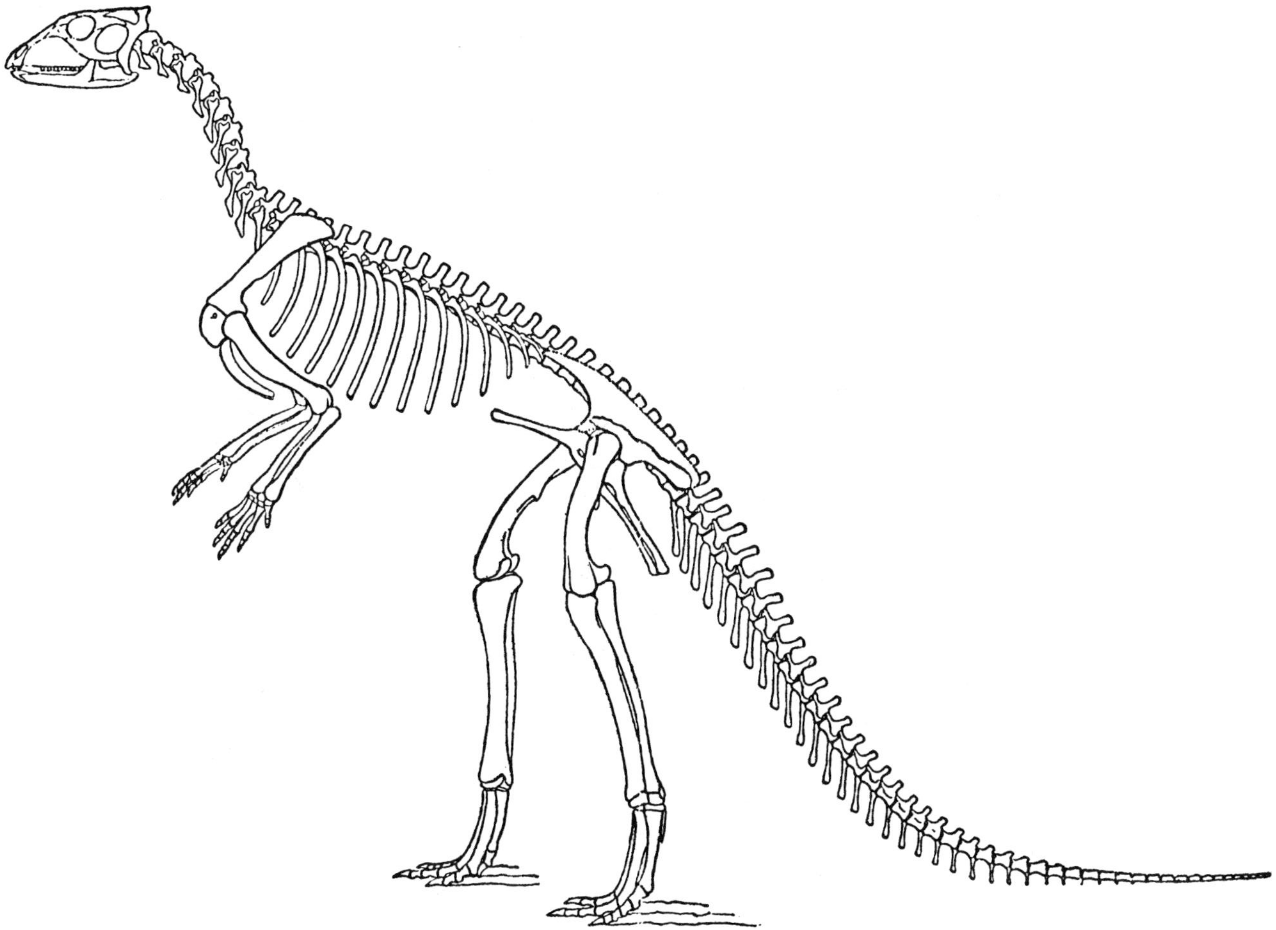

Plate LXXXIV

RESTORATION OF HYPSILOPHODON FOXII Huxley.
One-eighth natural size. Wealden, England.

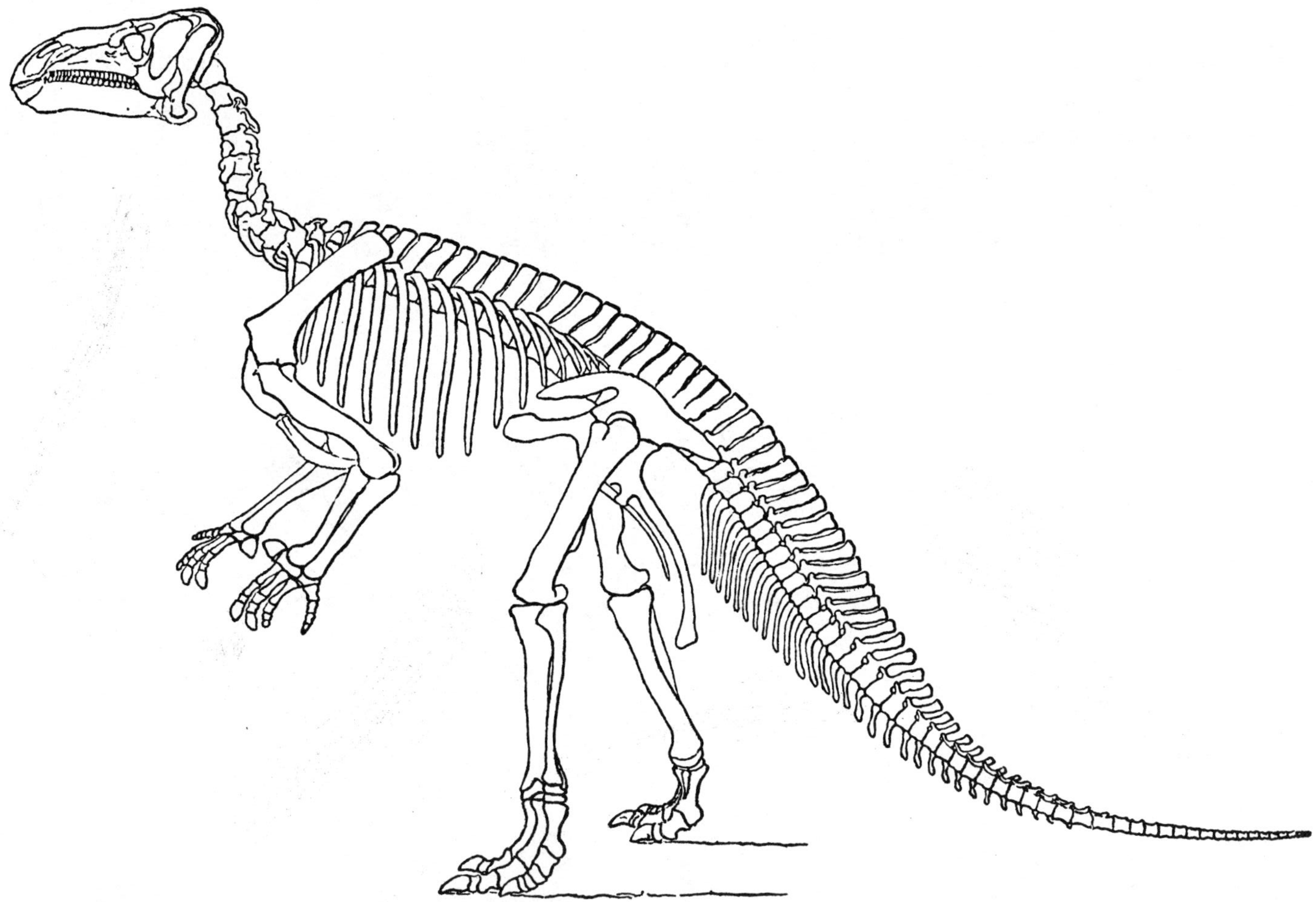

Plate LXXXV

RESTORATION OF IGUANODON BERNISSARTENSIS Boulenger.
One-fortieth natural size. Wealden, Belgium.

13

Reprinted from *N.Y. Acad. Sci. Ann.* **27**:31-38 (1916)

THEORIES OF THE ORIGIN OF BIRDS[1]

BY WILLIAM K. GREGORY

(*Presented before the Academy, 13 December, 1915, in connection with Mr. C. W. Beebe's paper on "A Tetrapteryx Stage in the Ancestry of Birds"*)

Comparative anatomists of the nineteenth century demonstrated that birds, in the entire ground plan of their brain, skeleton, reproductive organs and all other structures, as well as in their mode of development, are "glorified Reptiles," or "feathered saurians." In this instance the unanimous findings of comparative anatomy may be regarded as practically decisive.

But while all authorities agree that the assumed very remote ancestors of birds that lived in the Carboniferous and Permian periods of the earth's history were very probably scaly, lizard-like reptiles, there is no such unanimity regarding the structure and habits of the more immediate ancestors of birds, during the ages when scales were gradually transforming into feathers and the art of flying was still in its earliest stages. Professor Osborn, in 1900, after reviewing the evidence for the well-known view that birds and dinosaurs had been derived from a common ancestral stock that lived during the Permian period, said:[2] "In the origin of the birds we have to imagine, first, a terrestrial stage, in which bipedal was gradually substituted for quadrupedal progression; it would appear probable that the bipedal progression was first acquired during a terrestrial stage because the foot of birds is primarily a walking, and not a climbing, organ; second, a cursorial bipedal or, more probably, an arboreal stage, in which both fore limb and tail enjoyed a change of function contemporaneous with the acquisition of feathers."

In 1906 Mr. W. P. Pycraft, of the British Museum, argued that[3] in the stage preceding *Archæopteryx* (the oldest known fossil bird, of the Jurassic period) the ancestral birds probably lived in the trees, leaping from branch to branch and from tree to tree. "In these movements," he

---

[1] Manuscript received by the Editor 24 February, 1916.

[2] "Reconsideration of the Evidence for a Common Dinosaur-Avian Stem in the Permian," The American Naturalist, Vol. XXXIV, No. 406. 1900.

[3] "The Origin of Birds," Knowledge and Scientific News. September, 1906, pp. 531-532.

continues, "we may reasonably suppose the fore limbs were used for grasping at the end of the leap. The use of the fore limb for this work would naturally throw more work upon the inner digits—1-3—so that the work of selection would rapidly tend to the increased development of these, and the gradual decrease of the two outer and now useless members. Correlated with this trend in the evolution, the axillary membrane—the skin between the inner border of the arm and the body—became drawn out into a fold, while a similar fold came to extend from the shoulder to the wrist, as the fore limb, in adaptation to this new function, became more and more flexed. While the fingers, upon which safety now depended, were increasing in length, and growing more and more efficient, they were, at the same time, losing the power of lateral extension and becoming more and more flexed upon the fore-arm. And the growth in this direction was probably accompanied by the development of connective tissue and membrane along the hinder, post-axial border of the whole limb, tending to increase the breadth of the limb when extended preparatory to parachuting through space from one tree to another, long claws being used to effect a hold at the end of the leap.

"The hind limbs, though to a less extent, were also affected by the leaping motion, resulting in the reduction of the toes to four, and the lengthening and approximation of the metatarsals 2-4 to form a 'cannon' bone.

"The body clothing at this time was probably scale-like, the scales being of relatively large size and probably having a medium ridge, or keel, recalling the keeled scales of many living reptiles. Those covering the incipient wing, growing longer, would still retain their original overlapping arrangement, and hence those along the hinder border of the wing would, in their arrangement, simulate in appearance and function the quill feathers of their later descendants. As by selection their length increased, so also they probably became fimbriated and more and more efficient in the work of carrying the body through space.

"There is less of imagination than might be supposed in this attempt at reconstructing the primitive feather, inasmuch as there is a stage in the development of the highly complex feather of to-day which may well represent the first stage in this process of evolution. Creatures such as are here conjured up would bear a somewhat close resemblance to *Archæopteryx,* and it is contended that the discovery of earlier phases of avian development, phases preceding *Archæopteryx,* will show that this forecast was well founded. But in *Archæopteryx,* it is to be noted, the feathers differ in no way from the most perfectly developed feathers known to us."

Mr. Pycraft summed up this theory in his restoration entitled "One of the 'Pro-Aves.'" This hypothetical animal, as thus represented, stands about half way between a normal lizard-like reptile on the one hand and *Archæopteryx* on the other. It is represented as volplaning down from the trees, with arms outstretched. It is covered with scales, which on the back of the arm and sides of the tail have begun to lengthen out and transform into feathers.

The next year (1907) Baron Francis Nopcsa, in the Proceedings of the Zoölogical Society of London,[4] after reviewing the many resemblances between birds and running dinosaurs and showing that birds both in their mode of flight and in their limb structure differ in many important respects from bats, flying squirrels and other primarily tree-living animals that are provided with webs of skin for volplaning, came to the following conclusions:

"If we, after these preliminaries, now suppose that Birds, before attaining the *Archæopteryx*-state, originated from quadrupedal arboreal animals and only after having learnt to fly became bipedal, it is difficult to understand why they in general show Dinosaurian affinities, why they did not use both hind and fore limbs to the same extent for flight as they would have done for arboreal locomotion, why the bones of the pectoral region and of the wings show more primitive traces than the hind parts of the body, and why they did not, like all other quadrupedal flying animals, develop a patagium; whereas, if we consider that in *Archæopteryx* the anterior extremities, though bearing the most important ectodermal pinions, are less modified than the posterior extremities, which are already perfectly bird-like, and if we then suppose that Birds originated from bipedal Dinosaur-like Reptiles, it is easy to understand what induced the Birds to attain an *Archæopteryx*-like stage of evolution, for at first a certain amount of bipedal, and only afterwards a volant, modification would be required.

"While we can safely state that a bipedal animal never could or did develop a patagium without giving up bipedalism, this cannot be said of feather-bearing forms, for we may quite well suppose that *birds originated from bipedal long-tailed cursorial reptiles which during running oared along in the air by flapping their free anterior extremities.* . . . A double running and flapping action would—somewhat in accordance with Pycraft's views on this subject—subsequently easily lead to an enlargement of the posterior marginal scales of the antibrachium, and at the same time produce a certain amount of bipedal specialization

[4] "Ideas on the Origin of Flight." Proceedings of the Zoölogical Society of London, June 12, 1907.

"By gradually increasing in size, the enlarged but perhaps still horny hypothetical scales of the antibrachial margin would in time enable the yet carnivorous and cursorial ancestor of Birds to take long strides or leaps, much in the manner of a domesticated Goose or of a Stork when starting, and ultimately develop to actual feathers; this epidermic cover would also raise the temperature of the body, and thus help to increase the mental and bodily activity of these rapacious forms."

In 1913 Doctor Robert Broom described the skeleton of a small fossil reptile from the Upper Triassic beds of South Africa, which he named *Euparkeria capensis.* This animal, the type of which is in this museum, belonged to an ancient group of reptiles called "Pseudosuchia," some of which are found in the Triassic of Connecticut and others in the Upper Triassic of Scotland, Germany and South Africa. This group is of exceptional interest to palæontologists because of the largely primitive character of certain of its genera, which show marked evidences of affinity with such diverse later groups as Dinosaurs, Pterosaurs and Crocodilians. In discussing the affinities of this group Dr. Broom[5] said:

"There is still another group to which some Pseudosuchian has probably been ancestral, namely, the Birds. For a time one or other of the Dinosaurs was regarded as near the avian ancestor. The resemblance of the hind limb and pelvis seemed to make this extremely probable, and Huxley, Marsh, Cope, and others have all favored this view. Others, however, were more impressed by the apparently avian characters in the skeleton of the Pterodactyls, and especially in the striking avian appearances in the brain, and have argued in favor of a close affinity between the Birds and the Pterodactyls. Osborn, while recognizing the affinities to both groups, and especially to the Dinosaurs, believed that the Birds and the Dinosaurs had a common ancestor, probably in the Permian. Seven years ago, when describing the skeletogenesis of the Ostrich, I argued that the bird had come from a group immediately ancestral to the Theropodous Dinosaurs. The Pseudosuchia, now that it is better known, proves to be just such a group as is required. In those points where we find the Dinosaur too specialized we see the Pseudosuchian still primitive enough. The bird pelvis has probably developed from a type like that of *Ornithosuchus* by the pubis turning further back and the symphysis becoming lost. Whether the union of the metatarsals is a primary or a secondary character is a debatable point. The question is really whether the bird ancestor was a hopping bipedal animal before it flew, or if it only hopped after the wing had become specialized. I am

[5] Proc. Zoöl. Soc., 1913, pp. 631-632.

strongly of the opinion that it was a hopping animal first, and that the metatarsus became strengthened to support the weight of the body entirely borne by the hind feet. It is easy to understand a hopping animal taking to an arboreal life and ultimately developing a wing out of a four-toed hand, while it seems unlikely that the hind foot could ever have developed by arboreal habits. It is interesting to note that while the ancestor of the Pterodactyls had four toes in the manus, there is very clear evidence from the skeletogenesis of the bird that the latter also had a four-toed ancestor.

"A Pseudosuchian which through a bipedal habit had developed a strengthened ankle-joint and a firm metatarsus, and had lost the 5th digit from the manus would meet all the requirements of the avian ancestor."

The theory that the immediate ancestors of birds were arboreal animals has also been supported by Professor Abel,[6] of Vienna, who maintained that the birds and carnivorous dinosaurs arose from a common arboreal, ancestral group with climbing feet. The carnivorous dinosaurs soon reverted to terrestrial habits, while the birds, remaining arboreal, only returned to terrestrial life long after the acquisition of flight. The cleft between birds and carnivorous dinosaurs runs back perhaps to the beginning of the Trias. Professor Abel's conclusions rest partly on the facts that both the hands and the feet of the smaller dinosaurs show marked resemblances to those of the arboreal bird *Archæopteryx.*

Mr. D. M. S. Watson,[7] on the other hand, holds that the backwardly directed first toe of some of the earliest dinosaurs was not a perching adaptation, but served as a strut, or prop, for the support of an animal in the early stages of walking upright.

Quite recently Professor S. W. Williston[8] has expressed the belief (*in litteris*) that the consolidation of the instep bones in the oldest birds was an adaptation to digitigradism, or the habit of raising the body upon the toes, and that flight originated by leaping from below upward, not by gliding downward.

Still more recently Mr. C. W. Beebe, of the New York Zoölogical Park, has made certain discoveries which lend additional evidence for the view that the immediate ancestors of the birds were arboreal animals. In a paper entitled "A Tetrapteryx Stage in the Ancestry of Birds"[9] Mr.

[6] Grundzüge der Paläobiologie der Wirbeltiere. Stuttgart, 1912.

[7] "The Cheirotherium," Geol. Mag., Decade VI, Vol. I, No. 603, September, 1914, pp. 395-398.

[8] "Trimerorhachis, a Permian Temnospondyl Amphibian," Journ. Geol., Vol. XXIII, No. 3, April-May, 1915.

[9] Zoologica, Scientific Contributions of the New York Zoölogical Society, Vol. II, No. 2, November, 1915.

Beebe found that in the young of various species of doves, pigeons, jacanas and owls there is a reduced pelvic wing, consisting of a row of degenerate flight feathers and a second overlapping row of "coverts," all stretched upon the patagium, or skin-fold, behind the femur. Owing to the lateness of the season, it was impossible at this time to extend further his observations on the "pelvic wing" of birds; but he states that "judging merely from the pterylosis of the adult, many species of Coraciiformes, Scansores and Piciformes should show most interesting developments of this tract in the young birds." Mr. Beebe was naturally delighted to find that in the Berlin specimen of *Archæopteryx* there were some strongly marked impressions of feathers on both sides of the tibia and of still larger feathers lying between the pelvis and the bent back head, which he interprets as evidence of the "pelvic wing" in this oldest known bird. Partly from these data Mr. Beebe and Mr. Dwight Franklin have made their series of restorations to illustrate the evolution of birds from a "tetrapteryx stage," with four wings and a long segmental tail, to the modernized two-winged stage, with a normal fan-tail.

Thus Mr. Beebe, along with Pycraft, Abel and others, conceives the immediate ancestors of birds as arboreal animals with the habit of scaling downwards through the air after the fashion of flying squirrels; to this theory his discovery of a vestigial pelvic wing in modern birds lends obvious support.

For some years past the present writer has taken special interest in the problem of the origin of birds, partly for the reason that the subject forms one of the major problems in the Columbia University graduate course on the evolution of the vertebrates. Each year we hold a seminar on this subject, in which the various theories of the origin of the birds are duly advocated by graduate students, and the rival claims of the dinosaurs and other reptilian groups to close kinship with the birds are considered. From this annual review arises the impression that from the clash of conflicting hypotheses the following approximation to the facts may, from present evidence, be provisionally made out:

Far back in the Carboniferous ages the remote common ancestors of birds, dinosaurs, pterosaurs and other reptilian groups were very primitive lizard-like reptiles with extremely small brains, comparatively sluggish habits and a highly variable body temperature. This general description would doubtless fit many of the already known reptiles from the Carboniferous and Permian, but no one of these can yet be recognized as a direct ancestor to the later types.

In the harsh arid ages of the Permian and Trias were evolved hardier and more active carnivorous saurians, represented by the earliest dinosaurs, and especially by the "Pseudosuchians," such as *Ornithosuchus, Euparkeria.* Some of these small reptiles may have reared up on their hind legs in running, as certain lizards do; indeed the hind feet of *Euparkeria,* according to Broom, exhibit incipient adaptations to bipedal progression, while others were in all probability actively hopping types. But none of these animals show any very pronounced bird-like characters in the skeleton. Their structure, however, was, on the whole, of so generalized a type that the diverse peculiarities of the birds, pterosaurs and other groups could readily be derived from this source. In some of the Pseudosuchians the body is known to have been covered with horny plates, but others may have been clothed with the overlapping scales which must have preceded the evolution of feathers.

Thus the paleontological record as to the immediate ancestry of the birds is regrettably indecisive, but the principles of comparative anatomy appear to lead to some pretty safe inferences, as follows:

The transformation of long overlapping scales into feathers, whenever it did occur, was one of the critical steps in the evolution of birds. This transformation took place apparently over the whole surface of the body and along the legs and arms. No doubt it was correlated with the higher vitality and improved circulation of the pro-avis as compared with the ancestral reptile; no doubt also the air-retaining mesh of the feathery covering not only prevented undue radiation of heat from within during cold periods, but also screened off some of the burning heat of the sun. Possibly this feather-armor would also protect the ambitious aviator in his many falls from the branches. The pro-Aves were surely quick runners, both on the ground and in the trees, but it is not yet clear whether the upright position was first attained upon the ground or in the trees. They very early acquired the habit of perching upright on the branches, as shown by the consolidated instep bones, grasping first digit and strong claws of *Archæopteryx.* Their slender arms ended in three long fingers provided with large claws which were at first doubtless used in climbing.

These active pro-Aves contrasted widely in habits with their sluggish remote reptilian forebears. In the pursuit of their prey they jumped lightly from branch to branch and finally from tree to tree, partly sustained by the folds of skin on their arms and legs and later by the long scale-feathers of the pectoral and pelvic "wings" and tail. That they held the arms perfectly still throughout the gliding leap still appears doubtful, for all recent animals that do that have never attained true flight. I cannot avoid the impression that a vigorous downward flap of the arms, even before they became efficient wings, would assist in the "take off" for the leap, and that another flap just before landing would check the speed and assist in landing.

In correlation with this active arboreal life the brain became highly developed; the olfactory lobes grew smaller, the cerebra, optic lobes and cerebellum attained large size and high development. The brain-case was correspondingly expanded, while the skull as a whole acquired a very light construction. The earliest birds retained the sharp conical teeth implanted in distinct sockets which were characteristic of their "thecodont" ancestors (such as *Euparkeria*). The teeth and jaws were adapted for quickly snapping at living prey, perhaps insects. These entirely reptilian teeth, which in the case of Marsh's "toothed birds" bore an extraordinary resemblance to the teeth of mosasaurs, were retained long after the main adaptations for flight were established. But perhaps during the Cretaceous period the ancestors of modern birds lost their teeth as the horny beak at the front of the jaws grew backward.

The skull of birds is of a modified reptilian type and has no doubt been derived simply by the loss of the upper temporal bar, by the inturning of the pterygoid bones and by the enlargement of the internal nares. In short, the whole architecture of the bird skeleton, as indeed the whole internal anatomy, are unquestionably a modification of a primitive reptilian type. The consensus of opinion is that this ancestral type was nearly related to the primitive Archosauria (Diapsida), or two-arched reptiles, and was very widely removed from the mammals, mammal-reptiles, turtles, plesiosaurs and ichthyosaurs.

The hypothesis that the ratite birds have come off from some group of reptiles other than that which gave rise to the carinate birds is, in the writer's judgment, entirely untenable: first, because an examination of the skulls of various ratites and carinates shows agreement in fundamental plan, with minor, though well-marked, differences in details; secondly, because the entire organization of the ratites indicates merely a cursorial readaptation of carinate types; thirdly, because the tinamous are in many characters allied both to the ratites and to the carinates, and thus strengthen the conservative view that the class Aves is broadly monophyletic in origin.

14

Reprinted from *N.Y. Acad. Sci. Ann.* **23:**302-306 (1913)

A SUMMARY OF "THE ORIGIN OF MAMMALS" BY R. BROOM

W. K. Gregory

10 November, 1913

Section met at 8:15 p. m., Vice-President W. D. Matthew presiding.

The minutes of the last meeting of the Section were read and approved.

The following nomination of officer for the year 1914 was made and approved for transmission to the Council:

For Vice-President of the Academy and Chairman of the Section: Professor Raymond C. Osburn.

Dr. William K. Gregory was elected Secretary for the year 1914.

The Secretary read a letter from the Recording Secretary, stating that the Council had voted to assign to any section requesting it, a certain volume of the Annals, to contain papers emanating from the members of that section, the control of publication, however, remaining as heretofore with the Council.

It was moved and seconded that a committee be appointed by the Chairman to consider the publication of a separate volume of the Annals to contain papers read at the meetings of the section. Motion carried. The Chairman stated that he would appoint such a committee.

It was moved and seconded that the next meeting of the section, on 8 December, 1913, should be devoted to a conference on Unit Characters.

By request of the Secretary, it was voted that a Committee on Arrangements should be appointed by the Chairman to assist the officers of the Section in carrying on the work of the Section and in extending its membership.

The following lecture was then offered:

Dr. Robert Broom, The Origin of Mammals.

Summary of Paper

Dr. **Broom,** formerly Professor of Geology and Zoölogy in Victoria College, Stellenbosch, South Africa, illustrated his lecture with lantern slides, and especially with a series of South African mammal-like reptiles, and said in abstract: In 1876, Owen, in describing the fossil reptiles of

South Africa, pointed out numerous mammal-like characters seen in them and in 1880 definitely expressed the view that the primitive mammals living to-day in Australia are the direct descendants of a reptilian ancestor such as he had described. Huxley, on the other hand, favored the descent of the mammals from a salamander-like form, and the contest between those who believe they are descended from amphibians and those who look on reptiles as their ancestors has been urged ever since—sometimes rather vigorously.

When Cope, in 1880, studied the remarkable pelycosaurs, fin-backed reptiles found in the old Permian rocks of New Mexico and Texas, he came to the conclusion that he had found, if not the mammalian ancestors, at least forms allied to them, and in this I believe he was quite correct.

Between 1888 and 1905, Professor Osborn published a considerable number of papers dealing with the origin of mammals, in which he argued that the ancestor of the mammal was probably a member of that group of very mammal-like reptiles found in South Africa and called Cynodonts. This view of Osborn's seems at first sight opposed to that of Cope's, but in all probability both views were correct, the Pelycosaurs being a side branch from a direct line very near to the early mammalian ancestors, the Cynodonts being probably the immediate ancestors of the mammal.

Baur, who worked here in America and died some fifteen years ago, was in favor of the reptile origin. Seeley adopted a rather curious view. He believed that the egg-laying mammals came from reptiles but that other mammals arose from amphibians. On the whole, the Germans have favored the amphibians as ancestors, while English opinion, although somewhat divided, has mainly been in support of the reptilian theory. The majority of Americans, doubtless influenced by Cope and Osborn, have always favored the descent of the mammals from a reptilian ancestor.

I became interested in the question in 1885 and practically resolved then that I would contribute what I could to the solution of the problem. In 1892, I went to Australia and spent some years in studying the egg-laying mammals and marsupials. In 1897, I went to South Africa and have been working in that region for the last seventeen years. In these seventeen years, nearly every specimen that has been picked up there has passed through my hands.

After describing the Karroo formation in which these fossils are found, the speaker described some of the principal types of mammal-like reptiles as follows: The oldest animals we meet with in the Karroo formation in any number are of middle Permian age, shall we say of the year 18,000,000 B. C. These are of especial interest from the resemblance they bear to the American Permian reptiles from Texas and New Mexico. One of the largest and best-known animals is called *Pareiasaurus.* It is a large-limbed reptile, about nine feet in length and standing about three and one-half feet in height. In many points of its organization it shows affinities with the American reptile *Diadectes,* of which a mounted skeleton is to be seen in the American Museum. Another group of animals contemporaneous with *Pareiasaurus* is the reptilian group of Dinocephalians. These also were large reptiles with very powerful limbs. Although herbivorous and having no remarkable specialization of the spines of the vertebræ, they are nevertheless fairly closely allied to the very remarkable American fin-backed Pelycosaurs, of which skeletons are to be seen in the American Museum.

One of the most striking peculiarities of the Karroo reptiles is that almost all agree with *Pareiasaurus* and the Dinocephalians in having powerfully developed limbs. How these have been evolved is a matter of doubt, but there can be little question that it was this strengthening and lengthening of the limbs that started the evolution which ultimately resulted in the formation of the warm-blooded mammals.

The best-known, and in some respects the most remarkable, of the Karroo reptiles belong to a group named by Owen the Anomodonts, from their having horny beaks like the turtles or birds with, in addition in many forms, a pair of large walrus-like tusks. The first specimen was discovered as far back as 1844 and was called *Dicynodon,* but, although many skulls have been discovered, only three or four fairly good skeletons have been found. In limbs, shoulder and pelvic girdles and essential structure of the skull and in the number of joints of the toes, they strikingly resemble the mammals, and although the curious development of the beak obscures the mammal-like character of the skull, it is essentially built on the mammalian plan, and there is little doubt that although the Anomodonts are a side offshoot from the mammalian stem, they are closely allied to the mammalian ancestor.

A form nearly allied to *Dicynodon* is called *Endothiodon.* It has no large tusk but a number of small teeth. Although much rarer than *Dicynodon,* fortunately an almost complete skeleton has been discovered, which has recently been mounted in the Museum laboratories by Mr. Charles Falkenbach under my direction, and of which a photograph is given. The extremely mammal-like condition of the limbs is very manifest, and there is little doubt that the animal waddled about somewhat after the manner of the pigmy hippopotami of Liberia, seen at the New York Zoölogical Park. Attention may be called to the relatively enormous size of the skull and the curious way in which the long point of the lower jaw passes up into the groove in the upper.

We find many other mammal-like reptiles of which the Therocephalians, Dromasaurians and the Cynodonts are the most important. Although these insectivorous and carnivorous types are less mammal-like in some respects than the Anomodonts, they agree more closely with the mammals in the construction of the skull. They all have long, slender limbs adapted for running. The earlier members, such as the lower Therocephalians, have the number of toe joints as still found in the lizards and most reptiles, viz., 2, 3, 4, 5, 3; but the Anomodonts, the lower Dromasaurians and the higher Cynodonts have the same number of joints in the toes as is retained in ourselves, 2, 3, 3, 3, 3. It is rather interesting to look at one's hand and realize that the fingers have all these joints because a remote ancestor took to walking with the feet under the body, supporting it off the ground, rather than with the feet to the side as in the lizards and crocodiles.

The Cynodonts occur in the Triassic formation and a few survive into the Jurassic. In most points of structure, they are extremely mammal-like and it is frequently impossible, if the specimen is at all incomplete, to say whether we are dealing with one of the Cynodonts or a mammal. The lower jaw is almost entirely formed by a large single bone, the posterior bones being small, and the bone on which the jaw hinges is also small, thus foreshadowing the mammalian condition, the dentary bone, the angular, articular and surangular being quite small, as is also the quadrate bone. The teeth are in most forms of a carnivorous type, composed of sharp incisors, long sharp canines and cusped molars, the cusps being almost exactly like those of the carnivorous mammals.

A couple of months ago, I discovered that in the Cynodonts, the incisors, canines and premolars are preceded by an earlier set exactly as in ourselves. It would probably be inappropriate to call them milk teeth, as it is very unlikely that the Cynodonts provided their young with milk, but there can be no doubt that the young had a first temporary set of front teeth like most mammals.

Besides solving the question of the origin of the mammals, the Karroo fossil beds have thrown some light on the origin of birds. There has been considerable discussion as to whether birds were derived from flying bat-like reptiles called pterodactyls or from the dinosaurs. Some have even gone so far as to derive the flying birds from the pterodactyls and the running birds such as the ostrich from the dinosaurs. Dr. Lucas is one of those who favors a double origin for the birds. Professor Osborn some years ago argued in favor of the birds and dinosaurs having come from a common ancestor in Permian times. A few years ago I maintained, as the result of my studies on the development of the ostrich, that the ancestor of the bird though not a dinosaur was nearly a dinosaur, and that the bird and the carnivorous dinosaur were derived from a group of primitive dinosaur-like reptiles that were capable of running on their hind legs. A recent discovery in South Africa reveals just such a type as we required for the common ancestry of the birds and the dinosaurs, and this form is also not far removed from the ancestor of the pterodactyl. The birds, pterodactyls and carnivorous dinosaurs are all probably sprung from a small reptile such as the one recently discovered in South Africa and named by me *Euparkeria.*

Another interesting fact that seems to be brought out by our study of the South African fossil forms is that it was probably the development of the active Cynodonts that led to the development of the active reptiles such as *Euparkeria.* For possibly two million years, the carnivorous mammal-like reptiles had an abundant supply of food in the form of the small Anomodonts. In lower Triassic times, the smaller Anomodonts seem to have become extinct for some reason, and the carnivorous forms had to obtain a new diet, which was probably a little lizard-like animal called *Procolophon,* and possibly other small reptiles of a similar type. It was possibly this new activity that gave rise to the Cynodonts. In upper Triassic times, the Procolophons became extinct, and the small Cynodonts were driven to attacking the more active types like *Euparkeria.* The rivalry between these forms resulted in the greatly increased activity of both, the active four-footed forms becoming the primitive mammals and those which run on their hind legs gave rise to the theropodus dinosaurs and the ancestral birds. The further evolution of the bird was doubtless the result of its taking to an arboreal habit and developing feathers.

A fuller report of Dr. Broom's lecture is printed in the "American Museum Journal," December, 1913.

Professor Osborn spoke of the commanding importance of Dr. Broom's researches and discoveries and moved that the Secretary should be instructed to record on the minutes a cordial vote of thanks to Dr. Broom. The motion was unanimously carried.

The lecture was followed by a reception to Dr. Broom.

The Section then adjourned.

15

Reprinted from *Am. Naturalist* **22:**1067-1079 (1888)

THE EVOLUTION OF MAMMALIAN MOLARS TO AND FROM THE TRITUBERCULAR TYPE.[1]

BY HENRY FAIRFIELD OSBORN.

THE dentition in the recent Mammalia is so diverse that the most sanguine evolutionist of fifteen years ago could not have anticipated the discovery of a common type of molar, in both jaws, as universal among the Mmamalia of an early period as the pentadactyle foot, and as central in its capacity for development into the widely specialized recent types.

The tritubercular molar, discovered by Professor Cope in the Puerco, is exactly such a type, and may be considered with the pentadactyle foot as playing a somewhat analogous rôle in mammalian history, with this important difference—the unmodified pentadactyle foot was probably inherited direct from the reptiles, and its subsequent evolution, with a few exceptions, has been in the direction of the greater or less *reduction* of primitive elements towards special adaptation, as, to borrow an extreme illustration, in the transition from *Phenacodus* with 26 elements in the manus to *Equus* with only 12 such elements. On the other hand, the tritubercular tooth was not inherited, but in all probability developed within the mammalian stock, from a hypothetical form with almost, if not quite simple conical molars, implanted by single fangs, in a nearly homodont series.[2] No such primitive type of mammalian dentition is actually known, although *Dromotherium* approximates it; but the apparent reversion to this type among the *Cetacea*, and apparent retention of it in the *Edentata*,[3] support all the independent evidence upon this point derived from the Mesozoic Mammals. The principle of growth was the regular addition of new parts to the simple cone, not at random, but according to a certain definite

[1] Read in the geological section of the British Association at Bath, September, 1888. Read in abstract by Prof. Cope, National Academy is Sciences, at New Haven, Nov., 1888.

[2] See Author "Structure and Classification of the Mesozoic Mammalia." Jour. Phila. Academy, 1888, p. 240.

[3] See Oldfield Thomas, "The Homologies and Succession of the Teeth in the Dasyuridæ." Phil. Trans., 1887, p. 458.

order which apparently progressed independently in different phyla, through a series of sub-tritubercular stages until trituberculy [1] was attained.

The tritubercular molar consists essentially of three cusps, forming what may be called the primitive triangles, so disposed that the upper and lower molars alternate. This, when attained, formed a central stage from which the great majority of recent molar types have diverged by the addition, modification and reduction of cusps; we must except the Monotremes, the Edentates, and possibly the Cetaceans, although there is considerable evidence that the cetacean molars were once of the triconodont type.[2] Among extinct orders, the Multituberculata (*Plagiaulax*, *Tritylodon*, etc.) must also be excepted from this series and discussion.

The almost universal predominance of trituberculy in the early geological periods, is very significant of the uniformity of molar origin. Of twenty known Mesozoic genera,[3] all except three[4] show trituberculy in some of its stages. As to the Lower Eocene, eighty-two Puerco species, representing twenty-six genera and five orders (Creodonta, Tillodontia, Lemuroidea, Condylarthra, Amblypoda), only four species have quadritubercular teeth, all the remainder are tritubercular.[5] Prof. Rütimeyer has recently pointed out the predominance of this type in the nearly parallel Egerkingen beds The contemporary Cernaysien fauna in the collection of Dr. Lemoine at Rheims, recently examined by the writer, shows exclusively tritubercular molars or their derivatives. By the Middle Eocene the lines of divergence towards the existing types of molars were well advanced, but trituberculy persisted in the dentition of several orders, in which it is found to-day (Lemuroidea, Insectivora, Carnivora, and many Marsupialia).

[1] First employed by Rütimeyer, "Ueber Einige Bezi ehungen zwischen den Säugethierstämmen Alter und Neuer Welt." Abh. d. schweiz. pal. gesellsch., Vol. XV., 1888, p. 54.

[2] See Brandt, "Die Fossilen u. Subfoss. Cetacean Europas." Taf. XXXII., figs. 4-9.

[3] The list given by the writer (op. cit., p. 247) is found to contain several synonyms. See "Additional observations upon the Structure and Classification of the Mesozoic Mammalia." Proc. Phila. Acad., Nov., 1888, p. 292

[4] Dicrocynodon (Diplocynodon), Docodon, Enneodon, Marsh.

[5] Cope, "Synopsis of the Vertebrate Fauna of the Puerco Series," Am. Phil. Soc., 1888. p. 298.

It follows that it is quite as essential for the comparative anatomist to thoroughly grasp the meaning and history of each of the component cusps of the tritubercular molar and of their derivatives, as it is to perfectly understand the elements of the manus and pes. For, the homologies of the cusps can now be determined almost as certainly as those of the digits. Take a human molar, for example, every component tubercle has its pedigree, and it can be demonstrated, almost beyond a doubt, which of these tubercles is homologous with the single reptilian cone. The writer recently (op. cit., p. 242) proposed the adoption of a distinct nomenclature for the different cusps of the tritubercular molar, and offered a series of terms for the primary cusps based as far as possible upon the primitive position and order of development, and in most instances in accord with their secondary position. This nomenclature can be extended to the secondary cusps in the sextubercular superior, and quinquetubercular inferior molars. The terms now in general use are based, for the most part, upon the secondary or acquired position, and in no instance upon the homologies of the cusps in the upper and lower molars, or even in corresponding molars of different genera, thus involving much confusion. For example, the Antero-internal cusp of the lower molar of *Mioclænus* is not homologous with the antero-internal cusp of *Hyopsodus*, nor with the antero-internal cusps of the upper molar of either genus.

The present contribution is based principally upon the writer's studies among the Mesozoic Mammalia, and, with some additions, upon Prof. Cope's numerous essays upon the tritubercular type in the Tertiary Mammalia.[1]

Four propositions may be laid down for discussion:—

(1.) That tritubercnly was acquired during the Mesozoic period, in a series of stage, beginning with the single cone and attaining to the primitive sectorial type in the Jurassic period.

[1] Professor Cope's essays abound with discussions and notes upon the origin and succession of the tritubercular type. (See collection, in "Origin of the Fittest"). He has outlined the transition from the single cone to the tritubercular crown (p. 347); the tubercular sectorial (p. 246); the quadritubercular type (p. 245 and p. 359); the *Spalacotherium* molars as a transition to the *tritubercular* (p. 259). The acquisition of the superior and inferior quadritubercular molar (p. 361). The prediction of the discovery of Carnivora with triconodont molars (p. 365), and of the simple tritubercular type in both jaws (p. 362).

(2.) The majority of Mesozoic mammals showed trituberculy in some of its stages. Present evidence goes to show that the remaining, or aberrant types, if such existed, did not persist. The majority of the persisting forms of later periods were derived from the forms, with simple tritubercular molars, of earlier periods. It follows that trituberculy was an important factor in survival.

(3.) The definite homologies of the primary and to some degree of the secondary cusps in the upper and lower molars can be established.

(4.) The mode of succession of tooth forms favors the kinetogenesis theory advanced by Ryder and Coope.

There are three general observations to be made:—

First.—In attempting to complete the history of each of the cusps, we naturally find that the palæontological record is not sufficiently perfect to admit of our following a certain type along a single phylum back to the primitive type. We must at the outset proceed upon the principle of similar effects, similar causes. For example, since the history of the development of the intermediate tubercles in the superior molars of the Lemuroidea (Pseudolemuroidea, Schlosser) is perfectly clear during the Wasatch and Bridger epochs—it is safe to infer that the intermediate tubercles of the Ungulate molars, which are fully developed in the underlying Puerco, had the same history. *Second.*—There are in each period *Aberrant types* which embrace either incomplete or degenerate tritubercular stages, *i.e.*, a high specialization in which the past record is obliterated, or, finally, stages in non tritubercular lines of development. *Third.*—In the parallel evolution of trituberculy in different phyla we find that the progression is by no means uniform. In every geological period in which the fauna is well known we observe *progressive* genera which outstrip the others in reaching a certain stage of molar development, contrasted with *persistent* types which represent arrested lower stages of development, while between them are the *central* types which represent the degree of evolution attained by the majority of genera. The latter may be said to constitute the stage which is characteristic of the period.

The Stages of trituberculy may now be defined as seen in different types in their order of succession:—

I. *Haplodont* Type (Cope).[1] A simple conical crown. The fang usually single and not distinguished from the crown. This type has not as yet been discovered among the primitive Mammalia.

A *Protodont* Tub. Type.[2] The crown with one main cone, and lateral accessory Cuspules; the fang grooved. There is some question as to the advantage of distinguishing this as a type, for it stands intermediate between types I. and III. Example, *Dromotherium* of the American Triassic.

II. *Triconodont* Type (Osborn, op. cit., p. 242). The crown elongate, trifid, with one central cone and two distinct lateral cones The fang double. Example, *Triconodon.*

III. *Tritubercular* (Cope)[3]. The crown triangular, surmounted by three main cusps, the central cone placed internally in the upper mollars and externally in the lower molars. Example, the lower molars of *Spalacotherium* and *Asthenodon.* This type is rare in its primitive condition as above defined.

The upper and lower molars are alike in types I. and II.; in type III. they have a similar pattern but with the arrangement of the homologous cusps reversed. These types are all primitive. In the following sub types, the primitive triangle forms the main portion of the crown, to which other "secondary" cusps are added, the homologies of which in the upper and lower molars are somewhat doubtful. Parallel and with an intimate relation to the addition of the secondary cusps, is the division of the tritubercular into a secodont and bunodont series, according to the assumtion of a purely cutting or crushing function. In departing from the primitive type, the upper and lower molars diverge in structure, and the homologies of the secondary cusps in each are somewhat doubtful.

LOWER MOLARS.

A. *Tubercular Sectorial,* sub type (Cope). *a.* The primitive

[1] "The Homologies and Origin of the Types of Molar Teeth in the Mammalia Educabilia." Journ. Phila. Acad., 1874. The term *Homodont* was previously applied to this type by Rütimeyer, "Odontographie der Hufthiere, etc." Verh. d. Naturforsch. Gesellsch. in Basel, Band. III., 1863, p. 563. In the writer's opinion this term has acquired a special significance as applied to a whole series of teeth, viz., the reverse of "heterodont," and may well be retained in this sense.

[2] Osborn, op. cit., p. 222.

triangle elevated and its cusps connected by cutting crests; a low posterior heel. *b.* This type embraces a *quinquetubercular* form in which the heel consists of two cusps, an internal and external.[1] *c.* In the Bunodont series it develops into the *quadritubercular* form, by the loss of one of the primitive cusps.

UPPER MOLARS.

B. *Tritubercular. a.* The primitive triangle in the secodont series purely tricupsid. *b.* This embraces a *quinquetubercular* form in which "intermediate" tubercles are developed, both in the Secodont and Bunodont series. *c.* In the Bunodont series a postero-internal cusp is added, forming the *sextubercular* molar.

NOMENCLATURE OF THE CUSPS—As above stated, there is no doubt about the homologies of the three "primary" cusps (protocone, paracone, metacone) in the upper and lower molars They may be given the same terms, with the arbitrary suffix *id*, to distinguish the lower cusps. The first "secondary" cusps (hypocone-hypoconid), added to the upper and lower molars of the primitive triangle, modify the crown from a triangular to a quadrangular shape, and hence may be considered homologous. The three additional secondary cusps (protoconule, metaconule, entoconid) evidently have no homology with each other.

TERMS NOW IN USE.	PROPOSED TERMS.[2]	ABBREV.
Upper Molars.		
Antero-internal cusp	Protocone.	pr.
Postero- " " or 6th cusp	Hypocone.	h.
Antero-external "	Paracone.	p.
Postero- " "	Metacone.	m.
Anterior Intermediate cusp	Protoconule.	pl.
Posterior " "	Metaconule.	ml.
Lower Molars.		
Antero-external cusp	Protoconid.	pr.
Hostero- " "	Hypoconid.	h.
Antero-internal cusp, or 5th cusp	Parrconid.	p.
Intermediate or antero-internal cusp (in quadritubercular molars)	Metaconid.	m.
Postero-internal cusp	Eutoconid.	e.

Evolution of the Cusps. The cusp evolution in the Mesozoic period has been fully discussed by the writer (op. cit., pp.240-4)

[1] American Naturalist, April, 1883, p. 407.

[2] I am much indebted to my colleagues Professors Macloskie and Winans for tie assistance in the selection of these terms.

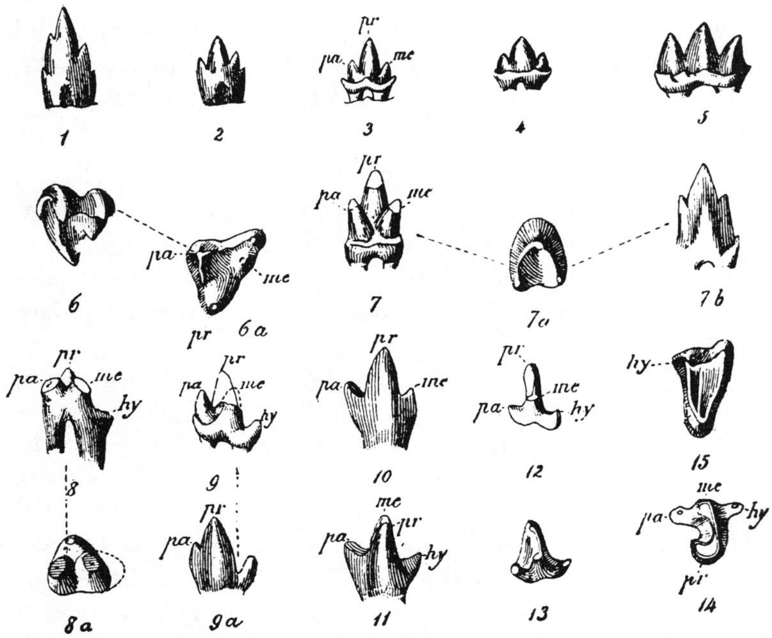

FIG. 1.

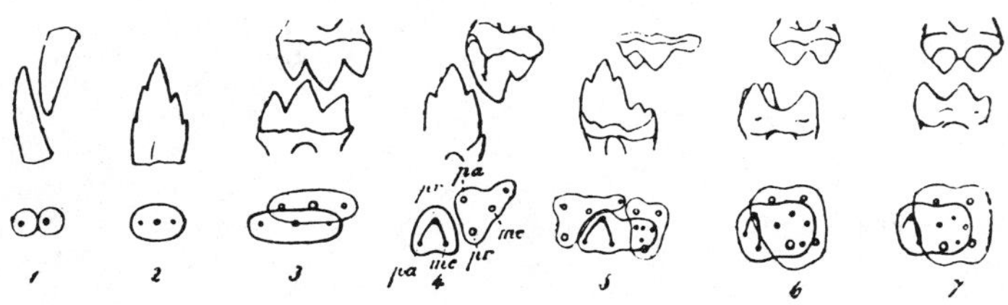

FIG. 2.

FIG. 1.—Molar teeth of Mesozoic Mammalia.
FIG. 2.—Molars of opposite jaws in normal mutual relation.

and in the Tertiary period, by Professor Cope, so that only a brief résumé is necessary here. In *Dromotherium* (fig. 1), from the upper Triassic, the oldest mammalian type known, with the exception

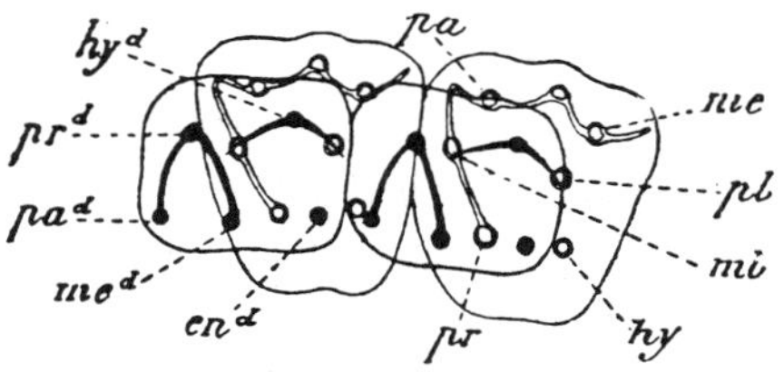

FIG. 1.—Diagram of quadritubercular molars of both jaws in normal mutual relation; the superior cusps double lines; the inferior black.

of *Microlestes*, the molars have a main protoconid with several minute lateral cuspules, differing in size in the different teeth, but in general giving a trifid appearance to the crown. The molars of the contemporary *Microconodon* (fig. 2) also have unpaired fangs, but distinctly trifid crowns, with the anterior and the posterior cusps, or para and metaconids, upon the slopes of the protoconid. This Triconodont type reappears, with the addition of a cingulum and paired fangs, in *Amphilestes* (fig. 3) and *Phascolotherium* (fig. 4) of the lower Jurassic and persists in *Tricondon* (fig. 5) of the upper Jurassic. In this succession we observe especially the relative subsidence of the protoconid and upgrowth of the para- and metaconids. Contemporary with *Amphilestes* is the classical genus *Amphitherium* (fig. 6). A recent examination of the type specimen by the writer revealed the very interesting fact that the molars of this genus are probably of the primitive tubercular- sectorial types,—the oldest known example. Only the paracone and metaconids and hypoconids have been observed heretofore, but one can see the tip of the main external cusps between the former pair. This pattern is repeated, with a considerable elevation of the heel, in *Peramus* of the upper Jurassic.[1] Neither of the two foregoing are of the primitive heelless tritubercular type which is apparently found in *Spalacotherium* also upper Jurassic, and in the nearly related if not synonymous *Peralestes*, Plate XXV. Contemporary with the above, are numerous genera of the *Stylodon* order; among these, *Asthenodon* is of

[1] This genus includes also *Septocladus dubius* Owen, and *Spalacotherium minus* Owen. See Proc. Phila. Acad., Nov. 1888, p. 292.

the primitive tritubercular type without the hypoconid, all the remainder present various modifications of the tubercular sectorial

This covers our knowledge of trituberculy in the Mesozoic period. No bnunodont forms are known—they were probably developed during the Cretaceous, for a few are found well developed in the Puerco. In the Sectorial series many of the types do not widely depart from those seen in the Jurassic, but the Bunodont series are universally characterized by the initial or advanced development of the proto- and metaconules in the upper molars and the apperance of the *Entoconid* upon the inner side of the hypoconid below.

The principles governing cusp development.—It is remarkable to note in how many particulars the actual succession of molar development in the Mesozoic period coincides with the theoretical scheme of origin of trituberculy proposed by Cope[1] and supported by Wortman[2] several years ago. At that time *Spalacotherium* and the genera now embraced under the *Triconodontidæ* were the only Mesozoic mammals whose molar structure was fully known, and the views of these authors were partly speculative and partly deductive from recent dental anatomy.

Two hypotheses may be advanced to explain the evolution of the ritubercular type. The first is that the type has been acquired by the selection of accidental variations in the production of new cusps and modelling of old ones. The second is, that the interaction of the upper and lower molars in the movements of the jaws has resulted in local increase of growth at certain points, resulting first in new cusps, then in a change of position and of form in the cusps. Both hypotheses are open to numerous objections and are by no means mutually exclusive, but the whole subject is so complicated as to require a separate treatment. The balance of evidence in tritubercular evolution seems to favor the second or kinetogenesis theory—as apparently witnessed in two laws of cusp development.

I. The primary cusps first appear as cuspules, or minute cones,

[1] "The Evolution of the Vertebrata Progressive and Retrogressive," American Naturalist, April, 1885, p. 350.

[2] "The Comparative Anatomy of the Teeth of the Vertebrata," 1886 p. 418.

at the first points of contact between the upper and lower molars in the vertical motions of the jaws.

II. The modelling of the cusps into new forms, and the acquisition of secondary position, is a concomitant of interference in the horizontal motions of the jaws.

The second law applies especially to the evolution of the molars after the acquisition of the tritubercular stage, and has been ably proposed and supported by Ryder,[1] principally in its application to recent types of teeth. The first, although not heretofore distinctly formulated, is partly founded upon facts and principles advanced by Cope, and applies chiefly to the stages which have been discussed in this essay.

During the Homodont mammalian or sub-mammalian molar stage, the jaws were probably isognathous and the simple cones alternated as in the *Delphinidæ* (fig. 1). The first additions to the protocone appeared upon its anterior and posterior surfaces. The growth of the para- and metaconids involved anisognathism,[2] for we find in the later triconodonts that the lower molars closed inside of the upper (*Triconodon*, fig. 2). There are several transition forms such as *Tinodon* and *Menacodon* between the primitive triconodont type and *Spalacotherium*, and it has been assumed by Cope and the writer (op. cit., p. 243) that the para- and metaconids were first formed upon the anterior and posterior slopes of the protoconid and then rotated inwards, but it is also possible that they were originally formed upon the inner slopes. In the complemental formation of the upper and lower triangles the jaws remained nearly isognathous (fig. 4). There is no evidence as to the origin of the hypoconid, which as a rule preceded the hypocone, as it was developed very early. In the *Stylacodontidæ*, *Phascolestes*, *Amblo-*

[1] "On the Mechanical Genesis of Tooth Forms." Proc. Phila. Acad., 1878, p. 45.

[2] As employed by Ryder (op. cit., p. 45). "So as not only to indicate respectively parity and disparity in transverse diameter of the crowns of the upper and lower molars, but also the parity or disparity in width transversely, from outside to outside," etc.

It is clear that in the homodont condition, with the teeth simply piercing the food, the greatest comminution (of the food) is effected by isognathism; in the triconodont stage, the jaws must be anisognathous to close upon each other, but the tritubercular stage admits a return to isognathism by the alternation of the triangles.

therium, etc., the crowns rapidly increased in transverse diameter (fig. 7) and, in some genera, they (Kurtodon) so far lost the tritubercular aspect that, but for the connecting form *Asthenodon* (fig. 6), we might hesitate to place them in this series. The key to the further evolution of the crown is seen in the bunodont series during the lower Eocene period.

The superposition of the lower and upper molar patterns brings out many interesting facts. First, even in the complex crowns of the bunodont molars the primitive triangles retain their primitive alternating arrangement. Second, the jaws are somewhat anisognathous. Third, in support of the first law of cusp development, we observe that the protoconule and metaconule are developed at the points of contact with the ridges which extend from the hypoconid, and, secondly, that the hypocone appears at the point where the paraconid abuts against the protocone. It follows from a comparison of numerous species of *Pelycodus* and *Mioclœnus* that as the hypocone develops, the paraconid recedes, as first observed by Cope; a fact difficult to reconcile with the kinetogenesis theory. In this manner the inferior primitive triangle is broken, as the upper molars develop into the sextubercular and the lower into the quadritubercular type. The complemental development of the upper and lower molars in the known genera of successive horizons is approximately displayed in the subjoined table.

The Eocene list of genera will be greatly reduced, especially in the Tritub.-tuberc-.sectorial type, when the upper and lower jaws are found associated, and it must be clearly understood that the sub-types a, b, c, in the above table, are very closely related by transition forms. In fact, in the carnivorous forms, the extreme secodont and bunodont types are frequently seen side by side, as in the first and second inferior molars of *Didymictis*. The chief distinction between these two series is the greater development of the secondary cusps and the almost invariable loss of the paraconid in the latter; this is effected by the broader surfaces of contact in the bunodont crowns. In the secodont series, on the other hand, the development of the secondary cusps is subordinated, and the metaconid is almost invariably suppressed.[1]

[1] See Cope: "Origin of the specialized Teeth of the Carnivora." Am. Naturalist, March, 1879.

PLATE XXVI.

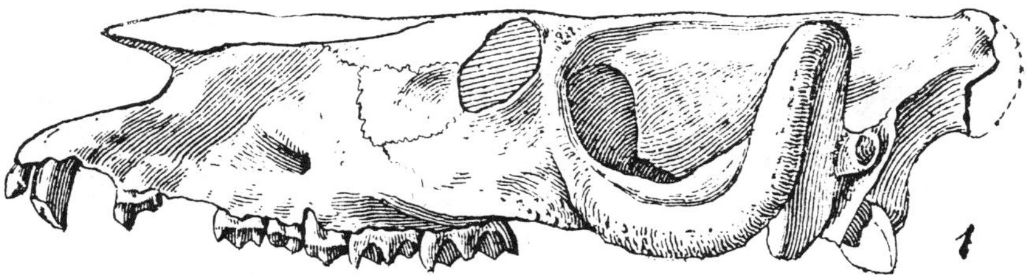

1

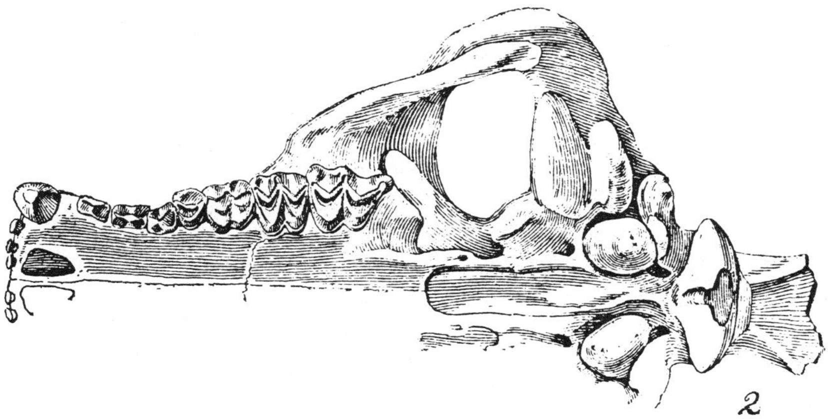

2

FIG. 1.—*Merycochœrus macrostegus* Cope. ¼
FIG. 2.—*Merycochœrus superbus* Leidy. ¼

		Secodont Series.			Bunodont Series.				
	I. A.	II.	III.	a.	a. c.	b.	c.		
Molars, Upper	....	Tricono-dont.	Trituber-cular.	Trituber-cular.	Trituber-cular.	Quinque-tuberc.	Sextuber-cular.	Aberrant.	Lopho-dont.
Molars, Lower	Protodont	Tricono-dont.	Trituber-cular.	Tuberc. Sectorial.	Quadritu-bercular.	Quinque-tuberc.	Quadritu-bercular.		Symbo-rodont.
Middle Eocene,* Bridger	....	....	....	22	4	7	9	1	11
Lower Eocene,** Puerco	....	....	....	6	2	2	1	....	....
Upper Jurassic***	....	3	2	5	....	....	....	4	....
Lower Jurassic	....	3	....	1	....	....	....	....	....
Triassic	1	1	....	....	....	....	....	....	....

* From Prof. Marsh's list of Bridger genera, excluding the Rodentia (see various papers). The Insectivora and Creodonta belong to sub-type a; the Lemuroidea, to sub-types b and c.
** From Prof. Cope's list of Puerco genera (see "Synopsis, etc.").
*** From Author's list of Mesozoic genera (see op. cit., p. 247).

Adapis and *Anaptomorphus* are examples of Sub-types a, c, associated; for it frequently happens that the paraconid atrophies without a complete enlargement of the hypocone. A study of the diagram demonstrates, however, that the association of Sub-types b and c is impossible. The recent monkeys *Tarsius* and *Loris* afford a good illustration of the association of

tritubercular
quinquetubercular
sextubercular
and
quadritubercular

molars.

The subsequent evolution of the molars, in different orders, was characterized, first, by the loss of the primary cusps, *e.g.*, the metaconid in the Carnivora,[1] the paraconid in the Ungulata. Second, by the loss of some of the secondary cusps, *e.g.*, the proto- and metaconules in the Artiodactyla.[1] Third, by the metamorphosis in the form of the cusps. This subject has been fully

[1] Schlosser: "Beitrage zur Kenntniss der Stammgeschichte der Hufthiere," Morph. Jahrb., 1886, p. 123, has especially drawn attention to the probability that the Artiodactyla were derived from sexitubercular forms.

treated by Rütimeyer, Kowalevsky, Cope, Schlosser and others.

The Relation of Trituberculy to the Persistence of Mammalian Phyla.—The above table shows somewhat indefinitely, but none the less positively, the general progression of the Mammalia, to and from the primitive tritubercular type. As already stated, even with our present very limited knowledge, certain stages appear to have been characteristic of certain periods, as follows: the triconodont in the lower Jurassic; the primitive tritubercular and tubercular sectorial in the upper Jurassic; the secodont and bunodont sub-types of trituberculy, predominated in the Puerco; in the Bridger, the Perissodactyl ungulates had mostly passed beyond into the lophodont and symborodont types, and the Artiodactyls were approximately in the stage of sub-types c; but the Lemuroidea, Creodonta, Insectivora, etc., were, almost without exception, tritubercular.

There can be little doubt that, parallel with the tritubercular forms, in each period, there were aberrant or degenerate types, but it is difficult to determine which these are. Many Mesozoic types, which the writer formerly considered aberrant, have now proven to be tritubercular.[3] The upper Jurassic genera included under the *Dicrocynodontidæ* (see Marsh, *Amer. Journ. Sc.*, April, 1887, p. 338) are apparently aberrant. There are several degenerate types among the Puerco and Wasatch Creodonts, such as *Dissacus* and *Mesonyx*. But there is a striking proof of the superiority of the tritubercular molar in the fact that, according to our present knowledge at least, the Jurassic mammals possessing aberrant or degenerate molar types did not persist into the Puerco, nor did such types in the Puerco persist into the Bridger. There is some doubt as to the persistence of the sub-tritubercular stage; the writer previously considered the *Thylacinus* molars as triconodont; but Mr. Lydekker has called attention to the probability that the metaconid has disappeared and been replaced by a heel as in the sectorial teeth of the Carnivora. The disappearance of the degenerate types may be attributed to the general principle that rapid specialization and loss of parts leads ultimately to extinction, by depriving the animal of the means of adaptation to new conditions, or surroundings.

[3] See "Additional Observations upon the Structure and Classification of the Mesozoic Mammalia." Proc. Phila. Acad., Nov., 1888.

The mechanical superiority of the tritubercular type, over every other has been repeatedly demonstrated in its plastic capacity of adaptation to the most extreme trenchant and crushing functions.

ERRATA

Page 1067, paragraph 1, line 4 should read "among the Mammalia of . . . ." Page 1070, line 11 from the top of the page should read "advanced by Ryder and Cope." Page 1071, line 4 from the top of the page should read "II. *Protodont* Type . . . ." Page 1071, line 14 from the top of the page should read "molars and externally . . . ." Page 1072, line 5 from the bottom of the table should read "Postero-." Page 1072, line 4 from the bottom of the table should read "Paraconid." Page 1072, line 1 from the bottom of the table should read "Entoconid." Page 1074, line 22 from the top of the page should read "tritubercular type." Page 1077, lines 17 and 18 from the top of the page are transposed.

16

Reprinted by permission from *Nature* **115:**195–199 (1925)

Australopithecus africanus: The Man-Ape of South Africa.

By Prof. Raymond A. Dart, University of the Witwatersrand, Johannesburg, South Africa.

TOWARDS the close of 1924, Miss Josephine Salmons, student demonstrator of anatomy in the University of the Witwatersrand, brought to me the fossilised skull of a cercopithecid monkey which, through her instrumentality, was very generously loaned to the Department for description by its owner, Mr. E. G. Izod, of the Rand Mines Limited. I learned that this valuable fossil had been blasted out of the limestone cliff formation—at a vertical depth of 50 feet and a horizontal depth of 200 feet—at Taungs, which lies 80 miles north of Kimberley on the main line to Rhodesia, in Bechuanaland, by operatives of the Northern Lime Company. Important stratigraphical evidence has been forthcoming recently from this district concerning the succession of stone ages in South Africa (Neville Jones, Jour. Roy. Anthrop. Inst., 1920), and the feeling was entertained that this lime deposit, like that of Broken Hill in Rhodesia, might contain fossil remains of primitive man.

I immediately consulted Dr. R. B. Young, professor of geology in the University of the Witwatersrand, about the discovery, and he, by a fortunate coincidence, was called down to Taungs almost synchronously to investigate geologically the lime deposits of an adjacent farm. During his visit to Taungs, Prof. Young was enabled, through the courtesy of Mr. A. F. Campbell, general manager of the Northern Lime Company, to inspect the site of the discovery and to select further samples of fossil material for me from the same formation. These included a natural cercopithecid endocranial cast, a second and larger cast, and some rock fragments disclosing portions of bone. Finally, Dr. Gordon D. Laing, senior lecturer in anatomy, obtained news, through his friend Mr. Ridley Hendry, of another primate skull from the same cliff. This cercopithecid skull, the possession of Mr. De Wet, of the Langlaagte Deep Mine, has also been liberally entrusted by him to the Department for scientific investigation.

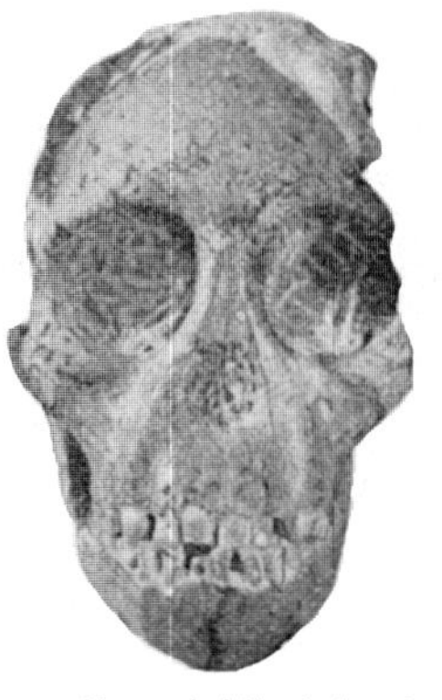

Fig. 1.—Norma facialis of *Australopithecus africanus* aligned on the Frankfort horizontal.

The cercopithecid remains placed at our disposal certainly represent more than one species of catarrhine ape. The discovery of Cercopithecidæ in this area is not novel, for I have been informed that Mr. S. Haughton has in the press a paper discussing at least one species of baboon from this same spot (Royal Society of South Africa). It is of importance that, outside of the famous Fayüm area, primate deposits have been found on the African mainland at Oldaway (Hans Reck, *Sitzungsbericht der Gesellsch. Naturforsch. Freunde*, 1914), on the shores of Victoria Nyanza (C. W. Andrews, *Ann. Mag. Nat. Hist.*, 1916), and in Bechuanaland, for these discoveries lend promise to the expectation that a tolerably complete story of higher primate evolution in Africa will yet be wrested from our rocks.

In manipulating the pieces of rock brought back by Prof. Young, I found that the larger natural endocranial cast articulated exactly by its fractured frontal extremity with another piece of rock in which the broken lower and posterior margin of the left side of a mandible was visible. After cleaning the rock mass, the outline of the hinder and lower part of the facial skeleton came into view. Careful development of the solid limestone in which it was embedded finally revealed the almost entire face depicted in the accompanying photographs.

It was apparent when the larger endocranial cast was first observed that it was specially important, for its size and sulcal pattern revealed sufficient similarity with those of the chimpanzee and gorilla to demonstrate that one was handling in this instance an anthropoid and not a cercopithecid ape. Fossil anthropoids have not hitherto been recorded south of the Fayüm in Egypt, and living anthropoids have not been discovered in recent times south of Lake Kivu region in Belgian Congo, nearly 2000 miles to the north, as the crow flies.

All fossil anthropoids found hitherto have been known only from mandibular or maxillary fragments, so far as crania are concerned, and so the general appearance of the types they represented has been unknown; consequently, a condition of affairs where virtually the whole face and lower jaw, replete with teeth, together with the major portion of the brain pattern, have been preserved, constitutes a specimen of unusual value in fossil anthropoid discovery. Here, as in *Homo rhodesiensis*, Southern Africa has provided documents of higher primate evolution that are amongst the most complete extant.

Apart from this evidential completeness, the specimen is of importance because it exhibits an extinct race of apes *intermediate between living anthropoids and man*.

In the first place, the whole cranium displays *humanoid* rather than anthropoid lineaments. It is markedly dolichocephalic and leptoprosopic, and manifests in a striking degree the *harmonious relation* of calvaria to face emphasised by Pruner-Bey. As Topinard says, "A cranium elongated from before backwards, and at the same time elevated, is already in harmony by itself; but if the face, on the other hand, is elongated from above downwards, and narrows, the harmony is complete." I have assessed roughly the difference in the relationship of the glabella-gnathion facial length to the glabella-inion calvarial length in recent African anthropoids of an age comparable with that of this specimen (depicted in Duckworth's "Anthropology and Morphology," second edition, vol. i.), and find that, if the glabella-inion length be regarded in all three as 100, then the glabella-gnathion length in the young chimpanzee is approximately 88, in the young gorilla 80, and in this fossil 70, which proportion suitably demonstrates the enhanced relationship of cerebral length to facial length in the fossil (Fig. 2).

The glabella is tolerably pronounced, but any traces of the salient supra-orbital ridges, which are present even in immature living anthropoids, are here entirely absent. Thus the relatively increased glabella-inion measurement is due to brain and not to bone. Allowing 4 mm. for the bone thickness in the inion region, that measurement in the fossil is 127 mm. ; *i.e.* 4 mm. less than the same measurement in an adult chimpanzee in the Anatomy Museum at the University of the Witwatersrand. The orbits are not in any sense detached from the forehead, which rises steadily from their margins in a fashion amazingly human. The interorbital width is very small (13 mm.) and the ethmoids are not blown out laterally as in modern African anthropoids. This lack of ethmoidal expansion causes the lacrimal fossæ to face posteriorly and to lie relatively far back in the orbits, as in man. The orbits, instead of being subquadrate as in anthropoids, are almost circular, furnishing an orbital index of 100, which is well within the range of human variation (Topinard, " Anthropology "). The malars, zygomatic arches, maxillæ, and mandible all betray a delicate and humanoid character. The facial prognathism is relatively slight, the gnathic index of Flower giving a value of 109, which is scarcely greater than that of certain Bushmen (Strandloopers) examined by Shrubsall. The nasal bones are not prolonged below the level of the lower orbital margins, as in anthropoids, but end above these, as in man, and are incompletely fused together in their lower half. Their maximum length (17 mm.) is not so great as that of the nasals in *Eoanthropus dawsoni*. They are depressed in the median line, as in the chimpanzee, in their lower half, but it seems probable that this depression has occurred post-mortem, for the upper half of each bone is arched forwards (Fig. 1). The nasal aperture is small and is just wider than it is high (17 mm. × 16 mm.). There is no nasal spine, the floor of the nasal cavity being continuous with the anterior aspect of the alveolar portions of the maxillæ, after the fashion of the chimpanzee and of certain New Caledonians and negroes (Topinard, *loc. cit.*).

In the second place, the dentition is *humanoid* rather than anthropoid. The specimen is juvenile, for the

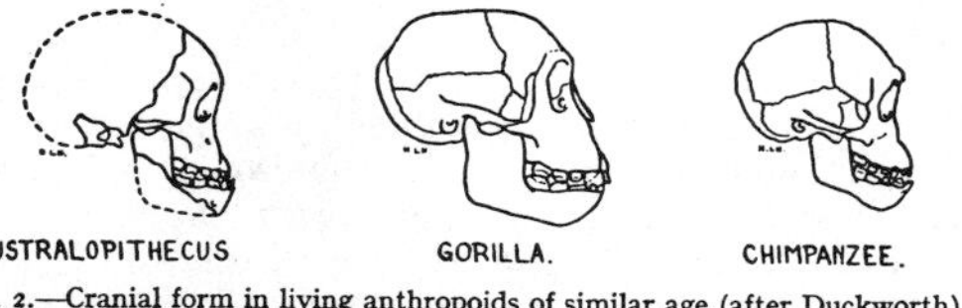

Fig. 2.—Cranial form in living anthropoids of similar age (after Duckworth) and in the new fossil. For this comparison, the fossil is regarded as having the same calvarial length as the gorilla.

first permanent molar tooth only has erupted in both jaws on both sides of the face ; *i.e.* it corresponds anatomically with a human child of six years of age. Observations upon the milk dentition of living primates are few, and only one molar tooth of the deciduous dentition in one fossil anthropoid is known (Gregory, " The Origin and Evolution of the Human Dentition," 1920). Hence the data for the necessary comparisons are meagre, but certain striking features of the milk dentition of this creature may be mentioned. The tips of the canine teeth transgress very slightly (0·5-0·75 mm.) the general margin of the teeth in each jaw, *i.e.* very little more than does the human milk canine. There is no diastema whatever between the premolars and canines on either side of the lower jaw, such as is present in the deciduous dentition of living anthropoids ; but the canines in this jaw come, as in the human jaw, into alignment with the incisors (Gregory, *loc. cit.*). There is a diastema (2 mm. on the right side, and 3 mm. on the left side) between the canines and lateral incisors of the upper jaw ; but seeing, first, that the incisors are narrow, and, secondly, that diastemata (1 mm.-1·5 mm.) occur between the central incisors of the upper jaw and between the medial and lateral incisors of both sides in the lower jaw, and, thirdly, that some separation of the milk teeth takes place even in mankind (Tomes, " Dental Anatomy," seventh edition) during the establishment of the permanent dentition, it is evident that the diastemata which occur in the upper jaw are small. The lower canines, nevertheless, show wearing facets both for the upper canines and for the upper lateral incisors.

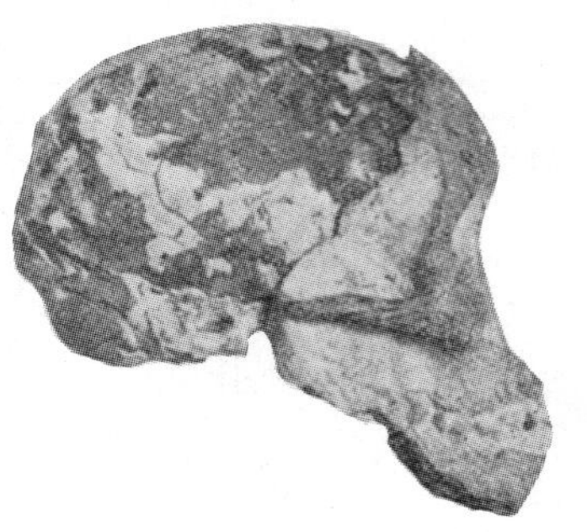

Fig. 3.—Norma lateralis of *Australopithecus africanus* aligned on the Frankfort horizontal.

The incisors as a group are irregular in size, tend to overlap one another, and are almost vertical, as in man ; they are not symmetrical and well spaced, and do not project forwards markedly, as in anthropoids. The upper lateral incisors do project forwards to some extent and perhaps also do the upper central incisors very slightly, but the lateral lower incisors betray no evidence of forward projection, and the central lower incisors are not even vertical as in most races of mankind, but are directed slightly backwards, as *sometimes* occurs in man. Owing to these remarkably human characters displayed by the deciduous dentition, when contour tracings of the upper jaw are made, it is found that the jaw and the teeth, as a whole, take up a parabolic arrangement comparable only with that presented by mankind amongst the higher primates. These facts, together with the more minute anatomy of the teeth, will be illustrated and discussed in the memoir which is in the process of elaboration concerning the fossil remains.

In the third place, the mandible itself is *humanoid* rather than anthropoid. Its ramus is, on the whole, short and slender as compared with that of anthropoids, but the bone itself is more massive than that of a human being of the same age. Its symphyseal region is virtually complete and reveals anteriorly a more vertical outline than is found in anthropoids or even in the jaw of Piltdown man. The anterior symphyseal surface is scarcely less vertical than that of Heidelberg man. The posterior symphyseal surface in living anthropoids differs from that of modern man in possessing a pronounced posterior prolongation of the lower border, which joins together the two halves of the mandible, and so forms the well-known *simian shelf* and above it a deep genial impression for the attachment of the tongue musculature. In this character, *Eoanthropus dawsoni* scarcely differs from the anthropoids, especially the chimpanzee; but this new fossil betrays no evidence of such a shelf, the lower border of the mandible having been massive and rounded after the fashion of the mandible of *Homo heidelbergensis*.

FIG. 4.—Norma basalis of *Australopithecus africanus* aligned on the Frankfort horizontal.

That hominid characters were not restricted to the face in this extinct primate group is borne out by the relatively forward situation of the foramen magnum. The position of the basion can be assessed within a few millimetres of error, because a portion of the right exoccipital is present alongside the cast of the basal aspect of the cerebellum. Its position is such that the basi-prosthion measurement is 89 mm., while the basi-inion measurement is at least 54 mm. This relationship may be expressed in the form of a " head-balancing " index of 60·7. The same index in a baboon provides a value of 41·3, in an adult chimpanzee 50·7, in Rhodesian man 83·7, in a dolichocephalic European 90·9, and in a brachycephalic European 105·8. It is significant that this index, which indicates in a measure the poise of the skull upon the vertebral column, points to the assumption by this fossil group of an attitude appreciably more erect than that of modern anthropoids. The improved poise of the head, and the better posture of the whole body framework which accompanied this alteration in the angle at which its dominant member was supported, is of great significance. It means that a greater reliance was being placed by this group upon the feet as organs of progression, and that the hands were being freed from their more primitive function of accessory organs of locomotion. Bipedal animals, their hands were assuming a higher evolutionary rôle not only as delicate tactual, examining organs which were adding copiously to the animal's knowledge of its physical environment, but also as instruments of the growing intelligence in carrying out more elaborate, purposeful, and skilled movements, and as organs of offence and defence. The latter is rendered the more probable, in view, first, of their failure to develop massive canines and hideous features, and, secondly, of the fact that even living baboons and anthropoid apes can and do use sticks and stones as implements and as weapons of offence (" Descent of Man," p. 81 *et seq.*).

Lastly, there remains a consideration of the endocranial cast which was responsible for the discovery of the face. The cast comprises the right cerebral and cerebellar hemispheres (both of which fortunately meet the median line throughout their entire dorsal length) and the anterior portion of the left cerebral hemisphere. The remainder of the cranial cavity seems to have been empty, for the left face of the cast is clothed with a picturesque lime crystal deposit; the vacuity in the left half of the cranial cavity was probably responsible for the fragmentation of the specimen during the blasting. The cranial capacity of the specimen may best be appreciated by the statement that the length of the cavity could not have been less than 114 mm., which is 3 mm. greater than that of an adult chimpanzee in the Museum of the Anatomy Department in the University of the Witwatersrand, and only 14 mm. less than the greatest length of the cast of the endocranium of a gorilla chosen for casting on account of its great size. Few data are available concerning the expansion of brain matter which takes place in the living anthropoid brain between the time of eruption of the first permanent molars and the time of their becoming adult. So far as man is concerned, Owen (" Anatomy of Vertebrates," vol. iii.) tells us that " The brain has advanced to near its term of size at about ten years, but it does not usually obtain its full development till between twenty and thirty years of age." R. Boyd (1860) discovered an increase in weight of nearly 250 grams in the brains of male human beings after they had reached the age of seven years. It is therefore reasonable to believe that the adult forms typified by our present specimen possessed brains which were larger than that of this juvenile specimen, and equalled, if they did not actually supersede, that of the gorilla in absolute size.

Whatever the total dimensions of the adult brain may have been, there are not lacking evidences that the brain in this group of fossil forms was distinctive in type and was an instrument of greater intelligence that that of living anthropoids. The face of the endocranial cast is scarred unfortunately in several places (cross-hatched in the dioptographic tracing—see Fig. 5). It is evident that the relative proportion

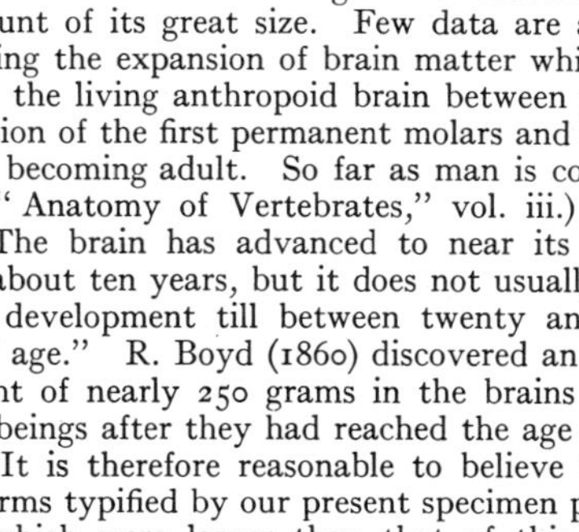

FIG. 5.—Dioptographic tracing of *Australopithecus africanus* (right side), × ⅓.

of cerebral to cerebellar matter in this brain was greater than in the gorilla's. The brain does not show that general pre- and post-Rolandic flattening characteristic of the living anthropoids, but presents a rounded and well-filled-out contour, which points to a symmetrical and balanced development of the faculties of associative memory and intelligent activity. The pithecoid type of parallel sulcus is preserved, but the sulcus lunatus has been thrust backwards towards the occipital pole by a pronounced general bulging of the parieto-temporo-occipital association areas.

To emphasise this matter, I have reproduced (Fig. 6) superimposed coronal contour tracings taken at the widest part of the parietal region in the gorilla endocranial cast and in this fossil. Nothing could illustrate better the mental gap that exists between living anthropoid apes and the group of creatures which the fossil represents than the flattened atrophic appearance of the parietal region of the brain (which lies between the visual field on one hand, and the tactile and auditory fields on the other) in the former and its surgent vertical and dorso-lateral expansion in the latter. The expansion in this area of the brain is the more significant in that it explains the posterior *humanoid* situation of the sulcus lunatus. It indicates (together with the narrow interorbital interval and human characters of the orbit) the fact that this group of beings, having acquired the faculty of stereoscopic vision, had profited beyond living anthropoids by setting aside a relatively much larger area of the cerebral cortex to

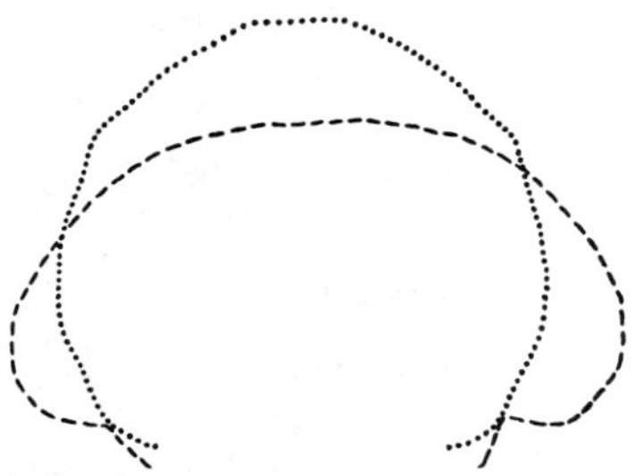

FIG. 6.—Contour tracings of coronal sections through the widest part of the parietal region of the endocranial casts in Australopithecus ...... and in a gorilla - - - -.

serve as a storehouse of information concerning their objective environment as its details were simultaneously revealed to the senses of vision and touch, and also of hearing. They possessed to a degree unappreciated by living anthropoids the use of their hands and ears and the consequent faculty of associating with the colour, form, and general appearance of objects, their weight, texture, resilience, and flexibility, as well as the significance of sounds emitted by them. In other words, their eyes saw, their ears heard, and their hands handled objects with greater meaning and to fuller purpose than the corresponding organs in recent apes. They had laid down the foundations of that discriminative knowledge of the appearance, feeling, and sound of things that was a necessary milestone in the acquisition of articulate speech.

There is, therefore, an ultra-simian quality of the brain depicted in this immature endocranial cast which harmonises with the ultra-simian features revealed by the entire cranial topography and corroborates the various inferences drawn therefrom. The two thousand miles of territory which separate this creature from its nearest living anthropoid cousins is indirect testimony to its increased intelligence and mastery of its environment. It is manifest that we are in the presence here of a pre-human stock, neither chimpanzee nor gorilla, which possesses a series of differential characters not encountered hitherto in any anthropoid stock. This complex of characters exhibited is such that it cannot be interpreted as belonging to a form ancestral to any living anthropoid. For this reason, we may be equally confident that there can be no question here of a primitive anthropoid stock such as has been recovered from the Egyptian Fayüm. Fossil anthropoids, varieties of Dryopithecus, have been retrieved in many parts of Europe, Northern Africa, and Northern India, but the present specimen, despite its youth, cannot be confused with anthropoids having the dryopithecid dentition. Other fossil anthropoids from the Siwalik hills in India (Miocene and Pliocene) are known which, according to certain observers, may be ancestral to modern anthropoids and even to man.

Whether our present fossil is to be correlated with the discoveries made in India is not yet apparent; that question can only be solved by a careful comparison of the permanent molar teeth from both localities. It is obvious, meanwhile, that it represents a fossil group distinctly advanced beyond living anthropoids in those two dominantly human characters of facial and dental recession on one hand, and improved quality of the brain on the other. Unlike Pithecanthropus, it does not represent an ape-like man, a caricature of precocious hominid failure, but a creature well advanced beyond modern anthropoids in just those characters, facial and cerebral, which are to be anticipated in an extinct link between man and his simian ancestor. At the same time, it is equally evident that a creature with anthropoid brain capacity, and lacking the distinctive, localised temporal expansions which appear to be concomitant with and necessary to articulate man, is no true man. It is therefore logically regarded as a man-like ape. I propose tentatively, then, that a new family of *Homo-simiadæ* be created for the reception of the group of individuals which it represents, and that the first known species of the group be designated *Australopithecus africanus*, in commemoration, first, of the extreme southern and unexpected horizon of its discovery, and secondly, of the continent in which so many new and important discoveries connected with the early history of man have recently been made, thus vindicating the Darwinian claim that Africa would prove to be the cradle of mankind.

It will appear to many a remarkable fact that an ultra-simian and pre-human stock should be discovered, in the first place, at this extreme southern point in Africa, and, secondly, in Bechuanaland, for one does not associate with the present climatic conditions obtaining on the eastern fringe of the Kalahari desert an environment favourable to higher primate life. It is generally believed by geologists (*vide* A. W. Rogers, "Post-Cretaceous Climates of South Africa," *South African Journal of Science*, vol. xix., 1922) that the climate has fluctuated within exceedingly narrow limits in this country since Cretaceous times. We must therefore conclude that it was only the enhanced cerebral powers possessed by this group which made their existence possible in this untoward environment.

In anticipating the discovery of the true links between the apes and man in tropical countries, there has been a tendency to overlook the fact that, in the luxuriant forests of the tropical belts, Nature was supplying with profligate and lavish hand an easy and sluggish solution, by adaptive specialisation, of the problem of existence in creatures so well equipped mentally as living anthropoids are. For the production of man a different apprenticeship was needed to sharpen the wits and quicken the higher manifestations of intellect—a more open veldt country where competition was keener between swiftness and stealth, and where adroitness of thinking and movement played a preponderating rôle in the preservation of the species. Darwin has said, " no country in the world abounds in a greater degree with dangerous beasts than Southern Africa," and, in my opinion, Southern Africa, by providing a vast open country with occasional wooded belts and a relative scarcity of water, together with a fierce and bitter mammalian competition, furnished a laboratory such as was essential to this penultimate phase of human evolution.

In Southern Africa, where climatic conditions appear to have fluctuated little since Cretaceous times, and where ample dolomitic formations have provided innumerable refuges during life, and burial-places after death, for our troglodytic forefathers, we may confidently anticipate many complementary discoveries concerning this period in our evolution.

In conclusion, I desire to place on record my indebtedness to Miss Salmons, Prof. Young, and Mr. Campbell, without whose aid the discovery would not have been made ; to Mr. Len Richardson for providing the photographs ; to Dr. Laing and my laboratory staff for their willing assistance ; and particularly to Mr. H. Le Helloco, student demonstrator in the Anatomy Department, who has prepared the illustrations for this preliminary statement.

[*Editor's Note:* The University of Montana Library loaned the *Nature* volume from which camera copy of this article was made.]

Part IV

FUNCTIONAL MORPHOLOGY, PALEOBIOLOGY, AND PALEOECOLOGY

Editor's Comments on Papers 17 Through 22

The papers reprinted in Part IV deal with fossil vertebrates as living, breathing animals. Although most of these papers date from the present century, this is not a new subject. Indeed, one of Cuvier's prime concerns was to restore the appearance and habits of extinct vertebrates (Papers 1A and 1B). In the twentieth century, however, there has been progressively more interest in reconstructing the biology (paleoecology) of fossil forms. Two basic approaches have been taken to achieve this goal: the autecological approach, concentrating on the functional anatomy and morphology of the individual (Papers 17, 18, and 19), and the synecological approach, involving the study of the interrelations of different individuals and species in a community, fauna, or temporal succession of faunas (Papers 20, 21, and 22). Papers 21 and 22 also address the eternal question of why some species and groups of species become extinct. While most vertebrate paleontologists have traditionally favored gradualistic explanations to account for extinctions, recently there has been quite a bit of interest in the possibility that some major, worldwide extinctions (such as is seen at the Cretaceous-Tertiary boundary) may have been due to catastrophic extraterrestrial causes, such as the impact of a large asteroid or comet (Silver and Schultz, 1982).

Sir Richard Owen (1804–1892) was born in Lancaster, England, and studied medicine and surgery at Edinburgh University and St. Bartholomew's Hospital, London. During his long career he was at the Hunterian Museum of the Royal College of Surgeons; the British Museum (1856), where he was superintendent of the Natural History departments; and the British Museum (Natural History), where he became the first director in 1881. Owen was inspired by Cuvier (Papers 1A and 1B) and as a young man attended some of Cuvier's lectures in Paris. Owen's researches and writings cover the whole spectrum of vertebrate anatomy from fishes to mammals, both living and extinct forms. He is especially well known for his work on Mesozoic reptiles, naming the group Dinosauria in 1841. One of Owen's important contributions to theoretical comparative anatomy is his clear separation of analogous organs (parts or organs with the same function in different organisms) from homologous organs (the same parts or organs in different animals despite their form or function).

Paper 17 is one of Owen's early works. In this paper he attempts to reconstruct the form of the tail of the Mesozoic ichthyosaurs (swimming reptiles somewhat analogous to extant dolphins and porpoises). The paper also represents an extremely early study in taphonomy (Behrensmeyer and Hill, 1982; Efremov, 1940); that is, Owen tries to figure out the processes that the bodies of the ichthyosaurs underwent from death to final burial and preservation as fossils. Owen based his analysis on what he considered to be a downward break or dislocation found consistently in the preserved tail vertebrae of ichthyosaurs. From this he deduced that the ichthyosaurs possessed a large, fleshy caudal fin. From the morphology of the caudal vertebrae, Owen deduced that this fin was not horizontal, as in porpoises, dolphins, and whales, but vertical. Subsequent studies and finds of ichthyosaurs have confirmed that these animals possessed vertical caudal fins. However, it is now known that the supposed dislocation or break was actually natural and that the ichthyosaurs bore reversed heterocercal tails (Huene, 1922; Romer, 1956). As a side note, it is interesting that, according to the inscription on the first page, the copy of Paper 17 reproduced here (now housed in the Reprint Collection of the Division of Vertebrate Paleontology, Peabody Museum of Natural History, Yale University) was originally presented by Owen to Gideon Algernon Mantell (1790–1852). Mantell was the original describer, in 1825, of the second known dinosaur, *Iguanodon* (the first was *Megalosaurus,* described by William Buckland [1784-1856] in 1824). For a reconstruction of a species of *Iguanodon* by Marsh, see plate 85 of Paper 12.

In Paper 18, Gregory (mentioned previously, see Paper 13) presents a clear and explicit, if perhaps somewhat simplified, mechanical analysis of quadrupedal locomotion in hoofed mammals. Such studies enabled him (Paper 18 and Gregory, 1929) to reconstruct the locomotor habits of extinct species. Gregory's studies along these lines have inspired, both directly and indirectly, a long series of mechanically oriented studies of vertebrate functional anatomy, especially tetrapod functional anatomy (e.g., Becht, 1953; Bramble, 1978; Gambaryan, 1974; Herring and Herring, 1974; Howell, 1944; Smith and Savage, 1956, 1959; Slijper, 1946).

Alexei Alexievich Borissiak (1872-1944) was one of the early leading figures in vertebrate paleontology in the Union of Soviet Socialist Republics (Sokolov, 1972). Trained as a mining engineer, Borissiak was a teacher and administrator, and also made original contributions in tectonics, stratigraphy, history of geology and paleontology, invertebrate paleontology and vertebrate paleontology. He founded the Paleontological Institute of the U.S.S.R. Academy of Sciences, and was its first director. Borissiak's professional research in vertebrate paleontology concentrated primarily on extinct rhinoceroses and chalicotheres (Perissodactyla). In Paper 19, Borissiak not only tries to bring the extinct chalicotheres back to life as living, functioning organisms, but he also digresses into a discussion of the origin and evolution of chalicotheres. In particular, Borissiak concludes, on the basis of shared-primitive (unspecialized or plesiomorphous) anatomical characters (Paper 19, p. 670), that chalicotheres have their closest relatives among titanotheres. Such methodology (using primitive characters to deduce relationships) was widely accepted at the time (1945) this paper was first published. Today, however, adherents of the cladistic school of phylogeny reconstruction argue that only shared and derived (specialized, advanced, homologous, evolved, or apomorphous) characters can be used to deduce phylogenetic relationships among organisms (see review in Cracraft, 1983; see also Schoch, 1983, who argues that the closest relatives of chalicotheres are found among ceratomorphs [tapirs and rhinos]). In consensus with modern opinion (Schoch, 1982), Borissiak (Paper 19, p. 678) did realize that, when hypothesizing a genealogical tree for the chalicotheres (or any group of extinct animals), it is virtually impossible to demonstrate direct ancestor-descendant relationships between known forms.

Paper 20, by the contemporary American Everett Claire Olson, is a classic paleoecological analysis that attempts to trace the effects of changing environmental conditions on successive vertebrate populations and communities within a limited geographical area through time. In Paper 21, the Americans Leigh Maiorana Van Valen and Robert Evan Sloan seek an understanding of the ecology of a particular taxonomic group, the extinct mammalian group Multituberculata. They chose to work primarily through analysis of the fossil assemblages in which multituberculates are found rather than through analysis of the functional morphology of individual animals or species (as exemplified by Papers 18 and 19). It should be noted that multituberculates are now known from the early Oligocene of North America (Krishtalka et al., 1982; Savage and Russell, 1983).

Paper 22 is by the contemporary Polish paleontologist Kazimierz Kowalski, who now resides in Algeria. In this contribution, Kowalski analyzes the mammalian fauna of an area (actually a small biogeographical province) that is known to have undergone definite environmental trauma: the Pleistocene fauna of Europe. The primary question that Kowalski addresses is why some mammalian species became extinct in this area during this period while others survived. Indeed, the extinction of many large terrestrial vertebrates on all continents (except Antarctica?) "during the late Pleistocene is one of the most striking phenomena in the history of life" (Van Valen, 1970, p. 469). Two general explanations have been suggested to explain these extinctions: (1) that man, either directly or indirectly, caused many of the extinctions; (2) that the unusually severe climatic changes are to be blamed. Van Valen (1970) reviewed the major relevant arguments and facts supporting each view, but came to no conclusions. Likewise, in Paper 22, Kowalski does not reach any simple, straightforward conclusions, but does greatly elucidate the problem.

In this selection Kowalski cites, but does not elaborate on, the conclusions of John E. Guilday concerning the differential extinction of mammalian species during late Pleistocene and recent times (Paper 22, p. 350). Guilday (1967, p. 121) concluded that in a "deteriorating ecosystem, large herbivores are more drastically affected than are small herbivores, by virtue of their greater demands upon the system for space, food, and cover" and that this can produce differential extinction in which large forms are eliminated. Guilday (1967) also suggested that our inability to account for both population and species extinctions and for continental extinction patterns during the Quaternary is not due to a lack of adequate factors or causes, but to a lack of evidence for singling out one or several of a great number of possible causes. Finally, Guilday (1967, p. 121) suggested that the great "prime mover" of extinctions during this time period was the "harshening of the environment during the Pleistocene, with sudden and great changes in the distribution of temperature and moisture throughout the world."

REFERENCES

Becht, G., 1953, Comparative Biologic-Anatomical Researches in Some Mammals, *Koninkl. Nederlandse Akad. Wetensch. Proc.*, ser. C, **56:**508-527.

Behrensmeyer, A. K., and A. P. Hill, 1980, *Fossils in the Making,* The University of Chicago Press, Chicago and London, 350p.

Bramble, D. M., 1978, Origin of the Mammalian Feeding Complex: Models and Mechanisms, *Paleobiology* **4:**271-301.

Cracraft, J., 1983, Cladistic Analysis and Vicariance Biogeography, *Am. Scientist* **71:**273-281.

Efremov, J. A., 1940, Taphonomy: A New Branch of Paleontology, *Pan-Am. Geol.* **74:**81-93.

Gambaryan, P. P., 1974, *How Mammals Run: Anatomical Adaptations,* John Wiley & Sons, New York, 382p.

Gregory, W. K., 1929, The Muscular Anatomy and the Restoration of the Titanotheres, and Mechanisms of Locomotion in the Evolution of Limb Structures as Bearing on the Form and Habits of the Titanotheres and the Related Odd-toed Ungulates, *U.S. Geol. Survey Monograph 55,* pp. 703-756.

Guilday, J. E., 1967, Differential Extinction During the Late Pleistocene and Recent Times, in *Pleistocene Extinctions: The Search for a Cause,* P. S. Martin and H. E. Wright, Jr., eds., Yale University Press, New Haven, Conn., and London, pp. 121-140.

Herring, S. W., and S. E. Herring, 1974, The Superficial Masseter and Gape in Mammals, *Am. Naturalist* **108:**561-576.

Howell, A. B., 1944, *Speed in Animals,* The University of Chicago Press, Chicago, 282p.

Huene, F. von, 1922, *Die Ichthyosaurier des Lias und ihre Zusammenhange,* Monographien zur Geologie und Palaeontologie 1, Verlag von Gebrüder Bornträger, Berlin, 122p.

Krishtalka, L., R. J. Emry, J. E. Storer, and J. F. Sutton, 1982, Oligocene Multituberculates (Mammalia: Allotheria): Youngest Known Record, *Jour. Paleontology* **56:**791-794.

Romer, A. S., 1956, *Osteology of the Reptiles,* The University of Chicago Press, Chicago and London, 794p.

Savage, D. E., and D. E. Russell, 1983, *Mammalian Paleofaunas of the World,* Addison-Wesley, Reading, Mass., 450p.

Schoch, R. M., 1982, Gaps in the Fossil Record: Fossils and Stratigraphy, *Nature* **299:**490.

Schoch, R. M., 1983, Relationships of the Earliest Perissodactyls (Mammalia, Eutheria), *Geol. Soc. America Abst. with Programs* **15:**144.

Silver, L. T., and P. H. Schultz, eds., 1982, *Geological Implications of Impacts of Large Asteroids and Comets on the Earth,* Geol. Soc. America Spec. Paper 190, Geological Society of America, Boulder, Colo., 548p.

Slijper, E. J., 1946, Comparative Biologic-Anatomical Investigations on the Vertebral Column and Spinal Musculature of Mammals, *Koninkl. Nederlandse Acad. Wetensch. Verh., Afd. Natuurk.* **42:**1–128.
Smith, J. M., and R. J. G. Savage, 1956, Some Locomotory Adaptations in Mammals, *Linnean Soc. London Jour. Zoology* **42:**603–622.
Smith, J. M., and R. J. G. Savage, 1959, The Mechanics of Mammalian Jaws, *School Sci. Rev.* **40:**289–301.
Sokolov, B. S., 1972, Academician A. A. Borisyak and the Development of Soviet Paleontology, *Paleontol. Jour.* **1972:**293–302.
Van Valen, L. M., 1970, Late Pleistocene Extinctions, *North American Paleontol. Conv., First, Proc.* **1:**469–485.

17

Reprinted from *Geol. Soc. London Trans.* **5**:511–514, Plate XLII (1840)

XXXVI.—*Note on the Dislocation of the Tail at a certain point observable in the Skeleton of many Ichthyosauri.*

By RICHARD OWEN, Esq., F.R.S., F.G.S., Hunterian Professor to the Royal College of Surgeons, London.

PLATE XLII.

IF the *Cetacea*, like the *Enaliosauria*, were known only by their fossilized skeletons, it can hardly be doubted but that their conjecturally-restored figures as when entire and alive, would have resembled, so far at least as regards the form of the tail, those which have been published of the *Ichthyosaurus*. It may, I think, be safely affirmed, that the depressed or flattened shape of the small vertebral centres which terminate the gradually tapering tail in the skeleton of the Dolphins and Whales, would never of themselves have suggested the existence, in the recent and entire animal, of so large and important an instrument of locomotion, due entirely to unossified and readily decomposable material, which the actual presence of these fish-like Mammalia in the existing state of things, places beyond the necessity of speculation and conjecture.

The relation, however, which the slight modification of the terminal caudal vertebræ above alluded to bears to the presence of a large, horizontal, ligamentous fin, induced me some time ago to examine such specimens of skeletons of *Ichthyosauri* and *Plesiosauri*, having the tail complete, as were accessible in London, with the view to obtain evidence of the possible existence or trace of a similar structure in these species; but I was then unable to obtain from the imbedded vertebræ satisfactory proof of their exchanging a compressed for a depressed form at this part, or deviating in any way (save in size, number of articular surfaces, and, in the *Plesiosauri*, in the greater excavation of their anterior and posterior surfaces of articulation) from the rest of the vertebræ of the tail. I concluded, therefore, that these air-breathing inhabitants of the ancient deep, partly from being cold-blooded, as the condition of the ossification of their skeleton proves, and therefore slow breathers, and more especially on account of the superaddition of posterior paddles, were devoid of any locomotive organ analogous to the tail-fin of the *Cetacea*; and that the hinder paddles principally served the purpose, in the absence of such a horizontal terminal fin, in bringing the head to the surface for the purpose of breathing.

Having recently examined many saurian skeletons now in London, the greater part of which have been disencumbered of their earthy shroud by the chisel of Mr. Hawkins, a condition of the tail which, on a former occasion, in a single instance had arrested my attention, but without calling up any theory to account for it, now more forcibly engaged my thoughts, from observing that it was repeated, with scarcely any variation, in five instances. The condition to which I allude is an abrupt bend or dislocation of the tail at about one-third of its whole length distant from the end; generally about the thirtieth caudal vertebra in the *Ichthyosaurus communis*; the terminal portion continuing, after the bend, almost as straight as the portion of the tail preceding it. In short, the appearance presented is precisely that of a stick which has been broken, and with the broken end still left attached, and depending at an open angle.

Now there is no modification of the vertebræ, where this bend takes place in the tail of the *Ichthyosauri* indicative of the tail having possessed, during the lifetime of the animal, greater mobility at this particular point than at any other; and it is scarcely possible to conceive how any force operating on the dead carcase at the bottom of the sea, or when it was imbedded in sedimentary deposits, could produce a fracture or dislocation of the tail at the precise point at which we find it in so many specimens, otherwise very dissimilar in regard to their general position and degree of perfection. I incline to believe, therefore, that the appearance is due to the operation of some force acting upon a ligamentous fin attached to that part of the extremity of the tail which is thus bent down; and that this force operated while the dead animal was floating and buoyed up by the gases generated by putrefaction, and when the ligaments and other connexions of the caudal vertebræ had been so far loosened as to allow the force which previously would have bent down the tail in a general curve like the top of a fishing-rod, to overcome the diminished resistance of the decomposing joints immediately proximal of its point of action, and dislocate the vertebræ at that part. The two portions of the tail, like the rest of the skeleton, would then continue to be held together by the common integument, until the rupture of the parietes of the abdomen would allow the pent-up gas to escape, and the body to sink to the bottom.

In thus descending through a sufficiently tranquil medium, the back being weighted with the heavy vertebral column, would be most likely the first part to touch the bottom; and if this consisted of soft clay or mud, a heavy trunk, like that of the *Ichthyosaurus,* would immediately subside therein back downwards, while the broken tail, supported by its flat surface, would afterwards be dragged down at an angle with the trunk, and with its extremity directed towards the ventral—now, as when the body floated, the uppermost side of the body: this is, in fact, the position in which the broken part of the tail is generally placed in the fossil skeletons.

The prevailing amount of dislocation which most of the skeletons of the *Ichthyosauri* present, is exactly such as might be conceived to take place in a dead animal body buoyed up sufficiently long to allow of a partial separation of the connecting ligaments and cartilages; the probable toughness of the integuments of the Enaliosaurians would tend to retain the gases generated by putrefaction sufficiently long to enable this extent of decomposition to take place while the animal floated, at least in regard to the ribs and bones of the pectoral and pelvic arches. The bones of the paddles on the contrary, and those of the tail, being connected together by denser ligaments and surrounded by a less proportion of soft animal matter, would be more likely to sink comparatively undisturbed, with the rest of the trunk, when the buoyant gases were ultimately set free.

The dead carcasses of the *Ichthyosauri* would of course be subject to such dislocations and separations as the attacks of predatory animals on a floating dead body would produce; but the appearance in the tail of the *Ichthyosaurus* here alluded to, is too uniform and common to be due entirely to an accidental and extrinsic cause. I am therefore disposed to attribute it to an influence connected with some structure of the recent animal; and most probably to the presence of a terminal tegumentary and ligamentous caudal fin, which, either by its weight, or by the force of the waves beating upon its extended surface, or by the action of predatory animals of strength sufficient to tug at without tearing it off, might, under the circumstances already mentioned, give rise to a dislocation of the caudal vertebræ immediately proximal of its attachment. We have evidence, however, in the form of the vertebræ that the supposed fin was not a horizontal one, and I have already observed that the superaddition of posterior paddles in these air-breathing marine animals is the compensation for the absence of such an organ; and, therefore, a tegumentary fin, if developed at the extremity of the tail, would probably be placed in the position best adapted to produce with rapidity those lateral movements of the head most needed in a short-necked predatory aquatic animal.

Since the preceding notes were written and read to the Society, I have had the opportunity of detaching some of the terminal caudal vertebræ of an *Ichthyosaurus communis* from the substance in which they were imbedded;

and have received, through the kindness of Sir Philip Grey Egerton, a separate terminal caudal vertebra of an *Ichthyosaurus* from Lyme Regis, all of which exhibit a compressed form nearly as well marked, in comparison to the vertebræ of the rest of the spine, as is the depressed form in the terminal vertebræ of the *Cetacea*. This structure offers additional and very satisfactory proof of the existence of a caudal tegumentary fin expanded in the vertical direction. Such a fin would be especially useful in the short- and stiff-necked Ichthyosaur; while in the Plesiosaur, where the length and structure of the neck are so favourable for rapid lateral inflections of the head, an instrument like that which produces the corresponding movements in fishes would be unnecessary. Now, it is an interesting fact, that in all those skeletons of Plesiosaurs in which the tail is perfect, it is straight; and there is no indication of the partial fracture or bend in them, which is so common in the tails of the Ichthyosaurs.

March 18th, 1838.

G. Mantell, D.C.L. &c. &c.
with the author's best
respects.

PLATE XLII.

Illustrates Mr. Owen's paper on the Dislocation of the Tail at a certain point, observable in the Skeleton of many *Ichthyosauri*: p. 511.

All the tails represented in the accompanying plate belong to skeletons of *Ichthyosauri* now in London and formerly in the collection of Mr. Hawkins: and these illustrations will not lose force from the fact of their being taken from figures of those skeletons published without any reference to the structure or hypothesis in question.

Upper right-hand figure from a specimen found in the lias at Walton in Somersetshire. (See Mr. Hawkins's work on *Ichthyosauri* and *Plesiosauri*: p. 32. Pl. XVII.)

Lower right-hand figure, from a specimen found in lias at Lyme Regis. (See Mr. Hawkins's work, p. 25. Pl. VII.)

Middle figure.
Left-hand upper figure. } See the Second Part of Mr. Hawkins's work.

Left-hand lower figure from a specimen found in lias at Walton in Somersetshire. (See Mr. Hawkins's work, p. 34. Pl. XXI.)

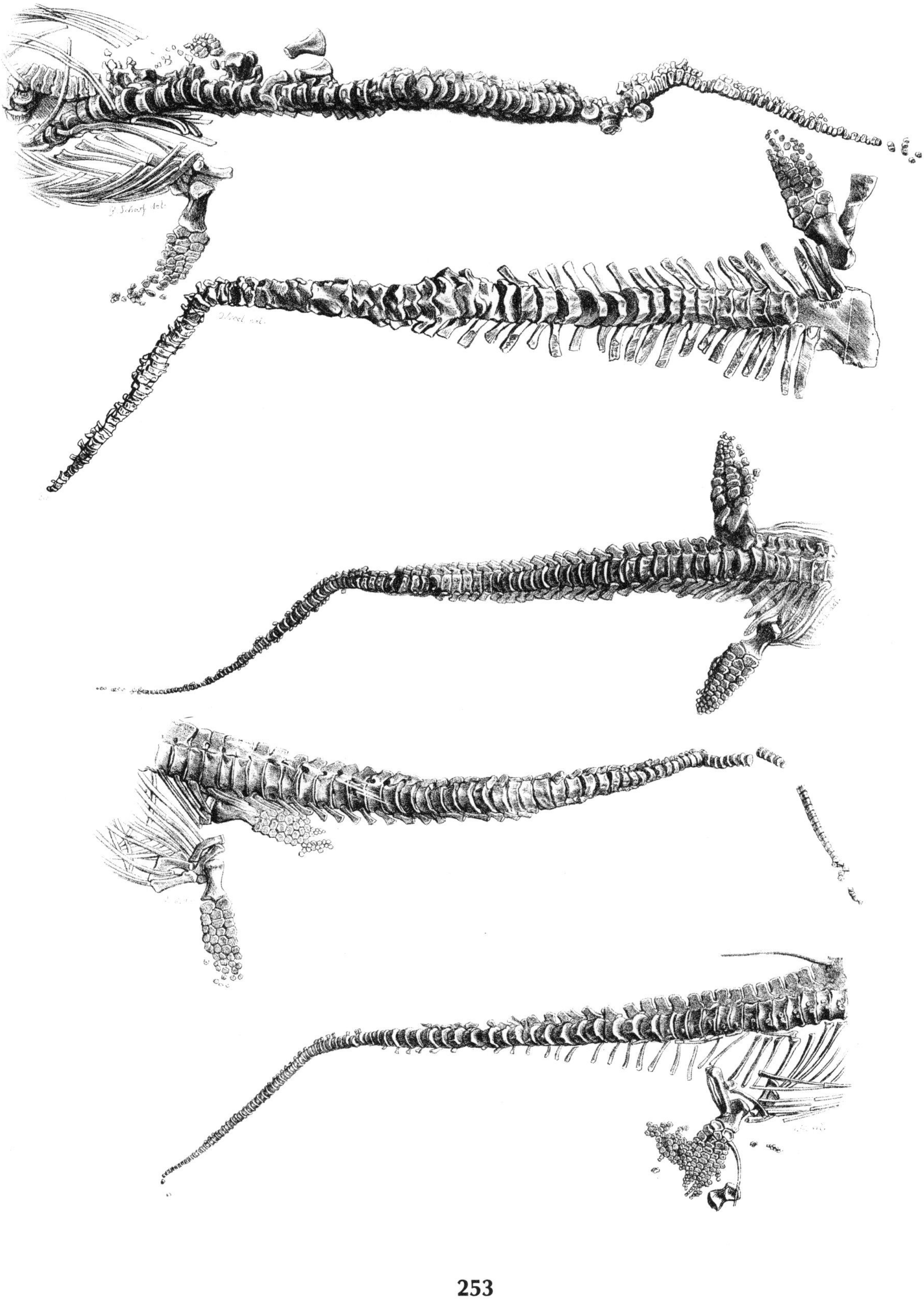

18

Reprinted from *N.Y. Acad. Sci. Ann.* **22**:267–294, Plate XXXIV (1912)

NOTES ON THE PRINCIPLES OF QUADRUPEDAL LOCOMOTION AND ON THE MECHANISM OF THE LIMBS IN HOOFED ANIMALS

BY WILLIAM K. GREGORY

CONTENTS

Introduction

The movements and locomotive mechanisms of animals were the subject of a classic work by Borelli in 1680, entitled "De Motu Animalium . . ." The chief pioneers of modern research were the brothers Weber (Eduard and Wilhelm), authors of "Die Mechanik der menschlichen Gehwerkzeuge," Göttingen, 1836. Marey, the author of the handbook on "Animal Mechanism"[1] (1874), invented elaborate apparatus for analyzing and graphically recording the movements of the limbs of men and animals, while more recently the mathematics of human locomotion have been developed by O. Fischer and others. General reviews of animal mechanics and of human locomotion are given by Marey,[2] Haycraft[3] and Luciani.[4] A very able analysis of the mechanism of locomotion in the horse is given in "The Horse in Motion," by J. D. Stillman,[5] while the extensive series of photographs by Muybridge[6] record the actual positions of the limbs and body assumed in motion by ungulates and other animals.

None of the above mentioned works considers the subject from the evolutionary point of view. Ryder,[7] and especially Cope,[8] pointed out certain adaptations in the feet of ungulates, such as the reduction of the digits in the "Diplarthra" and the so-called displacement of the metacarpals, and Cope used these observations in his argument for the hypothesis of the transmission of acquired characters. He also made the following very important observations:[9]

"In animals which leap, the distal segments of the limbs are elongated; in those which do not leap, but which merely run or walk, it is the proximal segments of the limbs which are elongated.

"Animals which run by leaping are divided into those which run and leap with all fours, as Diplarthra, and those which run and leap with the posterior limbs only, as the jerboas and kangaroos. In both types, the distal segments of the hind limbs are elongated, and in the Diplarthra, those of the fore limb also.

"Animals which do not leap in progression (elephants, Quadrumana, bears) are always plantigrade and have very short feet but elongate thighs and mostly tibias."

---

[1] 12mo. London, 1874.

[2] *Op. cit.*

[3] E. A. Schäfer: Text book of Physiology. Edinburgh and London. pp. 228-273, 1900.

[4] Physiologie des Menschen . . . Ins Deutsche übertragen and bearbeitet von Dr. Silvestro Baglioni und Dr. Hans Winterstein . . . Siebente Lieferung. Jena, 1906.

[5] Executed and published under the auspices of Leland Stanford, 4to, Boston, 1882.

[6] Animals in Motion . . . Third Impression 4to, London, 1907.

[7] Am. Nat., vol. 11, p. 607. 1877.

[8] "The Mechanical Causes of the Development of the Hard Parts of the Mammalia." Journ. of Morphology, vol. 3, pp. 137-290. 1889.

[9] *Ibid.*, p. 151.

Cope neglected to follow up the important mechanical and adaptive corollaries of these facts. He merely drew the very questionable inference that "those elements which receive the principal impact in progression are those which increase in length [in phylogeny]."[10]

In his paper on "The Angulation of the Limbs of Proboscidia, Dinocerata, and other Quadrupeds, in Adaptation to Weight," Osborn[11] concluded that

"The straightening of the limb [in the Dinocerata, Proboscidia, etc.] is an adaptation designed to transmit the increasing weight through a vertical shaft. Correlated with it are the shifting of the facets into the direct line of pressure and the alteration of their planes from an oblique to a right or horizontal angle with relation to the vertical shaft."

Gaudry,[12] in describing the limbs of extinct South American ungulates, endeavored to show how the pose of these animals could be inferred from a study of the facets,—an idea which had been previously advanced by Osborn.[13] Gaudry also designated as "rectigrade" the pose of elephants and similar heavy forms which stand with straightened limbs and toes, resting the weight chiefly on the pad.

The marked contrasts in the limbs and musculature between the slow-moving heavy-bodied ungulates and the slender swift-footed or cursorial types in various phyla, constitute a subject which will be discussed in considerable detail in the monograph on the Titanotheres by Professor Osborn. At his suggestion, the following notes, forming a part of the present writer's studies on this subject, are now published, together with some of the drawings which have been made by Mr. Erwin S. Christman, under the direction of the writer, for the monograph above mentioned. The writer is also indebted to Dr. W. D. Matthew for valuable criticisms and suggestions.

Limbs regarded as Compound Levers

The simple principle that the limbs of quadrupeds are compound levers and that the relative lengths of the upper, middle and lower segments are adapted to specific loads, muscular powers and speeds, although well understood by students of human and equine locomotion, has apparently not hitherto been applied to elucidate the adaptive contrasts between cursorial and "graviportal"[14] ungulates.

---

[10] *Loc cit.*

[11] Am. Nat., vol. xxxiv, pp. 89-94. 1900.

[12] "Fossiles de Patagonie. Les Attitudes de quelques Animaux." Ann. de Paléontologie, tomei.

[13] "Patriofelis and Oxyæna restudied as Terrestrial Creodonts." Bull. Amer. Mus. Nat. Hist., Vol. 13, pp. 270, 271. 1900.

[14] This word has been invented by Professor Osborn to describe the conditions in heavy-bodied animals with long proximal and short distal limb segments.

EVOLUTIONARY STAGES OF QUADRUPEDAL LOCOMOTION

In the Stegocephalian stage of quadrupedal locomotion, the short limbs were held widely outward from the body, the humerus and femur were very short and the feet were spreading and flat. Crawling was effected in part by a sharp downward pull of a proximal segment (humerus or femur), thus tilting the body upward on the same side and throwing the weight on the opposite foot. The long axis of the body was meanwhile thrown into alternate lateral curves, the advancing fore limb being on a convexity, the advancing hind limb on a concavity.

In the late reptilian or early mammalian stage, the feet were brought around partly under the body, the elbow and knee began to be drawn in, the scapula was rotated backward as the coracoid lost its connection with the sternum, and the body became well raised off the ground. According to a hypothesis advanced elsewhere by the writer,[15] this process was associated with the acquisition of climbing, or semi-arboreal, habits, structural vestiges of which remain in the partly divergent first digit and many other characters of early Eocene mammals.[16]

The Lower Eocene ancestors of the various orders of ungulates had probably all long since passed through these earlier stages of quadrupedal terrestrial locomotion, and at that time many of them had perhaps become more or less digitigrade. The primitive "Protungulates," *Meniscotherium, Periptychus, Pantolambda,* may give us some idea of what the several ancestors of the subungulate series (Hyracoidea, Embrithopoda, Proboscidea, Amblypoda) may have been like. They also preserve apparent traces of arboreal ancestry in the relatively short, spreading hands and feet, long limb bones, the humerus with large entocondyle and entepicondylar foramen, the undiminished power of pronation and supination of the forearm and many other characters. The Basal Eocene *Euprotogonia,* the ancestor of the Condylarth *Phenacodus,* with slender subunguligrade feet, represents a more advanced stage of evolution, in the direction of the Perissodactyls.

FACTORS OF LONG-DISTANCE TRAVELLING POWER IN UNGULATES

The primitive ungulates of the Lower Eocene were doubtless surrounded by environmental conditions which set the premium of survival upon improvements in long-distance travelling power and in speed. These improvements have been attained in various ways and in the most

[15] "The Orders of Mammals," Bull. Amer. Mus. Nat. Hist., Vol. 27, p. 226, 1910.

[16] MATTHEW: Arboreal Ancestry of the Mammalia, Amer. Nat., Vol. 38, 1904, pp. 811-818.

diverse lines of evolution,—in the elephant no less than in the antelope. The general factors of long-distance travelling power may be grouped broadly under the headings: (A) Endurance and (B) Speed.

A. ENDURANCE

Endurance may be measured either by (1) the length of time an ungulate can keep in motion without rest or refreshment, or by (2) the amount of reserve strength left after a stated expenditure of energy, or by (3) the relative quickness of recuperation. Endurance increases with practice. As metabolism increases, the muscles, lungs, heart and other organs of the thorax become stronger and larger.

More rapid metabolism requires more food and larger digestive apparatus. Although the Lamarckian hypothesis in its crude form is very probably untenable, it is a fact that herbivorous animals have longer and heavier digestive tracts than carnivores. Moreover, as fast as the dentition has become adapted for the harder, less nutritious kinds of food, the digestive apparatus must have become more complex and much heavier.

While the enlarging thorax and abdomen have made available a great increase in energy, they have caused an even more rapid increase in total weight.

With increasing total weight, the internal and external resistance to be overcome in locomotion also rises, and hence the margin between the total energy available and the energy required in progressing a given distance is lessened. In other words, endurance, the first great factor of long-distance travelling power, is directly proportional to the efficiency of the adaptations for minimizing the waste of energy.

ADAPTATIONS FOR MINIMIZING WASTE OF ENERGY

Higher efficiency in locomotion has doubtless been attained, first, by advantageous modifications of the organs of propulsion (such as are described below), secondly, by improvements in the supporting framework, thirdly, by improvements in the methods (*a*) of conserving the inertia of forward motion, (*b*) of taking up shock, (*c*) of preventing dislocation and (*d*) of minimizing lost motion.

Momentum.—The shocks and strains to which the locomotive apparatus is subject vary with the momentum of the body in motion. Hence as momentum is the product of mass by velocity, the shocks and strains experienced by heavy animals in rapid motion are very great, and devices for lessening them become conspicuous.

Propulsion and the center of gravity.—"Perfect quadrupedal locomotion," says Stillman,[17] "requires uniform support to the center of gravity (of the whole animal) and continuous propulsion by each extremity in turn." In walking and running, by the straightening or extension of the limbs, the center of gravity is raised and thrown in advance of the centers of support. The body thus falls forward, the center of gravity describing a curve of greater or less convexity, the forward motion being accelerated by the thrust of the propelling limb.

The first shock of the downward fall in the running horse is taken up by the forwardly stretched and slightly bent hind limb (Fig. 1) placed beneath or in advance of the center of gravity; the gradually stiffening muscles of the thigh and back checking the downward momentum (Stillman, p. 91). The rearing muscles thus come into play and serve to let the fore part of the body down gently.

Dislocation of the fully extended forelimb in landing is prevented partly by (*a*) the crutch-like action of the limb itself (which is slung from the converging fibers of the serratus magnus attached to the top of the scapula) and by (*b*) the contraction of certain muscles of the shoulder, neck and back (Stillman, pp. 61, 62, *et seq.*). To these arrangements and conditions, Stillman attributes the absence of the clavicles in the horse.

The center of gravity in the smoothly trotting horse describes a relatively flat trajectory, whereas in the "bounding" movement, or gallop, the center of gravity ricochets and the trajectory consists of a series of cycloids of marked convexity. This mode of locomotion, while very rapid for short distances, is too wasteful for heavy-bodied animals, which require a relatively flat trajectory and a maximum saving of inertia.

Sinuous movement of the body in running.—By the bending backward of the pelvis, first on one side and then on the other, the thrusts of the femora are brought more nearly into line with the anteroposterior axis, while wrenching of the pelvis is prevented by the contraction of the longissimus dorsi of the opposite side (Stillman, p. 36). By this means also, the length of the stride is directly increased. The same sinuous motion of the body is associated with the "figure-of-8" movement of the limbs noticed by Pettigrew.[18]

In this connection may be noted also the devices for avoiding "interference" of the limbs (*e. g.*, "stifle action" of the iliacus, preventing the knee from striking the abdomen; oblique trochlea of the astragalus carrying the advancing foot around its fellow of the opposite side). Dislocation

---

[17] *Op. cit.*, p. 87.
[18] Animal Locomotion, p. 39. 12mo. New York, 1874.

is provided against not only by the ligaments, but also by the metapodial keels, by the grooved trochlea of the astragalus, by the cnemial crest of the tibia, etc.

Spiral configuration of limb bones and of articular surfaces.—Goodsir, Pettigrew[19] and others have shown that the articular surfaces of the elbow, ankle and calcaneo-astragalar joints are spirally warped surfaces which act after the manner of screws. The limbs as a whole also are twisted levers with the ridges and muscles arranged spirally. "This arrangement," says Pettigrew, "enables the higher animals to apply their traveling surfaces to the media on which they are destined to operate at any degree of obliquity so as to obtain a maximum of support or propulsion with a minimum of slip. If the traveling surfaces of animals did not form screws structurally and functionally, they could neither seize nor let go the fulcra on which they act with the requisite rapidity to secure speed, particularly in water and air."

Lost motion.—Lost motion through backward slipping of the foot upon the ground is provided against in the horse by the form and details of the hoof, and in the elephant by the plantar pads.

Pendulum action of the limbs.—The brothers Weber held that in rapid locomotion the limbs swing freely as pendula, but Marey and later investigators, according to Luciani,[20] hold that the natural swing of the leg is very largely damped and controlled by the flexor muscles. In favor of the view that there is some measure of analogy to the pendulum, we observe that in the horse, the center of gravity of the limb, corresponding to the "bob" of a pendulum, is relatively proximal in position, and this is associated with rapid oscillation of the limb, whereas in the elephant, the center of gravity of the limb is farther down the shaft, and here we have a slower oscillation of the limb. It will be observed that while the body is moving forward, the propelling limb is moving backward, and its own backward momentum, due to weight alone and to the pull of the extensors, must be overcome by the forward pull of the flexor muscles and by the forward pull of the body as a whole. Hence the heavier the limb, the greater the force expended in overcoming and reversing the momentum of each limb at the end of each stride.

SPEED

The speed of a quadruped or biped in motion is measured by the product of the length of the stride into the rapidity of the stride.

[19] Animal Locomotion, pp. 23-24, 28, 29. 1874.
[20] *Op. cit.*, p. 126.

The diverse adaptations in the limbs, considered as compound levers, are related to either or both of the factors, "length of stride" and "rapidity of stride."

LENGTH OF STRIDE

Length of limb.—Length of limb is the first factor of length of stride. It is generally proportional to height at the shoulders and hips. Length of limb has been attained in cursorial animals by lengthening the lower and middle segments of the limb, in graviportal animals by lengthening especially the proximal segments of the limb.

Angle of stride.—Angle of stride is the second factor. It is measured by the arc described by the lower end of the femur or humerus in swinging from the position of extreme extension to that of extreme flexion. A wide angle of stride not only lengthens the stride, but also enables each limb, first, to be placed in turn below or beyond the center of gravity in order to secure more continuous support for the center of gravity, and, secondly, it enables the propelling limb to exert its propulsive effort for a relatively long period.

Acceleration increment of stride due to ballistic power of limbs.—Those portions of the stride that are due simply to the length of the limb and to the angle of the stride might, if determined, be illustrated by moving the inert limbs of a dead animal suspended in the air. In the slow walk of a biped, the successive positions of the legs might for our purposes be represented by a series of inverted V's (/\/\ /\ /\ /\ /\ ) with the lower ends touching. Each /\ represents a single step and two successive /\ 's represent a stride. In the rapidly moving animal, however, the stride receives a very considerable increment, due to the impetus imparted by the propelling limb and to the forward motion of the body as a whole, which carries the forwardly moving foot to a position far in advance of its own unaided reach. This "acceleration increment," as it may be called, increases with the velocity of the movement and is proportional to what we may designate the ballistic power of the limb. This ballistic power may be defined as excess propulsive power over and above that which is necessary to move the limbs as stilts and to support the weight of the body; it is expended in lengthening the stride. Ballistic power and the acceleration increment of the stride are measured by the length of time at least three of the feet are off the ground together during a single stride of a quadruped running at full speed. In Fig. 1, representing an elephant in rapid motion (ambling), it will be observed that three of the feet are never off the ground at the same instant; whereas Stillman's Figs. 2-10 show a galloping horse in which at least three of the feet are off the ground at once in seven out of nine phases of one stride. The same figures show that the elephant has three feet on the ground together during five-sixths of one stride, and during the remaining sixth the body is supported by one fore foot and the opposite hind foot, whereas the horse in question never has three feet on the ground together during the stride there pictured. Thus it will be seen that the "acceleration incre-

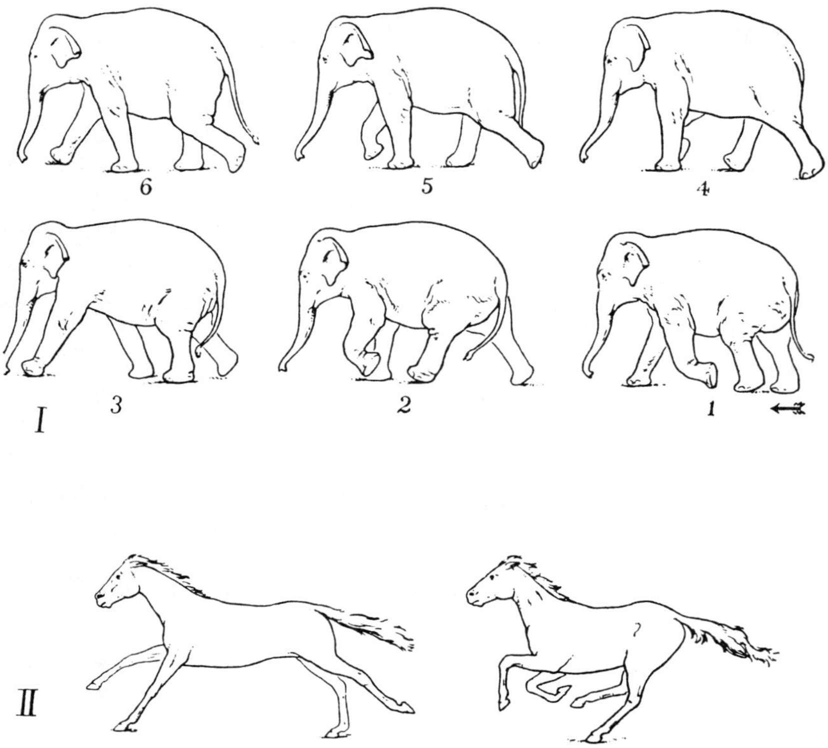

FIG. 1.—*Graviportal and cursorial modes of locomotion contrasted in the amble of the elephant (I) and gallop of the horse (II)*

I after Muybridge; II after Stillman

ment" of the stride and the ballistic power of the limbs are at a maximum in cursorial animals and at a minimum in graviportal animals. The "acceleration increment" will no doubt increase also with the potency of the psychic motive and of the neural stimulus (cf., p. 282).

In brief, the factors of length of stride are (1) length of limb, (2) angle of stride, (3) acceleration increment.

RAPIDITY OF STRIDE

Rapidity of stride, the second major factor of speed, is determined by the rate of oscillation of the limbs, especially of the proximal segments. The conditions determining rapidity of stride are discussed below (p. 281).

MECHANICAL AND PHYSIOLOGICAL RELATIONS OF POWER AND SPEED IN THE LOCOMOTIVE APPARATUS

CONTRACTILE FORCE OF LOCOMOTIVE MUSCLES

The contractile force of a muscle (*i. e.*, its ability to overcome inertia at a given instant) is proportional to the number of its contractile fibers, when these fibers are parallel to the direction of contraction. The force of such a muscle is therefore proportional to the sectional area of the muscle.[21] In the case of weights lifted vertically by isolated muscles, the work (W) performed is measured by the product of the muscular force (F), multiplied by the distance (D) through which the load is lifted[22] ($W = F \times D$). This distance is proportional to the length of the muscle,[23] for the "shortening" of a muscle is proportional to its length. Hence the total work performed will be proportional both to the length of the muscle and to its sectional area, and hence to the mass or the weight of the muscle.[24]

The work performed by a long muscle is greater than that of a shorter one of the same sectional area.[25] Long and slender muscles such as the sterno-mastoid and the sartorius of man exert a small power over a long range; short and thick muscles such as the pectoralis major, the glutæus maximus or the temporalis develop a relatively great power multiplied by a short range.[26] The contractile force of muscle per unit of sectional area is much less in cold-blooded than in warm-blooded animals.[27] It is lessened by disuse and extreme fatigue and is increased by exercise, and hence is dependent upon the nervous system and general systemic conditions.

The contractile force is inversely proportional to the number of connective tissue fibers mingled with the striped muscle fibers. Hence muscles grade into tendons and ligaments. When a muscle is stretched,

---

[21] HAYCRAFT in Schäfer's Text book of Physiology, p. 242.
[22] *Ibid.*, p. 245.
[23] *Ibid.*, p. 244.
[24] *Ibid.*, p. 246; also Marey, p. 62. 1874.
[25] HAYCRAFT, p. 246.
[26] MAREY, p. 62.
[27] HAYCRAFT, p. 243.

it serves partly as a ligament. All muscles *in situ* are stretched to a certain degree, and thus act as ligaments.[28] "Over extension" of the muscle is prevented by the inextensible connective tissue fibers.[29] According to Stillman,[30] the length of the muscles cannot be increased by exercise, otherwise the tension necessary to prompt action would be lost.

The contractile force is highest when a muscle is stretched to its full "physiological length" (that is the greatest length it ever assumes during life). As shortening takes place, the contractile force becomes less and less (Haycraft).[31]

Contractile force and speed of movement augmented by "hold and let go" arrangements.—Fick and Helmholtz showed[32] that the greatest force and velocity of contraction are developed when the movement of the muscle is checked during the initial stages and when the resistance is suddenly diminished.

Amount of contraction.—The shortening of individual muscles is in general proportional to their length when in repose, but different investigators give somewhat different estimates. "While Weber described a muscle as shortening 70 per cent. of its length, when unweighted, more recent observers incline to put the shortening at 20 to 30 per cent. of its length" (Haycraft).[33] Marey[34] estimates "the mean shortening of a muscle while contracting, when it is not detached from the animal," as "about a third of its length when in repose." Bishop's estimate is one-fourth (Stillman, p. 31). "When the fibers are not parallel, but obliquely set, as in the gastrocnemius, we have a greatly extended transverse area of muscular fibers, which act therefore very powerfully, though, on account of their short length, they can exercise their pull but a comparatively short distance" (Haycraft).[35]

THE "ANGLE OF INSERTION" AND THE PRINCIPLE OF THE PARALLELOGRAM OF FORCES

In Fig. 2 (I), let AC represent a rod free to rotate around the point A in the direction CC′; let BD represent a contractile spring fastened at D, inserted on AC at B and forming the angle ABD (a). Assume that the length of BD is proportional to its contractile force; then from

[28] *Ibid.*, p. 245.
[29] *Ibid.*, p. 242.
[30] *Op. cit.*, p. 32.
[31] *Op. cit.*, p. 242.
[32] HAYCRAFT, *op. cit.*, p. 248.
[33] *Op. cit.*, p. 244.
[34] *Op. cit.*, p. 62.
[35] *Op. cit.*, p. 242.

the principle of the parallelogram of forces, we may resolve BD into two forces, the first AB acting in the direction of the rod AC and tending to press AC against its fulcrum A, the second component BR acting at right angles to the first and tangent to the arc of rotation BB′. The first component AB may be called the "centripetal component," the second BR may be called the "rotation component." In Fig. 2 (II), the contractile spring *bd* is of the same length as before, but the angle of

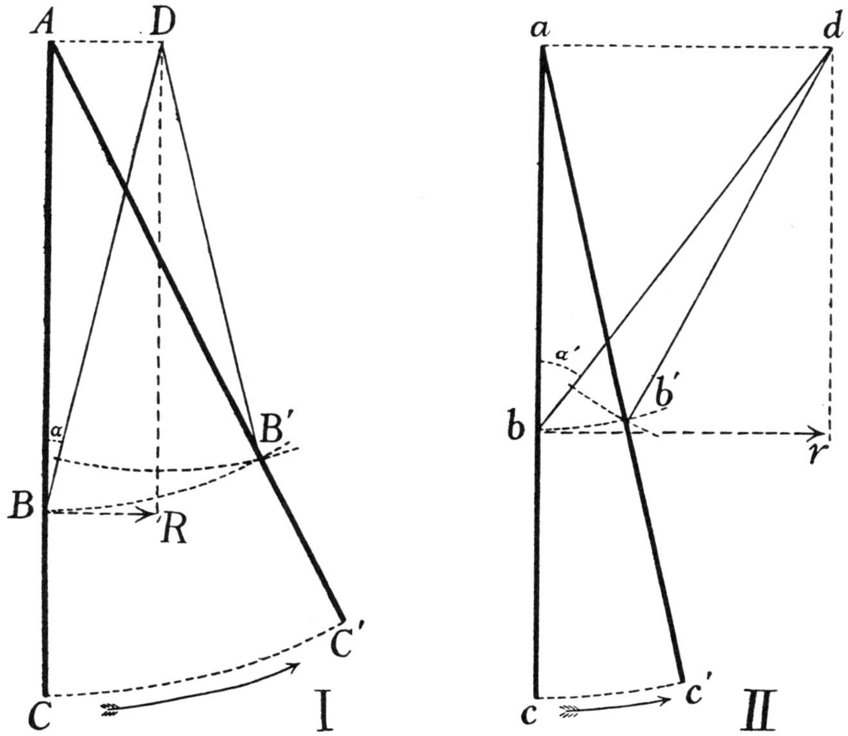

FIG. 2.—*Diagram illustrating the direct relation of the angle of insertion (a, a′) to the "rotation component" (BR, br) and the inverse relation of the angle of insertion (a, a′) to the "centripetal component" (AB, ab) and to the speed of the insertion point (proportional to BB′, bb′).*

insertion *abd* (*a*′) is increased; then the centripetal component *ab* will be less than AB, but the "rotation component" *br* will be greater than BR. Accordingly as the angle of insertion increases, the pull *across* the shaft becomes more direct, while the pull *along* the shaft decreases; in other words, the rotation component varies directly, the centripetal component inversely, with the angle of insertion.

Applying this to Fig. 6, II, we see that in the horse, the muscles figured are inserted at more open angles of insertion (α, β, γ) than in the mastodon or elephant and that their rotation components are therefore relatively greater, the centripetal components relatively less.

The foregoing principles were worked out independently by the writer, but Luciani[36] gives similar principles for human locomotion. He states that not all of the muscular force is available for the movement of the skeleton, that this is only the case when the insertion of the muscle on the bone reaches almost a right angle, as in the case of the masseters, which can exert their whole strength in pressing the lower jaw against the upper jaw. He says that generally, owing to the conditioning form of the skeleton, the muscle is attached more or less obliquely, so that the direction of its fibers makes a more or less acute angle with the long axis of the bone. In all these cases, a great part of the total force of the muscle is lost as regards movement. In every case, however, whatever the form of the muscle or the size of the angle of insertion may be, by resolving the total pull into its components, in accordance with the law of the parallelogram of forces, one can determine how much of the total pull is expended in the movement of the bone, assuming the other bones to be stationary. Luciani[37] also shows that the more acute the angle of insertion is, the smaller will be the component of rotation, and the nearer the angle of insertion approaches a right angle, the greater will be the component of rotation.

Relations of speed of movement and angle of insertion.—Returning to Fig. 2, we see that if DB contracts to DB′, the point of insertion will move from B to B′. If now the angle of insertion be increased to α' and *db* (equal to DB) contracts to *db′* (equal to DB′), then the point of insertion moves only through *bb′*, which is less than BB′. If the contraction time as well as the distance be constant, then B will move faster than *b*; that is, when the rate of contraction and length of muscle are constant the speed of the insertion point varies inversely with the angle of insertion.

It is also evident that if the angle of insertion and other factors remain constant the speed of the distal end of a long bone will increase as the point of insertion is moved toward the head of the bone. (Because BC will be larger.)

Variableness of the rotation component.—From Fig. 2, it will be seen that the angle *ab′d* is somewhat greater than *abd*, that is, both the angle of insertion and the rotation component increase as the muscle contracts.

---

36 Physiologie des Menschen, Siebente Lieferung, p. 115.

37 *Ibid.*, p 116.

"MOMENT OF RESISTANCE" AND "DIMINISHING LOAD"

In every lever, whether of the first, second or third order, the "power" and the "resistance," acting along parallel lines, but in opposite directions, are in equilibrium when the power multiplied by its effective distance from the fulcrum is equal to the resistance multiplied by its effective distance from the fulcrum. The "effective distance" is measured by a line passing through the fulcrum and perpendicular to the line of direction of the force. The product of a force multiplied by its effective distance from the fulcrum is called its "moment."

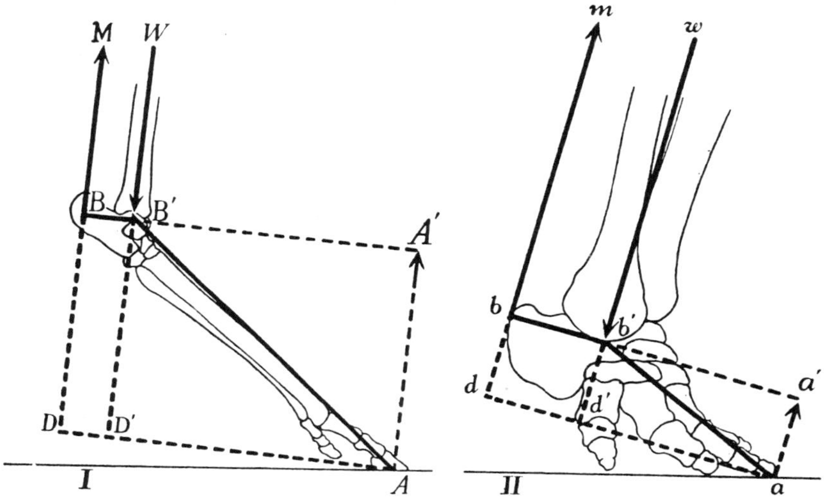

FIG. 3.—*I. Hindfoot of an extremely cursorial type (Neohipparion) showing at the instant of greatest extension of the foot a low moment of power of the calf muscles (M × BB') and a very high moment of resistance (W × B' A') of the pressure of the tibia upon the ankle.*
II. Hindfoot of an extremely graviportal type (Mastodon) showing at the instant of greatest extension of the foot a much higher moment of power of the calf muscles (m × bb') and a relatively lower moment of resistance (w × b' a') of the pressure of the tibia upon the ankle.

In Fig. 3, I, it would appear natural to assume that the point A, on the ground, is the fulcrum, and that the "resistance" is the pressure of the tibia upon the ankle joint at B', while the "power" is the contractile force of the muscles of the calf, applied at B. Similarly, Eduard Weber[38] described the human foot as a lever of the second order and gave for the relations of the forces and movements of the foot in raising the

[38] Cf. HAYCRAFT, *op. cit.*, p. 251.

weight of the body a formula which translated into the terms of our Fig. 3 would be as follows:

$$M \times BA' = W \times B'A'$$

But Knorz, Henke, Ewald and others, as quoted by Haycraft (*loc. cit.*), showed that the effective distance of the muscular force M is not BA', but BB', and that we should rather conceive the foot as a lever of the first order with the pivot at B', the "power" at B and the "resistance" (offered by the reaction of the ground upon the foot) at A. In that case, the moments around B' are as follows:

$$M \times BB' = W \times B'A'$$

Hence, other things being equal, the longer the foot, the greater will be the moment of resistance to be overcome by the muscles of the calf.

If the angle B'AD' be increased, as when the foot assumes a more vertical position, the effective distance B'A' decreases; that is, the moment of resistance decreases as the foot becomes more vertical. In other words, the "load" diminishes as the calf muscles contract. It has been shown by Fick and others (quoted by Haycraft, *loc. cit.*, p. 246) that when the force of muscular contraction is opposed to a diminishing moment of resistance, the muscle is capable of performing more total work (force $\times$ distance) than when the resistance is constant. Consequently, the diminishing resistance, conditioned by the raising of the foot, enables the calf muscles to perform their work under the most favorable conditions.

SUMMARY OF MECHANICAL AND PSYCHIC FACTORS IN POWER AND SPEED

The speed of the distal end of a "long bone" of the limbs will depend upon (1) the nearness of the point of insertion of the principal muscles to the joint or axle, (2) the smallness of the angle of insertion of the muscles, (3) the position of the muscle fibers with reference to the long axis of the muscle, and (4) the speed of contraction of the muscle itself. If a long muscle and a short muscle were isolated for experiment, it might prove that the short muscle would contract faster than the long one, but, in nature, a single movement of a long bone is produced by the simultaneous and coördinated action of muscles of varying length. Thus, in the act of extending the whole arm from the fully flexed position, the relatively short, broad thoraco-scapular and scapulo-humeral muscles contract in the same time as do the relatively long extensors of the forearm, irrespective of their lengths. The speed and force of contraction naturally depend partly upon the strength of the stimulus and partly on the resistance to be overcome.[39] The end result, as it were, determines the rate of contraction of the coördinated muscles. The effective regulation and correlation of muscular action is obviously an extremely complex function of the peripheral and central nervous systems. The force and speed of contraction of a given set of muscles in a living animal at a given moment are determined not only by many mechanical factors, of which a few have been mentioned above, but also by the whole psychic constitution of the animal and by the psychic effectiveness of the exciting "motive."

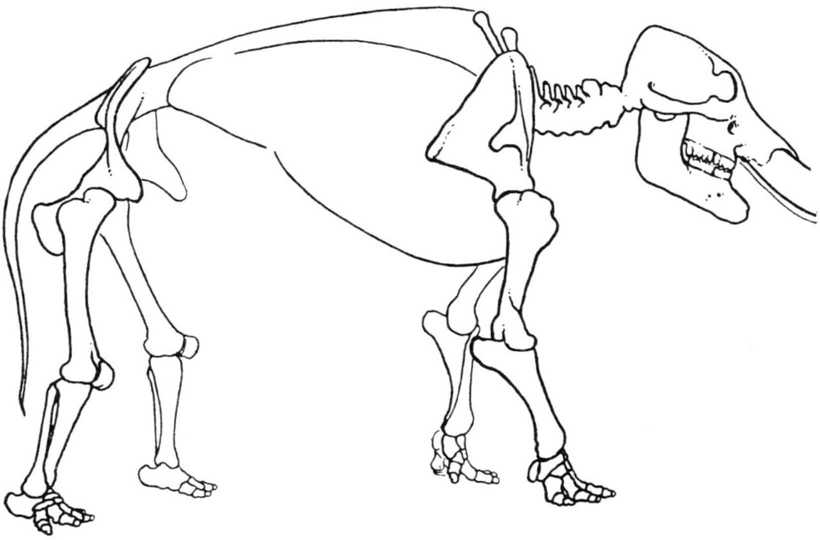

FIG. 4.—*"Graviportal" adaptations for the walk and amble in the Mastodon*

APPLICATION OF THE FOREGOING PRINCIPLES TO THE STUDY OF THE LIMBS OF UNGULATES

FUNCTIONAL SIGNIFICANCE OF THE ANGULATION OF THE LIMBS

As noted above (p. 269), the straightness of the limbs in the Proboscidea and similar heavy-bodied animals was interpreted by Osborn in 1900 as "an adaptation designed to transmit the increasing weight through a vertical shaft." While this is no doubt an incidental advantage of the straightness of the limbs, it is probably not the chief teleological "object." From a consideration of the mechanical principles governing the use of the limbs as compound levers (see pp. 278-281) and

[39] J. BURDON SANDERSON, in Schäfer's Text Book of Physiology, Vol. 2, p. 363, 1900.

from a comparison of the photographs (Fig. 1) of an ambling elephant and of a galloping horse, it seems probable that the straightness of the limbs in graviportal animals has been evolved *pari passu* with the short rectigrade feet and with an ambling even gait, in combination with a long stride of minimal acceleration increment (p. 274). Conversely, the bent or angulate character of the limbs in the horse and other cursorial animals is correlated in part with the very long, slender unguligrade feet and with a bounding galloping or trotting gait, in combination with a long, very rapid stride of maximal acceleration increment.

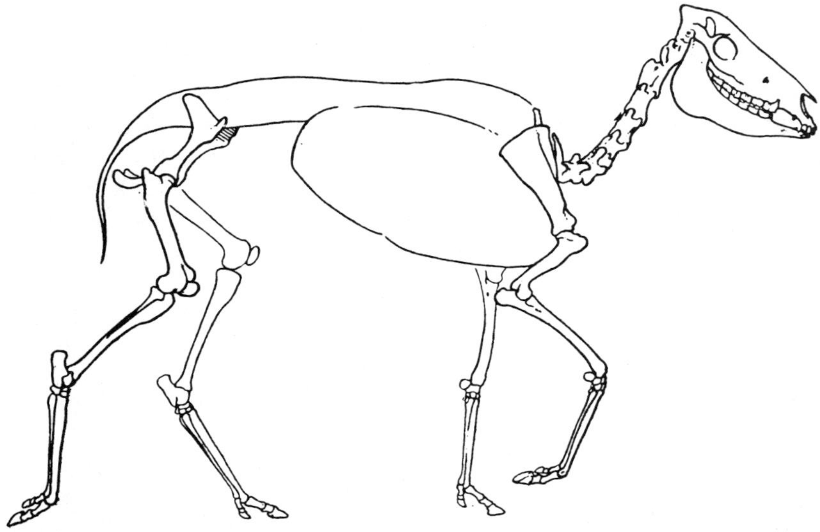

FIG. 5.—"*Cursorial*" *adaptations for the run, gallop, etc., in the Neohipparion*

In other words, the use and structure of the feet have been the teleological dominants which have determined the diverse modifications in the musculature, proportions and angulation of the proximal segments of the limbs, just as in early stages of aquatic adaptation in reptiles (*e. g.*, Thalattosuchia, Nothosauria, etc.) the aquatic habits are reflected more clearly in the feet or distal segments rather than in the proximal limb segments.

MECHANICS OF THE FOOT IN GRAVIPORTAL AND CURSORIAL FORMS

Comparing the structure and function of the graviportal and cursorial types of feet, we see (Figs. 3 and 4) that, in the elephant, the very massive gastrocnemius and soleus muscles are attached at a wide angle to the massive calcaneum, while the foot itself is very short. In the position shown in Fig. 3 (II) and as compared with conditions in the horse, this gives a relatively high moment of power (proportional to *bb'*) to the calf muscles and a relatively low "moment of resistance" (proportional to *b'a'*) to the great pressure of the shaft of the limb upon the astragalus. A considerable part of this "moment of resistance" is also subtracted by the supporting effect of the great pad of elastic tissue underneath the foot. No doubt the specimen from which Fig. 3, II, was drawn should have been mounted with the feet more nearly vertical; this would greatly shorten *b'a'* and further increase the advantage of $m \times bb'$. As the plantar pad is raised from the ground, more weight is thrown on the toes, but, at the same time, they are brought further back almost beneath the astragalus, thus reducing *b'a'* to a minimum, so that the load decreases as the muscles contract. This arrangement not only compensates for the fact that the greatest absolute force of a muscle is developed when it is stretched to its full psysiological length, but it also permits the muscle to perform a greater total quantity of work than would be the case if the load were increasing instead of diminishing (p. 280).

Similarly, in the horse, the greatest "moment of resistance" is when the foot is fully flexed forward (which is at the instant the foot touches the ground); the action of the extensor muscles is thus suddenly checked; this conditions physiologically a corresponding and sudden increase in the available energy (p. 277). By the raising of the heel the moment of resistance ($W \times BA'$), as in the case of the elephant, also decreases, *i. e.*, the load diminishes; but in the elephant, the motion of the foot is relatively slow and the acceleration increment of the stride (p. 275) is therefore slight, whereas in the horse the motion of the foot is very rapid, and the acceleration increment (through the high velocity imparted to the relatively light body by the spring-like extension or opening of the angles at the stifle, hock, fetlock and pastern) is very great, so much so that at least three of the feet are off the ground during a great portion of the time. In brief, short feet (as in the Proboscidea) slightly bending at the ankle, raise a heavy load through a short distance with a minimal acceleration increment of the stride; long feet (as in the horse), sharply bending at the ankle, throw a smaller load a long distance, with a maximal acceleration increment.

Metatarso-femoral ratios.[40]—That in cursorial animals the hind foot is long as compared with the femur, while in graviportal animals the

[40] The investigation of limb- and arch-form and proportions, and especially the establishment and significance of definite ratios between the limb segments, were suggested by Professor Osborn and taken up conjointly by him and the writer in the Eocene section of the Titanothere Monograph; the following observations on limb ratios are in part a preliminary publication of the joint results attained.

reverse is the case, is shown by the table of ratios, Plate XXXIV, and especially in the following examples:

$$\text{Ratio}: \frac{\text{Length of metatarsal III}}{\text{Length of femur}}$$

Graviportal		*Mediportal*		*Subcursorial*		*Cursorial*	
Coryphodon...	.14	Rhinoceros	.37	Eohippus...	.50	Equus......	.78
Uintatherium.	.10	Palæosyops	.21	Tragulus....	.56	Antilope....	1.00
Mastodon.....	.11			Mesohippus.	.57	Odocoileus..	1.00
Elephas.......	.13						
Brontotherium	.20						
Toxodon......	.17						

These ratios definitely prove the connection between the mode of locomotion and the length of the middle metatarsal as compared with the femur. The wide differences in the metatarso-femoral ratio, ranging from .10 in extreme graviportal forms to 1.00 and upward in cursorial forms, are partly bridged over in the mediportal and subcursorial types, and even more completely in a fifth group including certain primitive ungulates, the Condylarths, in which $\frac{\text{Mts. III.}}{\text{F.}}$ ranges from .43 to .31.

Perhaps the most important facts to keep in mind in comparing these and similar ratios (below) are that the ancestral Placentals probably had relatively short hands and feet and long limb bones (p. 270), but that there was doubtless a considerable range of variation in this respect even as far back as the Upper Cretaceous epoch. We are unfortunately unable to follow the ratios through approximate phyletic series except in a few cases (especially Titanotheres, Equidæ, Rhinocerotidæ), but, in every case, we can feel sure that the precise ratios attained in the end-forms are conditioned largely by the nature of the ratios in the stem-forms of each family.

Thus, the exceptional shortness of the feet in the Amblypoda is conditioned by the fact that this group, as represented by *Pantolambda*, had comparatively short feet before gigantism was developed. *Hyrax* is another example of a small form with very short feet, and from some such forms the Proboscidea probably arose.

Besides those phyla which had short feet in the ancestral forms and which merely emphasized this feature, there are many phyla which started from animals with feet of moderate length and later shortened up the feet to a considerable extent. Thus in the Titanotheres, the oldest and most primitive form (*Eotitanops*) has a metatarso-femoral ratio of about .34, which is not far from that of other early Perissodactyls, but in the collateral descendants of *Eotitanops*, we observe a relative shortening of the digits, correlated with increasing body size and straightening of the knees, so that $\frac{\text{Mts. III.}}{\text{F.}}$ drops from .34 through .31 and .30 in the Middle Eocene genera to .28, .26 and even .20 in the gigantic *Brontotherium*. Again, in the Rhinoceroses, the oldest, smallest and most primitive forms have a foot of moderate length ($\frac{\text{Mts. III.}}{\text{F.}}$ = .43-42), but by progressive relative shortening and broadening, the ratio drops to .24 in *Metamynodon*. In the Hippopotami, which are probably descended from animals proportioned about as in *Oreodon* (with an index of .38). gigantism and aquatic habits have brought about a reduction of the index to .26. Similarly in the Toxodonts, the smaller and more primitive forms had relative long, slender feet, while in the gigantic *Toxodon* $\frac{\text{Mts. III.}}{\text{F.}}$ falls to .17.

In those groups in which the most primitive known members had already attained a slender foot, with reduced side toes, gigantism is unable to effect a complete approximation to the graviportal type. Thus, in the bisons, which are undoubtedly descended from slender-footed forms having a metatarso-femoral ratio not less perhaps than .75, the sudden increase in size causes the ratio to fall but slightly (to .65). In the gigantic Irish elk, whose ancestors probably had a very high metatarso-femoral ratio (perhaps 1.00 or more), this ratio falls to .71.

The evolution of cursorial forms is also indicated by marked changes in the ratio under consideration. Thus, in the Equidæ, it rises from .53 in *Eohippus*, through .68 in *Mesohippus* to .99 in *Hypohippus*, reaching the extreme of 1.16 in the slender-limbed Upper Miocene horse, *Neohipparion*. In the relatively small and slender kiang, the ratio is still 1.00, but in the relatively heavy-bodied *Equus scotti* of the Pleistocene, the ratio is only .84, while in modern horses, we observe even in race horses a falling off of the index to .78, and in the stocky-limbed *Hippidion*, it drops to .72. Correlated with this fall in the length of the metatarsal III, we observe a straightening of the knee. These figures possibly may mean that the modern *Equus caballus*, on account of its great size, is somewhat less adapted to extreme cursorial locomotion than was the slender *Neohipparion* which closely paralleled the deer *Odocoileus*.[41] On the other hand, race horses seem to have comparatively long femora (*cf.* Stillman, p. 80) and cheetahs, hounds and other

[41] J. W. Gidley: Bull. Amer. Mus. Nat. Hist., Vol. 19, pp. 474-476. 1903.

forms that progress by bounding have quite long femora. In this connection, it must be remembered (p. 278) that a long femur, implying small angles of insertion of the principal extensors, gives relatively high speed of rotation of the insertion points, but low power for the rotation components. The long femur of graviportal forms has a wholly different meaning (p. 289).

Progressive reduction of the side toes in the Artiodactyla as in the Perissodactyla is accompanied by the elongation of the cannon bone (here represented by two coalesced digits, metatarsals III. and IV.) and by other cursorial adaptations. The primitive Artiodactyl foot with four complete side toes is represented in *Oreodon,* which has a ratio of .38. In *Sus* the ratio is .34, in the primitive four-toed camel *Eotylopus reedi* it rises to .52, in *Tragulus* to .66, whence it rises rapidly to 1.00 and more in the deer and antelopes, culminating in 1.35, an extreme figure, in the giraffe. In the giraffe, the development of great body size has not brought about any reduction in the length of the metacarpals, but, in correlation with the long neck, has even lengthened them.

Other adaptive contrasts in the feet.—There are other adaptive contrasts in the feet of graviportal and cursorial animals, as follows:

In graviportal forms, the astragalus is flattened down, for, when the weight of the body is raised, it is easier to force the tibia up a gentle slope than a steep one. No lateral keels are needed on the trochlear surface to prevent dislocation because of the breadth and spreading character of the tarsus and of the large size of the fibular malleolus. In cursorial animals, on the contrary, the curvature of the astragalar trochlear is steep and the range of movement wide; the trochlea keels help to keep the narrow tarsus in place.

With regard to the phalanges, in cursorial animals, the sharp bending at the fetlock and pasterns and the sudden straightening out of these joints under the pull of the powerful flexors of the foot greatly assists in projecting the body into the air (Stillman, *loc. cit.,* p. 89).

In graviportal animals, on the other hand, the terminal phalanges are reduced; the massive flexors of the foot raise the weight slowly and assist the animal in rolling from one foot to the other.

GRAVIPORTAL AND CURSORIAL TYPES OF TIBIA

Some of the tibio-femoral ratios $\left(\frac{T}{F}\right)$ given on Plate XXXIV may here be grouped as follows:

Graviportal

Uintatherium	.53
Coryphodon	.61
Pyrotherium	.56
Mastodon	.69
Elephas indicus	.60
Brontotherium	.54
Metamynodon	.58
Teleoceras	.57

Subcursorial or Primitive

Phenacodus primævus	.84
" wortmani	.97
Euprotogonia	1.01
Meniscotherium	.91
Procavia	.97
Hyrachyus	.95
Sus scrofa	.86
Eotylopus	.96

Mediportal

Palæosyops	.77
Rhinoceros indicus	.79
Tapirus	.80
Pantolambda	.76

Cursorial

Eohippus	1.00
Mesohippus	1.08
Neohipparion	1.17
Equus caballus	.92
" scotti	.88
Tragulus	1.09
Odocoileus	1.16
Gazella	1.25
Antilope	1.21
Antilocapra	1 23

These figures express the fact noted by Osborn, that, in general, graviportal forms have relatively a short tibia and long femur, while cursorial forms have a long tibia and short femur. It is highly probable that in the remote ancestors of all the Placental orders, the tibia was long, perhaps about as long as the femur. The primitive combination of long tibia and short feet is retained in *Meniscotherium, Hyrax* and to a less degree in *Euprotogonia, Phenacodus wortmani* and *Eohippus.* The steps through which the shortening of the tibia in graviportal animals has been attained may in some cases be discerned. Thus in the Amblypoda, the small and primitive *Pantolambda* has a tibio-femoral ratio of .76; in *Coryphodon,* this ratio falls to .61 and in the huge *Uintatherium* to .53; similarly in the titanotheres, the ratio drops from .77 in *Palæosyops* to .54 in the massive *Brontotherium.* In the Rhinoceroses, it falls from .95 (*Hyrachyus*) to .79 (*Rhinoceros*) and .58 (*Metamynodon*).

In the extremely cursorial lines, the long tibia of primitive mammals was usually lengthened slightly, but in the heavier members of cursorial phyla (*e. g., Equus, Bison*), we observe occasionally a falling off in the ratio.

The difference between the shortest graviportal tibia (*Uintatherium,* ratio .53) and the longest cursorial tibia (*Gazella,* 1.25) is much less extreme than in the metatarso-femoral ratios, which range from .11 to

1.35. There are also more ratios of intermediate type. Consequently, the tibio-femoral ratios taken alone do not always furnish a sure indication of the mode of locomotion.

From the viewpoint of adaptation, there are several plausible reasons why the tibia has not shortened to so extreme a degree as has the middle metatarsal. A short tibia implies a low knee joint and a long femur. A long femur, as stated below, is associated with a nearly vertical in-

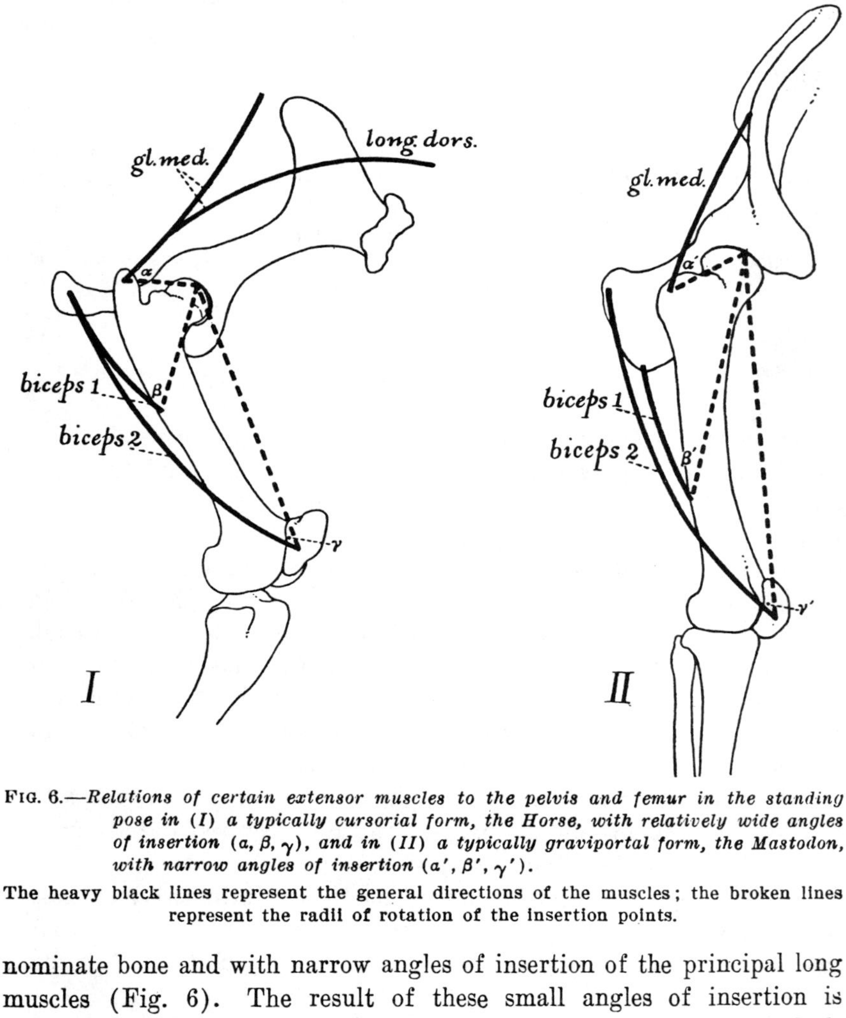

FIG. 6.—*Relations of certain extensor muscles to the pelvis and femur in the standing pose in (I) a typically cursorial form, the Horse, with relatively wide angles of insertion (α, β, γ), and in (II) a typically graviportal form, the Mastodon, with narrow angles of insertion (α', β', γ').*

The heavy black lines represent the general directions of the muscles; the broken lines represent the radii of rotation of the insertion points.

nominate bone and with narrow angles of insertion of the principal long muscles (Fig. 6). The result of these small angles of insertion is that the long muscles exert a powerful pull in the direction of the shaft of the femur (p. 278 and Fig. 2), an arrangement favorable to the lifting and support of great weight (p. 290). The short tibia and short foot, conditioning a low knee joint, also enable the very heavy animal to rise from the ground with comparative ease. At the same time, excessive shortening of the tibia (as in Ground Sloths) probably conditions very slow motion, which would be very disadvantageous to an animal that has to wander far in search of food.

GRAVIPORTAL AND CURSORIAL TYPES OF FEMUR

The femur of primitive animals (Creodonts, Condylarths, Stem Perissodactyls, etc.) is long, when compared with the size of the body. It has a prominent third trochanter situated in the upper fourth of the shaft and serving for the insertion of the glutæus superficials (seu maximus). This muscle arises from the enlarged tuber coxæ of the ilium and apparently functions as an adductor of the femur. The trochanter major projects high above the shaft.

These characters persist in the cursorial types, the chief difference being that here, as a rule, the femur is relatively short when compared with the size of the body. The short femur is associated with a subhorizontal innominate bone and with comparatively open angles of insertion of the long muscles (Fig. 6, I), which pull across the long axis of the femur and give high components of rotation and relatively low centripetal components (in the direction of the shaft of the bone). Such an arrangement, associated as it is with cursorial adaptations, seems less adapted for the support of a great dead weight than that noticed above in graviportal animals. In the standing pose, a sharp angulation at the knee conditions a tendency for the leg to collapse, which is counteracted by the stretching and tension of the opposing flexors and extensors of the thigh (Fig. 7). In the horse, these opposite tensions are transmitted by means of special tendons on opposite sides of the leg, which also serve to tie as it were the limb in position and thus relieve the muscles to a considerable extent. In graviportal animals, the long muscles are also stretched in standing and thus serve as ligaments, and since they are inserted at very narrow angles, their centripetal components are relatively high (p. 278).

In the femur of graviportal animals, the third trochanter is often reduced, absent or confluent with a long ridge running down from the great trochanter and situated well down the shaft. The functional meaning of this is possibly that the glutæus superficialis is less developed in graviportal animals, its place, perhaps, being usurped by the greatly enlarged glutæus medius. Again, it is conceivable that the third trochanter may have been crowded out, so to speak, by the enlargement of the powerful vastus muscles, which arise on the femur just above the third trochanter and which are one of the chief muscles used in lifting the body at the hips. The great trochanter of graviportal animals is very broad and heavy, but is more or less sessile, not jutting up so much

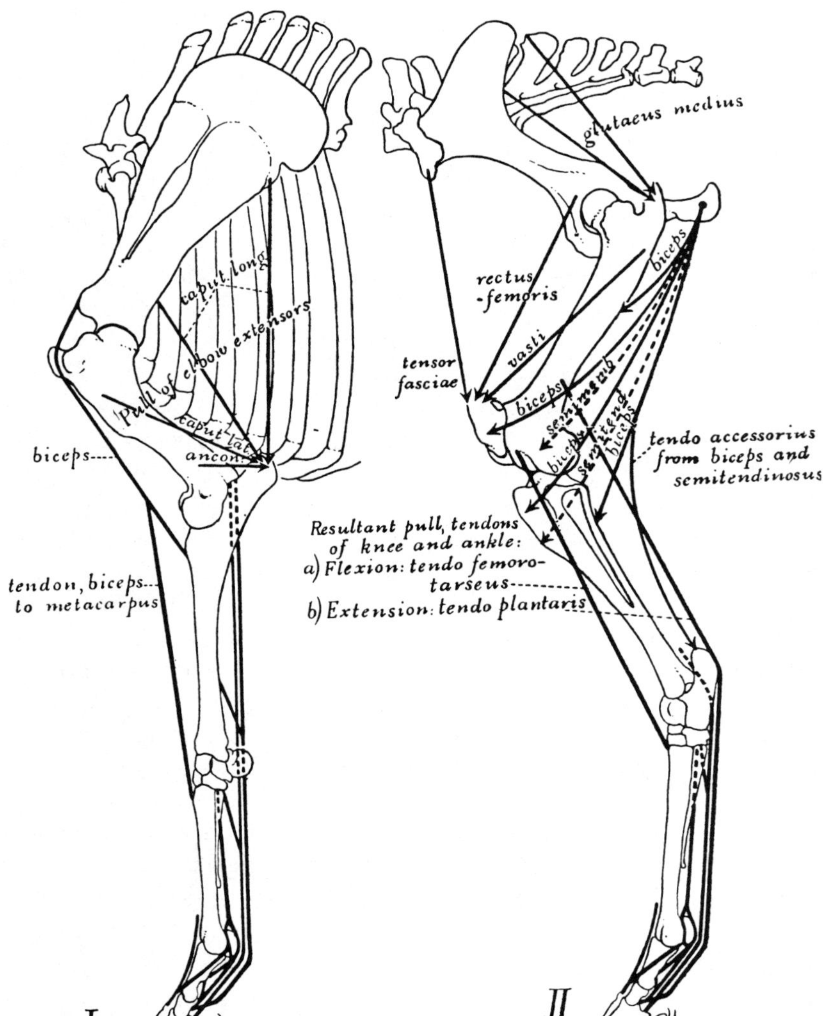

FIG. 7.—*The tendons and muscle-pulls which hold in place the fore- (I) and hind- (II) limbs of the standing horse*

After Schmalz

above the head of the bone. The reason is that the massive glutæus medius is inserted more upon the flat top and outer side of this projection than upon its anterointernal slope (Fig. 6, II), a fact correlated with the subvertical position of the ilium.

In brief, the long femur of graviportal animals serves for the attachment of long and very massive flexor and extensive muscles inserted at slight angles; it brings the knee joint well down below the belly, and it is further characterized by a very heavy sessile great trochanter and a more or less reduced third trochanter, located far down the shaft; the distal facets, as observed by Osborn, are more or less at right angles to the shaft, so that the femur rests subvertically upon the tibia.

GRAVIPORTAL AND CURSORIAL TYPES OF PELVIS

(Figs. 4-6)

The upper anterior border of the ilium in cursorial animals is concave, so that the anterior part of the glutæus medius is produced in front of the ilium and inserted into a long fossa in the posterior end of the fascia covering the longissimus dorsi.[42] The result is that these two muscles apparently pull in tandem, while the subhorizontal position of the pelvis permits the glutæus medius to be inserted on the great trochanter at an angle approaching a right angle. Thus the powerful and prolonged contraction of the combined glutæus medius and longissimus dorsi is exerted well across the long axis of the femur, producing a very powerful rotation component. Similarly the subhorizontal position of the ischium opens out the angles of insertion of the three branches of the biceps and of other extensors and flexors of the thigh.

In such progressively graviportal lines as the Amynodontinæ among the Rhinoceroses, we observe the transformation of the cursorial into the graviportal pelvis by the filling out of the anterosuperior concavity of the ilium, the broadening of the ilium and outward growth of the tuber coxæ and the increasing verticality of the whole pelvis. The result of these changes with regard to the muscles is that the glutæus medius is separated by a wall of bone from the longissimus dorsi, the insertion space of the glutæus medius and accessorius is greatly increased and the medius passes down subvertically on to the broad trochanter major of the femur; the angles of insertion of the flexors and extensors also lessen. Such graviportal adaptations have taken place independently in many lines of ungulates (Amblypoda, Proboscidea, Titanotheres, Amynodonts, Toxodonts) and also in the ground sloths.

---

[42] R. SCHMALTZ: "Atlas der Anatomie des Pferdes." Zweiter Teil: Topographische Myologie, Taf. 27 u. 28. 4to, Berlin, 1909.

GRAVIPORTAL AND CURSORIAL TYPES OF FORE LIMB

Much of what has been said above regarding the mechanical advantages of short hind feet in moving a heavy load or of long feet in throwing forward a light load applies also to the fore feet. The table of ratios (Plate XXXIV), however, shows that in both graviportal and cursorial types, the third metacarpal is usually longer in proportion to the radius than is the third metatarsal to the femur. The radius also is usually much longer in proportion to the humerus than is the tibia to the femur and the radiohumeral ratio responds less clearly to changes in mode of locomotion than does the tibiofemoral ratio. The adaptive reasons for these facts are not altogether clear, but they must be conditioned partly upon the marked differences in form and musculature between the shoulder girdle and the pelvis and upon the different situations of the fore and hind limbs with respect to the center of gravity of the animal. The relatively long metacarpals and long radius even of graviportal animals gives a very long reach to the fore arm, as shown in Fig. 1, I. In some graviportal animals, as in the Proboscidea, the humerus is much longer than the scapula, the elbow is widely exserted from the body and is turned outward in the back stroke, thus bringing the wrists well in. Interference is avoided by the sharp pronation of the radius, so that during flexion of the wrist the palm of the manus is turned partly outward. In this type of scapula, the backward extension of the posterior angle of the blade, as well as the length of the humerus, gives space for a very heavy caput longus of the triceps, which thus also secures a more direct upward pull on the olecranon. The postspinous fossa is located almost directly above the posterior part of the great tuberosity of the humerus, so that the pull of the infraspinatus muscle is nearly vertical. The tuber spinæ of the scapula is greatly enlarged, its upper border for the trapezius, its lower for the heavy deltoideus. In some other graviportal types (*e. g.*, *Toxodon*, *Rhinoceros*), the scapula instead of being broad and short is narrow and high, with vertically extended pre- and postspinous fossæ. In some cursorial types (*e. g.*, *Equus*), the scapula is also long, with rather narrow pre- and postspinous fossæ; in others (*e. g.*, Pecora), the scapula is fan-shaped, truncate at top, with prespinous fossa reduced and postspinous fossa much developed.

Regarding the relative power of the fore and hind limbs, many people think that in the horse, the hind limbs furnish much more than half of the locomotive power, but Stillman arrays cogent evidence[43] tending to

[43] *Op. cit.*, pp. 69, 79, 89.

show that the fore feet, drawn backward by the great pectoral triceps and powerful back muscles, contribute more than their share to the general result.

DIAGNOSTIC VS. CONVERGENT EVOLUTION VALUE OF LIMB RATIOS

From inspection of the table (Plate XXXIV), it will be seen that, allowing for imperfection of the fossil material and incompleteness of the series, the limb ratios have a certain degree of diagnostic value when taken in groups and that pure convergent evolution rarely brings about a close agreement in all four ratios at once.

The best cases of convergent evolution noted are as follows:

			Mts. III / F.	T. / F.	Mtc. III / H	R / H
Cursorial	Perissodactyla	Neohipparion	1.01	1.17	1.16	1.30
	Artiodactyla	Odocoileus	1.00	1.16	1.05	1.12
Traviportal	Edentata	Lestodon	.12	.51	.17	.60
	Amblypoda	Coryphodon	.14	.61	.19	.66
Mediportal	Titanotheriidæ	Brontops robustus	.26	.55	.37	.82
	Amynodontidæ	Metamynodon	.24	.58	.39	.81
	Rhinocerotidæ	Teleoceras	.25	.57	.37	.78
	Toxodontia	Toxodon	.17	.56	.38	.77

The greatest extremes are found in the following:

Traviportal	Amblypoda	Uintatherium	.10	.53	.19	.70
Cursorial	Giraffidæ	Giraffa	1.35	1.18	1.42	1.60

ERRATA

In the table on page 294, "Traviportal" should read "Graviportal." In Plate XXXIV, A.M. and A.M.N.H. stand for the American Museum of Natural History in New York.

Plate XXXIV

COMPARATIVE TABLE OF LIMB RATIOS IN THE HOOFED MAMMALS.

Species.	Femur (F)	Metatarsal III (Mts. III)	$\frac{\text{Mts. III}}{\text{F}}$ (F=1)	Tibia (T)	$\frac{\text{T}}{\text{F}}$ (F=1)	Humerus (H)	Metacarpal III (Mtc. III)	$\frac{\text{Mtc. III}}{\text{H}}$ (H=1)	Radius (R)	$\frac{\text{R}}{\text{H}}$ (H=1)
	mm.	mm.		mm.		mm.	mm.		mm.	
CONDYLARTHRA										
Euprotogonia puercensis	105	45	.43	107	1.01	......	......	......	......	......
Phenacodus wortmani	134	51	.38	{ 131 ? 138 r	.97 } 1.00	107	36	.65	89	.83
Phenacodus primævus	234	74	.31	198	.84	167	70	.42	146	.87
Meniscotherium terræubræ	100	29	.29	91	.91	82	22	.27	58	.70
AMBLYPODA										
Pantolambda bathmodon	149	36	.24	114	.76	124	31	.25	82	.66
Coryphodon lobatus	423	62	.14	260	.61	363	70	.19	240	.66
Uintatherium (Dinoceras) mirabile	692	70	.10	360	.53	540	106	.19	380	.70
PYROTHERIA										
Pyrotherium (figd. by Gaudry)	622	......	......	351	.56	452	......	......	238	.52
PROBOSCIDEA										
Mastodon americanus	1,020 e	117	.11	705	.69	885	165	.18	670	.75
Elephas indicus	1,020	138	.13	618	.60	810	183	.22	685	.80
E. (Loxodonta) africanus	1,050	144	.13	755	.71	1,000	205	.20	870	.87
HYRACOIDEA										
Procavia sp.	71	19	.26	69	.97	69	16	.23	46	.66
TOXODONTIA										
Toxodon sp.	577	101	.17	325	.56	387	147	.38	298	.77
EDENTATA GRAVIGRADA										
Hapalops sp.	150	18	.12	108	.72	130	22	.16	106	.80
Lestodon armatus	640	78	.12	330	.51	530	94	.17	325	.60
PERISSODACTYLA TAPIROIDEA										
Heptodon calciculus	175	75 e	.43	175	1.00	115	67	.58	114	.99
Tapirus americanus	262	108	.41	208	.79	205	106	.50	177	.86
Tapirus indicus	320	120	.37	258	.80	250	120	.48	228	.91
PERISSODACTYLA RHINOCEROTOIDEA										
Hyrachyus agrarius	254	110	.43	243	.95	197	93	.47	197	1.00
Hyracodon nebrascensis	267	114	.42	220	.82	202	114	.56	210	1.03
Rhinoceros indicus	495	180	.37	395	.79	385	186	.48	385	1.00
Teleoceras fossiger	408	105	.25	233	.57	305 e	114	.37	238	.78
Metamynodon planifrons	480	118	.24	280	.58	393	153	.39	320	.81
PERISSODACTYLA TITANOTHEROIDEA										
Eotitanops borealis	250 e	? 86	? .34	......	......	203	85	.41	......	......
Palæosyops major	433	137	.31	332	.77	......	......	......	......	......
Palæosyops leidyi	370	110	.30	290	.78	325	? 113	? .34	235	.72
Palæosyops sp. (A. M. N. H. No. 12205	......	......	......	......	......	? 340	106	.30	237	? .69
Limnohyops sp. (A. M. N. H. No. 11689	355 e	111	.31	285	.79	293	109	.37	228	.77
" " " " No. 11690	387	123	.31	283	.73	......	......	......	......	......
Manteoceras manteoceras	? 390	......	......	272	.69	......	......	......	......	......
Mesatirhinus petersoni, No. 11659	358	118	.33	283	.79	......	......	......	......	......
Dolichorhinus hyognathus, No. 13164	386	119 e	.30	......	......	315 e	......	......	284	? .81
Titanotherium trigonoceros	770	220 e	.28	430	.55	620	240	.38	520	.83
Brontops robustus	812	212 e	.26	448	.55	608	230	.37	504	.82
Brontotherium gigas ♀ (518 A. M.)	780	200	.20	427	.54	528	214	.40	478	.90
PERISSODACTYLA HIPPOIDEA										
Eohippus sp.	162	82 e	.50	162	1.00	121	64	.53	110	.90
Mesohippus sp.	178	121	.68	193	1.08	136	92	.68	136 e	1.00
Hypohippus osborni	278	218	.78	277	1.00	205	203	.99	260	1.27
Neohipparion whitneyi	249	252	1.01	293	1.17	187	218	1.16	244	1.30
Equus kiang	313	277	.88	310	.99	237	238	1.00	302	1.27
Equus scotti	370	263	.71	330	.88	289	243	.84	342	1.18
Equus caballus (race horse)	392	288	.73	363	.92	305	240	.78	363	1.19
Hippidion neogæum	340	214	.62	305	.89	273	198	.72	287	1.05
ARTIODACTYLA										
Oreodontidæ										
Oredon culbertsoni	161	62	.38	142	.88	138	57	.41	113	.81
Suidæ										
Sus scrofa	248	86	.34	215	.86	208	77	.37	168	.80
Hippopotamidæ										
Hippotamus amphibius	498	130	.26	332	.67	395	152	.38	270	.68
Camelidæ										
Eotylopus reedi	148	78	.52	142	.96	118	68	.57	100	.84
Camelus arabicus	470	325	.60	400	.80	363	330	.90	455	1.25
Tragulidæ										
Tragulus napu	94	62	.66	103	1.09	74	42	.56	62	.83
Cervidæ										
Odocoileus hemionus	253	255	1.00	295	1.16	198	208	1.05	223	1.12
Cervus megaceros	430	350	.71	454	1.05	334	342	1.02	358	1.07
Bovidæ										
Gazella dorcas juv.	140	132	.81	176	1.25	93	134	1.44	118	1.26
Antilope cervicapra	183	183	1.00	223	1.21	133	180	1.35	168	1.26
Bison bison	369	243	.65	355	.96	290	198	.60	293	1.01
Antilocapridæ										
Antilocapra americana	210	218	1.03	260	1.23	164	213	1.30	202	1.23
Giraffidæ										
Giraffa sp.	466	630	1.35	550	1.18	435	618	1.42	698	1.60

e—Estimated.

19

Reprinted from *Am. Jour. Sci.* **243:**667-679 (1945)

THE CHALICOTHERES AS A BIOLOGICAL TYPE.

A. BORISSIAK.

TRANSLATED BY I. P. TOLMACHOFF.

EXTINCT animals occur in the strata of the earth's crust only as skeletons, often fragmentary. The restoration of an animal on the basis of such remains is a difficult task requiring great erudition in the student. It is still more difficult to write the history of the group to which the animal belonged, and the difficulty increases in proportion to the differences of the extinct animal from animals still living. Extinct animals with skeletons very unlike those of recent animals are often a real enigma.

For a long time when only limb bones of *Chalicotherium* were known, the animal was placed among Edentata because its powerful claws suggest those of a sloth. When other parts of the skeleton were found, *Chalicotherium* was placed among the Ungulata in the Perissodactyla, although it differed from all other fossil and living Ungulata [then known] in having claws, not hoofs.

Holland and Peterson (1913) in their monograph bring together a complete list of literature on chalicotheres for the period of 1825-1913. Their detailed quotations give a good idea of previous knowledge of this group of animals. In following years the Chalicotherioidea continued to be studied chiefly by Americans (e.g. Matthew, 1929; Colbert, 1935a, b). Russian literature on Chalicotherioidea is very small in amount for until lately only a few chalicothere bones had been found in Russia. Quite recently, however, the bones of chalicotheres have been found in large numbers in the Tertiary deposits of the Golodnaya Steppe in Southern Kazakhstan. Here the Lower and Upper Tertiary strata are separated by a layer of conglomerate with abundant bones and teeth of mammals. The great majority of these bones belong to a large chalicothere. It has been possible to assemble an almost complete although composite skeleton, to study its elements in detail, and to come to conclusions about the habits of the animal and its phylogenetic relations (Borissiak, in press).

In the skeleton of chalicotheres (Fig. 1), striking features are the rather long neck, the small skull, the comparatively long and thin fore limbs, and the short, massive hind limbs. The feet were digitigrade and the replacement of hoofs by claws was especially well developed on the fore limbs.

In some species of chalicotheres the skull is long and low (like the skull of the horse), but in other species it is short, resembling, in these instances, the skull of a bear. The skull

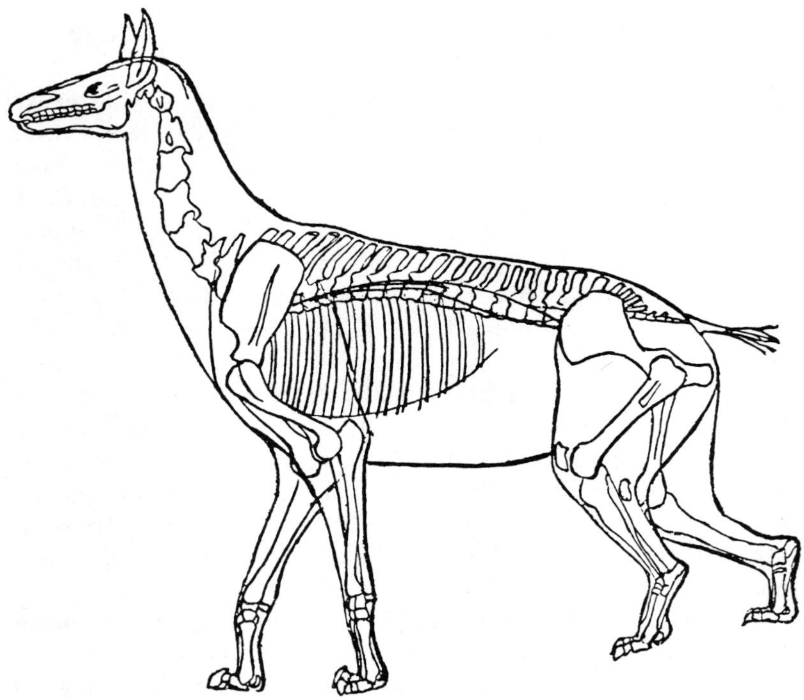

Fig. 1. Restored skeleton of *Phyllotillon betpakdalensis* (Flerov) from the Tertiary of the Golodnaya Steppe.

has characters of primitive Perissodactyla (tapirs, early titanotheres) but also a few characters known only in chalicotheres (e.g. two deep impressions on the sides of the presphenoid). The structure of the cervical section of the vertebral column was very peculiar. It was rather long and at the same time very massive, especially in comparison with the small skull. The vertebral centra were reduced relative to the well developed, flattened neural arches, which bore strong zygapophyses. Such a structure shows that the dorsal muscles of the neck were very strong in comparison with the relatively weak ventral muscles—relations opposite to those observed in typical Perissodactyla.

The structure of the fore limbs, particularly of the manus, was very peculiar and unlike that known in any other mammals. In the most specialized forms, only three of the four digits articulated with the carpus, the fourth and smallest digit being joined to the proximal part of the neighboring digit. The chalicothere carpus was low, with a small, narrow os magnum The last character distinguishes chalicotheres from horses, the os magnum of which is large, low, and flat, but in this respect chalicotheres are closer to titanotheres. The articulation of the carpus with the fore arm was a transverse arc which per-

Fig. 2. The second digit of the manus of *Phyllotillon betpakdalensis* (Flerov).

mitted considerable swinging of the manus to both sides. Articulation of carpus with metacarpus was very abrupt. On the anterior side of the distal bones of the carpus and on the proximal ends of the metapodials there are rugose swellings for the attachment of strong extensors, while in perissodactyls and in mammals in general limb flexors are usually more strongly developed than extensors. A walking or running animal pushes itself from the ground by flexing its limbs while extensors are used only to bring the limbs to a new place ahead, which is much easier work. The anterior muscles of the chalicothere scapula were also stronger than the posterior. The anterior faces of the metacarpals were evenly convex in all directions, as in some carnivores. The structure of the phalanges was particularly remarkable (Fig. 2). The second digit was stronger than the others (absence of tridactylism). The metacarpal facet of its first phalanx was not on the proximal end, but had moved to the anterior (or dorsal) face of the bone. In other digits this displacement was not so great; the facets here only cut off the anterior edge of the proximal end of phalanges. In other words, the metacarpal and first phalanx of the second digit were joined together at a right angle, not in a straight line. The first and second phalanges were often coössified. The ungual phalanx, again, met the second phalanx at a right angle. In this way was constructed a strong, flat, triangular claw with only slight movements in its parts. The whole digit was bent twice at right angles, forming a very strong hook. This structure was similar in the other digits but not so well developed as in the second. Separate segments of digits were only slightly movable but the digits as a whole had a wide movement owing to the spherical shape of the distal ends of the metacarpals.

The hind feet had in general the same structure but it was not so well developed as in the fore feet: the metatarsals had no distal swellings; the distal facets were not all spherical in shape; etc. The feet were tridactyl with a long middle digit, the metatarsals were more or less shortened, and the whole specialization of the phalanges was less advanced than in the fore feet. The hind feet were more massive than the fore feet; the calcaneum was low; the astragalus was straight-sided and almost without a neck; etc. On the whole, this is a rare case in which the specialization of fore and hind limbs has gone in different directions.

As in all animals, one group of anatomical characters changed constantly as the result of growing specialization in adaptation to the environment, while other inherited characters did not change and were common to all chalicotheres. The latter show the systematic position of the animals and in this case demonstrate that the closest relationship of the chalicotheres is with the titanotheres.

This relationship is particularly shown by the structure of the teeth of chalicotheres, which is similar to that of early and primitive titanotheres in which the upper molar crowns still had transverse lophs connecting the ectoloph with the lingual tubercles (protocone and hypocone). In the later titanotheres, the upper molars lost their transverse lophs. In chalicotheres the upper molars remained the same with surprising conservativeness. The only difference between the earlier and later forms is that the molars were more hypsodont in the latter. In their shape the teeth of chalicotheres represent the best developed browsing type. They were not good for hard grasses, much less for food with remnants of soil, such as bulbs, etc., but were exclusively adapted for soft leaves and branches. The cutting sides of the upper and lower molars worked like sharp scissors, while the strong inner tubercules were for grinding branches and leaves. The structure of the skull was also very similar to that of the early hornless titanotheres, although horns were so typical for the later titanotheres. In the skeleton there are also a great many characters suggestive of titanotheres. In spite of the great specialization of the cervical section of the vertebral column of chalicotheres, the atlas had exactly the same shape as in titanotheres. There is a great similarity in the structure of scapula, humerus, radius, and ulna. The structure of carpus and manus is particularly striking. The small, narrow but high magnum, the very peculiar scaphoid with a small, downwardly directed projection, the abrupt articulation of the phalanges with the carpus, the dilated proximal end of the ulna, the arcuate articulation between the fore arm and the manus—all these characters leave no doubt as to the relationship of the two groups, but weigh against any connection with horses, with which chalicotheres have occasionally been compared because of a certain similarity in the shape of their skulls. (In some restorations, chalicotheres were pictured with horse-like heads and long, curly manes.) The femur had a symmetrical vertical trochlea for the patella. There were also a great many resemblances in the bones of the tarsus. In general the chalicothere pes was more like the titanotheres than was the manus, as a result of the greater specialization of the fore limbs than of the hind limbs in chalicotheres. For example, the pes was typically tridactyl while tridactylism was lost in the manus.

The close relationship of chalicotheres with the early members of the titanothere family gives us an example of the application of Cope's law according to which the origin of a new group is usually from the less specialized forms of the old group.

We pass now to consideration of the characters of the skeleton of chalicotheres dependent on adaptation to their living conditions. New characters in the structure of the skeleton did not, of course, become fully developed at once. They are rather little developed in the Eocene chalicotheres of America and Asia. Even in the Oligocene forms of Europe and Asia they are still not completely expressed.

What were the living conditions that produced such a peculiar structure of the chalicothere skeleton? From the very beginning of the study of chalicotheres many attempts were made to answer this question. The structure of the feet was explained, for example, as an adaptation to digging, to climbing on trees, to catching tree branches in order to bend them to the ground, etc. (Abel, 1920; Gaudry, 1867; Koenigswald, 1932; Matthew, 1929; and others). Some of these ideas, often mutually exclusive, have been abandoned but a few are still accepted.

We have already seen that the neck and fore limbs of chalicotheres had a different structure from typical ungulates. In the neck the strongest muscles were dorsal, which shows that the animal pulled the head and neck upwards and backwards more often than down to the ground. In the fore feet extensors were stronger than flexors, which tells us that the fore feet were used less for walking than for some other work. This brings us to the conclusion that the normal position of the animal was standing on its hind feet while using the fore feet for clinging to the bark of trees when feeding on leaves and tender branches (Fig. 3). The recent goat climbs in this way when hunting after the leaves of large trees. Lifting the fore limbs for such climbing uses the extensors of the whole fore limb from the manus to the scapula. During feeding the head was moved backwards to pick up leaves distant from the trunk of the tree. The small size of the head was favorable for this operation of the neck. Chalicotheres did not embrace the tree trunk with their fore limbs but, in a manner of speaking, walked on the trunk. The manus was particularly well adapted for such "walking." The claw of the second digit was a strong hook which was driven into the bark of the tree more deeply as more pressure was placed on it.[1] Other digits helped the second very effectively, as each of them could change its place on the trunk with ease. Strong extensors not only raised the manus but also made it very stable. The arcuate articulation between the fore arm and the carpus made side movements of the whole manus very easy when the animal was looking for the best place on the trunk.

[1] This principle is used in hooks or hangers that are driven into a wall and support considerable weights.

The structure of the hind limb supplements the preceding picture. During feeding the animal stood on its hind feet only,

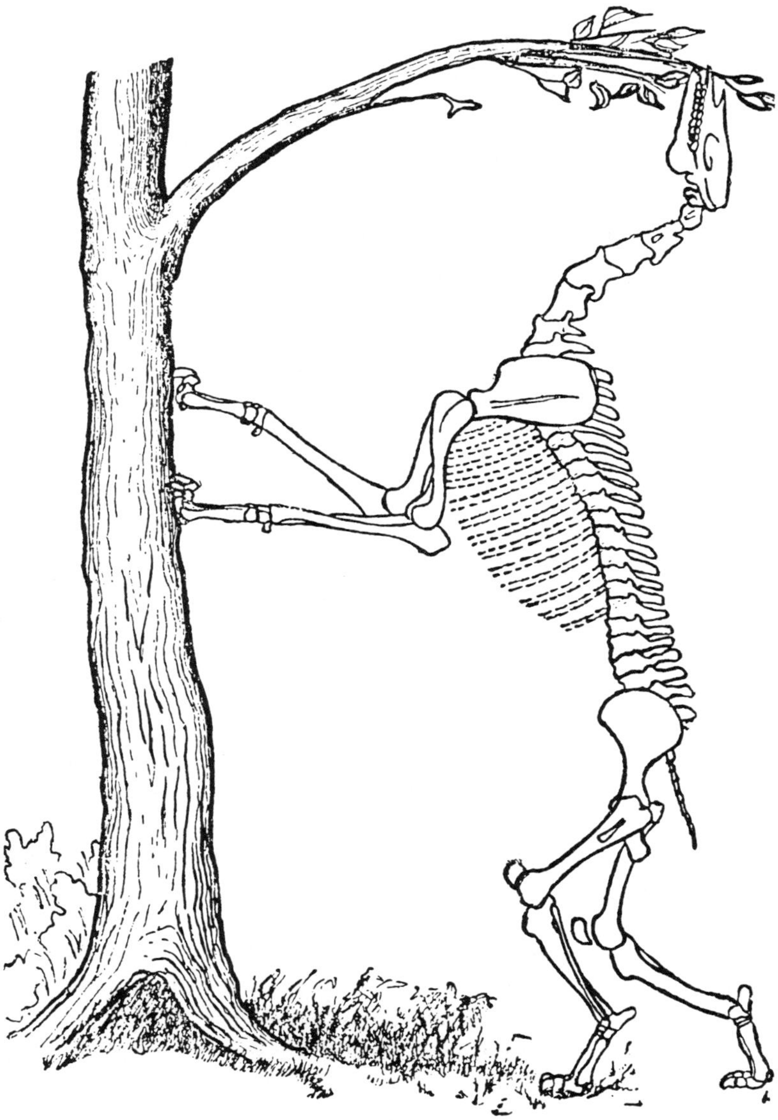

Fig. 3. The posture of the skeleton of *Phyllotillon betpakdalensis* (Flerov) when the animal was feeding.

and these were therefore constructed like the feet of heavy animals and were much heavier than the fore feet. In the forms of later geological age, the hind feet can be compared with the feet of mastodons for massiveness. The astragalus was flattened and moved on the cuboïd, the metatarsals were shortened, and the foot was less specialized than the manus. The claws were not so strong and tridactylism was well developed, while in the manus this development was hindered by the structure of the second digit.

Such is the biological type of the chalicotheres as disclosed by the study of their skeletons. This type arose as an adaptation to the environment. This adaptation is of a different degree in chalicotheres of different geological ages, becoming greater and greater with time.

It has been mentioned in the literature that the biological type of the chalicotheres resembles that of the giraffes. The specialization of the two animals was, however, in different directions. The whole skeleton of the giraffe was specialized for feeding on the leaves on the top of trees and did not progress any further. When the chalicotheres had their fore limbs on the ground, as they certainly sometimes did, they would become normal ungulates. Giraffes live in African savannas with scattered groups of trees. Chalicotheres apparently dwelt in dense forests where they had good protection and no competition because of their high adaptation to this environment. Chalicotheres survived through the whole Tertiary and died out only after that period in spite of their generally primitive inherited organization. Titanotheres, from which chalicotheres arose, had already become extinct in the Oligocene when there appeared the (better adapted) rhinoceroses.

The problem of the study of fossils is primarily a morpho-ecological analysis of the fossil skeleton. This analysis enables one to distinguish the rather stable inherited characters from those that were the result of adaptation to the environment. The latter characters permit reconstruction of the biological type of an animal and understanding of its ecological relations. This work is, however, only a preparation for another important task of the paleontologist—discovery of the phylogenetic relations of the given form. This can be done by the comparative study of restored representatives of the group. Such a study permits determination of the phylogenetic relationships of each form in reference to the other representatives of the group, in other words discovery of the place of each form in a common genealogical tree.

The peculiar specialization of the chalicotheres developed

gradually. The various representatives of this group, about twenty forms altogether of different geological ages, show different degrees of specialization, which was greater in the forms of later epochs, although we cannot establish an uninterrupted line with gradual modification. On the contrary, practically every form shows some peculiarities comprehension of which is rather difficult, chiefly because the material for study in most cases is very fragmentary. Among these peculiarities is the striking fact that the molar crowns are subquadrate in a few forms but elongated in others. Classification of chalicotheres has generally been based on this character. There have been attempts to bring these variations in the molars into correlation with differences in the structure of the skeletons. Relatively few skeletons are known, however, and associations with the teeth have been only occasionally established. New materials from the Tertiary of the Golodnaya Steppe have showed, however, that all Old World forms are very similar to each other in skeletal structure, whether the molar crowns are short or long. On the other hand, North American forms with long molar crowns, like those of some Asiatic and European forms, have skeletons of quite different type.

It is possible to think of the chalicotheres, after their separation from the titanotheres in the Eocene, as developing further into two separate lines, one in the Old World, and one in the New. Each line developed a number of branches. In Europe there evolved two well defined branches distinguished by the structure of their molars. This is all we positively know on the basis of the present material. Undoubtedly the history of the chalicotheres was much more complicated. A few observations demonstrate that there were subbranches, more especially in the beginning of chalicothere history, which became extinct without leaving any descendants. The genealogical tree of the chalicotheres is composed of several branches, the number of which will increase with discovery of new materials. Each branch of the genealogical tree of the chalicotheres represents a distinct direction of their evolution. The place of every form of chalicothere in this tree depends, first, on its pertinence to one or the other of the two main lines (according to which direction of evolution is typical for this form) and, second, on the degree of specialization. Direction and degree of specialization, like mathematical coördinates, determine the exact place of each form on the corresponding branch. To illustrate these statements we shall bring together information about the best known species of chalicotheres.

The most ancient representative of the Chalicotherioidea is *Eomoropus amarorum*, from the Middle Eocene of Northern America (Osborn, 1913). It was an animal of about the size

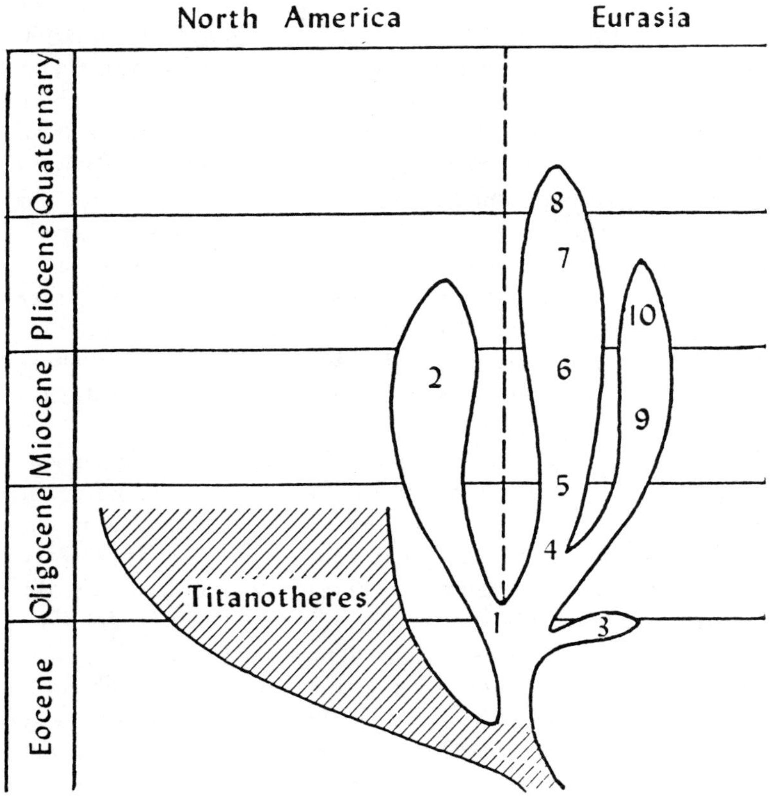

Fig. 4. Phylogenetic scheme of the Chalicotherioidea.
1, *Eomoropus*. 2, *Moropus*. 3, *Grangeria*. 4, *Schizotherium*. 5, *Phyllotillon*. 6, *Metaschizotherium*. 7, *Ancylotherium*. 8, *Postschizotherium*. 9, *Chalicotherium*. 10, *Nestoritherium*.

of the sheep with only insignificant specialization: the cervical vertebra did not show noticeable alterations but the metacarpals had already changed typically for the group in their general shape and joint structure, although they did not yet have callous swellings on the anterior side of the proximal ends. The specialization of the hind feet was still less advanced.

From the Asiatic Eocene isolated teeth are known about which it is not possible to say anything definite. They might belong to *Eomoropus* or to some new genus. At the end of Eocene in Mongolia lived *Grangeria*, a very peculiar form in the structure of its dental apparatus and limbs. *Grangeria* represented an early and short-lived side branch of chalicotheres.

In the Oligocene the European line is represented by several poorly known species of *Schizotherium*, still small animals. A few forms (*S. modicum*) had long molar crowns. The separation of the Chalicotherioidea into those with short and those with long molars had thus begun at this time.

At the end of the Oligocene or beginning of the Miocene in Middle and Southern Asia there lived large chalicotheres with long molars belonging to the genus *Phyllotillon* (Pilgrim, 1912). To this genus belongs *P. betpakdalensis* (See Borissiak, in press; Flerov, 1938), from the Golodnaya Steppe. Owing to the abundant material it is one of the best known chalicotheres. The specialized structure of the skeleton of *P. betpakdalensis* has already been described above. A few bones of *S. turgaicum* from the Middle Miocene of Turgai show remarkable similarity to the bones from the Golodnaya Steppe, but belong to an animal of a smaller size. In this case we have forms of the same branch but with different degrees of specialization determining their respective places on the branch. Another species of *Phyllotillon* lived simultaneously in the valley of the Indus River (Pilgrim, 1912).

In the Miocene also, but a little later than *Phyllotillon betpakdalensis*, there lived in North America *Moropus elatus* (see Holland and Peterson, 1913), a species with long molars very similar to those of *Phyllotillon* and similar also in degree of specialization, but different in the structure of various parts of the skeleton. *Moropus* is a representative of the American chalicothere line.

In the Miocene of Europe lived *Macrotherium*, known from fairly complete fossil remains. *Macrotherium* represents the short-molar branch of Eurasiatic chalicotheres. In spite of its short molars, *Macrotherium* was closer in skeletal structure to *Phyllotillon* than to the American *Moropus* with its long teeth. *Macrotherium* and *Phyllotillon* thus belonged to the same European trunk although to different branches.

In the Pliocene both branches of the European trunk were

represented by the largest species of chalicotheres. The form with long teeth is *Ancylotherium* found in the Pikermi Fauna of Greece (Gaudry, 1867). Its hind limbs resemble the limbs of a mastodon in massiveness. The Quaternary representative, *Postschizotherium*, of the same trunk had very strongly developed hypsodont teeth. The branch with short molars was also represented in the Pliocene and Quaternary by big animals.

Such is the history of the chalicotheres as far as it can be represented at the present time. Remains of chalicotheres are still very scarce and it is therefore impossible to make a complete genealogical tree with all branches and subbranches. Altogether we know about fifteen forms which can be placed at different points along a few branches. The task of the paleontologist is to find the position of these points on the branches on the basis of the coördinates as worked out above, and in this way to establish the phylogenetic relationships of the known forms.

We have, however, no basis for connecting these points by lines. This would give the idea of the direct descent of one form from another, which as a rule cannot be proved. In only one case is it possible to show with certainty the pertinence of two forms of different specialization to the same branch. This refers to *Schizotherium turgaicum* and *Phyllotillon betpakdalensis*. This does not mean, however, that one of these forms developed directly from the other and that the two can be connected by a straight line. We prefer to represent the branches as elongated leaves, on which are shown the different forms of chalicotheres according to their mutual relationships.

The study of chalicotheres shows us what paleontology can do for the history of the organic world. We can establish the ecological characters of animals and reconstruct the whole animals. The study of the genera of chalicotheres permits establishment of their phylogenetic relations.

Fossil remains are not complete and paleontology alone cannot decipher the complete history of animals and plants. It needs support from the other biological sciences dealing with recent organisms. But since paleontology alone has all the historical data, its conclusions check the deductions of other biological sciences.

Paleontology has not, however, said its last word. It still lags behind other biological sciences. Not only before but even since Darwin it has usually been in the service of geology, which has always made special demands for paleontological data. Only now are the possibilities of paleontology for the elucidation of many biological problems becoming clear. In the achievement of independence from geology and in becoming a great biological science, Soviet paleontology occupies a leading position.

REFERENCES.

Abel, O.; 1920, Studien über die Lebensweise von *Chalicotherium.* Acta Zool., **1**, 21-60.

Borissiak, A.; (In Press), A new representative of the chalicotheres from the Tertiary deposits of Kazakhstan. Mem. Pal. Inst. (Moscow), in press. (In Russian).

Colbert, E. H.; 1935a, Distributional and phylogenetic studies on Indian fossil mammals. III. A classification of the Chalicotherioidea. Amer. Mus. Novitates, No. 798, 1-16.

———; 1935b, Siwalik mammals in the American Museum of Natural History. Trans. Amer. Phil. Soc., N.S., **36**, i-x, 1-401.

Flerov, K. K.; 1938, Remains of Ungulata from Bet-pak-dala. Comptes Rendus Acad. Sci. (Moscow), **21**, 94-96.

Gaudry, A.; 1867, Animaux fossiles et géologie de l'Attique. Paris, 1862-1867.

Holland, W. J., and Peterson, O. A.; 1913, The osteology of the Chalicotheroidea; with special reference to a mounted skeleton of *Moropus elatus* Marsh, now installed in the Carnegie Museum. Mem. Carnegie Mus., **3**, 189-406.

Koenigswald, G. H. R. von; 1932, *Metaschizotherium fraasi,* N. G., N. Sp., ein neuer Chalicotheriide aus dem Obermiocän von Steinheim a. Albuch. Palaeontographica, Sup. Bd. 8, 1-24.

Matthew, W. D.; 1929, Critical observations upon Siwalik mammals. Bull. Amer. Mus. Nat. Hist., **56**, 437-560.

Osborn, H. F.; 1913, *Eomoropus,* an American Eocene chalicothere. Bull. Amer. Mus. Nat. Hist., **32**, 261-274.

Pilgrim, G.; 1912, The vertebrate fauna of the Gaj Series in the Bugti Hills and the Punjab. Pal. Indica, N. S., **4**, No. 2, 1-83.

INSTITUTE OF PALEONTOLOGY,
U. S. S. R. ACADEMY OF SCIENCE,
Moscow, U. S. S. R.

20

Reprinted from *Evolution* **6:**181-196 (1952)

THE EVOLUTION OF A PERMIAN VERTEBRATE CHRONOFAUNA

EVERETT C. OLSON

INTRODUCTION

This paper is concerned with the movement of a complex of interacting vertebrate populations through time under the influence of changing environmental conditions. This complex maintained a basic structural pattern throughout its existence, although, of course, there were changes of some of its components. The maintenance of this structural pattern appears to have been related to the general tendency toward ecological stability of the deltaic environment in which the vertebrates lived. Major changes in equilibrium of the system that might have resulted from immigration tended to be reduced by the partial ecological and geographical isolation provided by the environment, and by the internal balance of the complex, which was strong throughout much of the time involved.

Partial discontinuities exist between ecological systems today and clearly have existed in the past under a wide variety of circumstances. The units thus formed have persisted, changing in varying degrees, but maintaining their fundamental characteristics for periods of time that have geological significance. This persistence is reflected, as in the case studied, in the continuity of the structure of the evolving animal complex that is a part of the total ecological system. At any given moment of time in the history of such a system, the animals properly may be considered to constitute a fauna. The scope of such a fauna is limited by the particular ecological system of which it is a part. Furthermore, for convenience in reference, a fauna may be restricted or qualified, as may be appropriate in a given case, to include only part of the animals, as, for example, in the present case, the vertebrates. Interaction of the animal populations that constitute fauna, either direct or indirect, is implied by the association of the animals, although it is not explicit in the definition of a fauna.

When the factor of duration through time is introduced, the term fauna is no longer strictly applicable and may lead to confusion. For this reason the term chronofauna [1] is proposed to differentiate faunal units in which time is an important factor from those in which it is not. A chronofauna is defined as a geographically restricted, natural assemblage of interacting animal populations that has maintained its basic structure over a geologically significant period of time.

Environment, both biological and physical, plays an important part in most evolutionary studies and interpretations of past environments necessarily are involved in an analysis of a chronofauna. Reconstructions of ancient environments are subject to various difficulties and conclusions are rarely beyond question. Furthermore, there always remain biological and physical factors that cannot be considered. It is possible, nevertheless, to gain information concerning both physical and biological aspects of environment adequate to provide a background for study of many phases of adaptive modifications of animals. Such information becomes especially significant when changes of environment can be related to patterns of stability and modification of the animals of a chronofauna.

Once the history of each species of an assemblage is known and its relationships

[1] This term was suggested by Mr. Bryan Patterson in the course of discussions at an annual meeting of the Society of Vertebrate Paleontology.

to the physical environment have been determined, it becomes possible to interrelate the histories of many species, each of which to a greater or lesser degree provides a part of the environment of others. All known biological and physical aspects of the environment may then be studied in their relationships to patterns of stability and change of the complex of animals. There emerges, when sufficient data are available, a coherent picture of the evolution of a chronofauna in which the course of evolution of the elements, the populations, takes on new meaning in terms of the whole.

The Permian assemblages considered in the following sections of this paper constitute a vertebrate chronofauna. Interpretations of these assemblages and their environments are based upon information accumulated during some 17 years of study of the fossils and their field relationships. The animals are distributed among fish, amphibians and reptiles, the time is Early Permian and the physical setting was a delta. The delta lay marginal to non-deltaic areas, from which immigrations of vertebrates took place to modify the deltaic assemblage to a degree noted later in this report.

The Permian Chronofauna and Its Environment

The Geographic and Geologic Setting

The map[2] (fig. 1) shows the area of north central Texas with which the study is concerned and certain gross features of the geology. The vertebrates have been collected from the Arroyo, Vale and Choza, successively younger formations of the Clear Fork group of Early Permian age. Somewhat similar assemblages are well known from the underlying formations of the Wichita group, but the discussions will be limited to the formations cited above. The beds strike NE–SW and dip approximately 50 feet per mile to the northwest. The section is about 2,200 feet thick and the time involved in deposition probably was somewhere between 3 and 10 million years, perhaps nearer the former than the latter.

Certain major environmental features of this area during the Early Permian are fairly evident. The sediments were deposited on a delta that was margined by the sea to the south and west and by land to the north and east. The delta appears to have been sufficiently limited so that climatic differences between one part and another would not have been great. The sea did not cover the area during the Clear Fork, although it had done so earlier in the Permian and transgressed again later. Temperatures appear to have been moderate, for the animals and plants could not have otherwise survived. Rainfall was sufficient to support ferns, seed ferns, equisitales and conifers as well as to supply streams and ponds in which various fish and amphibians lived. The climate may be roughly compared to that of the same area in north central Texas today with, apparently, less seasonality and greater rainfall during the early part of Clear Fork time.

Study of more subtle aspects of the environment and the changes that occurred with time entail detailed analyses of intricately interrelated physical and biological evidence. Space permits only a summary statement of conclusions. Four principal physical subzones, which provided habitats for vertebrate animals, may be recognized: streams, ponds and lakes (hereafter referred to as ponds), pond margins, and flood plains and divides. The last mentioned will be referred to as uplands for the sake of simplicity. These subzones are not, of course, to be considered as completely separate units, for each makes marginal contact with the others and animals certainly would have tended to pass beyond the limits of a single subzone. Rather, they should be thought of as the principal, but not necessarily limiting, habitats of the various animals.

[2] This map is reproduced from Olson (1951a, fig. 1) with permission of the University of Chicago Press.

Each of the subzones can be recognized in the field from physical criteria. These include both gross features of the sedimentary deposits such as attitudes of beds, lateral continuity, etc., and internal characteristics of beds such as the size and arrangement of grains and composition. Pond deposits, for example, are characterized by gently dipping marginal beds, generally ovoid or elliptical outline, small size of particles and uniformity of particle size in a given bed, as well as color differences from surrounding sediments that reflect, in part, compositional differences. Deposits laid down in other subzones have equally distinctive features.

A brief survey of the environmental changes during the Clear Fork is given in figure 2. The sedimentation in the Arroyo was dominantly on flood plains. A thin but extensive series of red shales, silts and sands, deposited near the end of the Arroyo, indicates a marked change in environment. These deposits appear to have been laid down when the delta was extremely low. Of the four subzones, only the ponds appear to have remained habitable during this time. It is probable

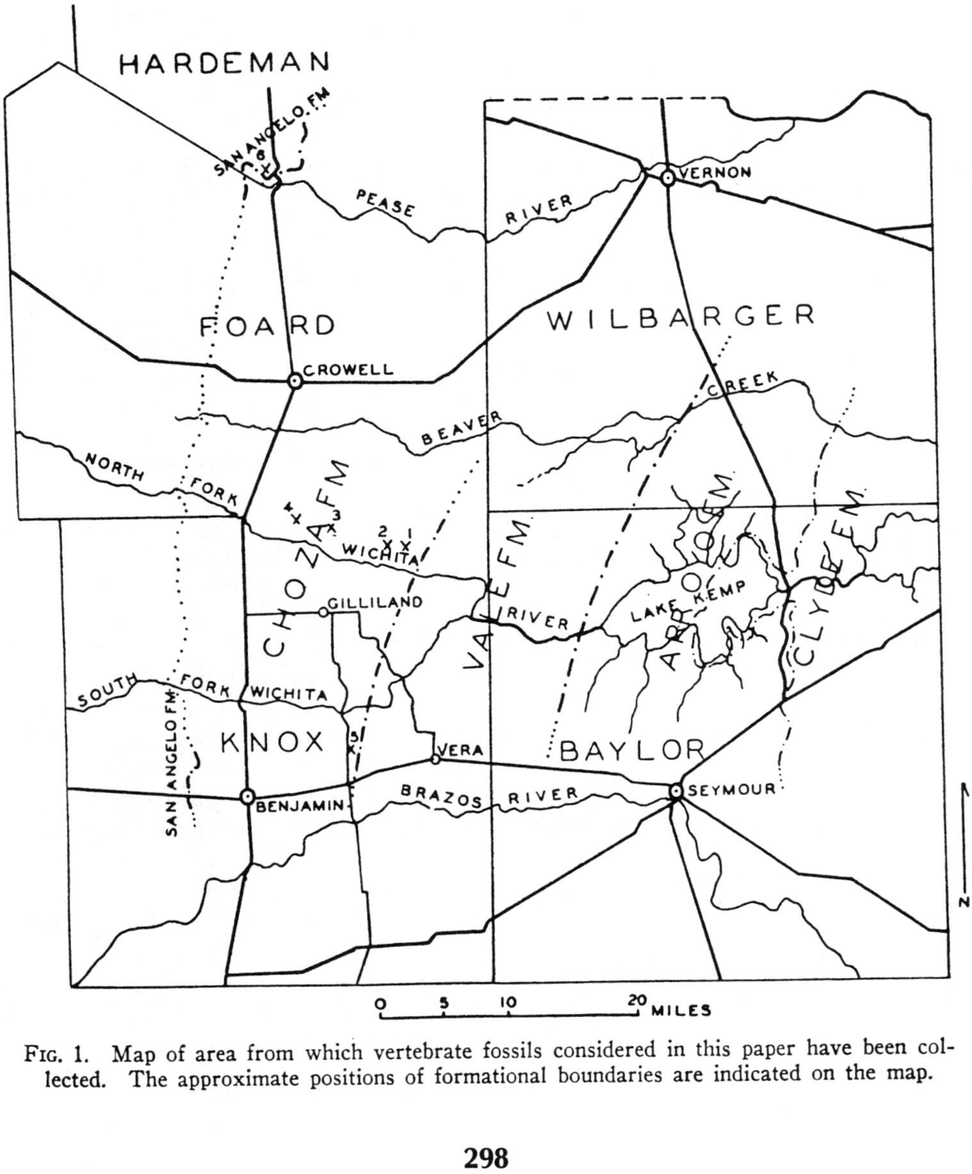

FIG. 1. Map of area from which vertebrate fossils considered in this paper have been collected. The approximate positions of formational boundaries are indicated on the map.

FM	Sedimentation	Topography	Rainfall
2200′			
	Extensive, even clay, sand and evaporite beds. Few channels and ponds. No known life record.	Same	Low rainfall
Choza	Predominantly flood plains. Channels small, few, widely spaced. Evaporite basins markedly increased. Some Playa, fresh water ponds.	Local relief low. Land near sea level.	Continued decrease in total rainfall.
	Few channels, not clay-pebble type. Initiation of evaporite beds. Scattered fresh water ponds.	Local relief moderately low. Basins with fresh water and evaporite deposits.	Decrease in total. Not torrential. Seasonality persistent.
1200′			
	Channel and lag deposits all of clay-pebble conglomerates, sediment locally derived. Ponds scattered, predominantly persistent. Ponds show seasonality in nature of deposition.	Local relief increased.	"Monsoonal" type fully developed. Torrential seasonal rains.
	Same, but beginning of clay-pebble conglomerates, derived from local sediments.	Local relief same. Source area high.	Initiation of "Monsoonal" type.
Vale	Marked channel development. Coarse, dipping, marginal flood plain deposits. Few but extensive ponds. Coarse materials of streams derived from source area.		
	Initiation of large stream channels with conglomeratic deposits. Ponds widely spaced.	Moderate local relief. Marked increase in relief in source area of sediments.	Increase in seasonality in source area of streams.
700′			
	Even red shale, with broad linear belts of silt and fine sand. Deposited by very slowly flowing water. No definite channels and few ponds.	Local relief very low. Area near sea level. Possibly tide-water estuaries.	Same.
Arroyo	Slight increase in ponding, especially temporary ponds.	Same	Slight increase in seasonality.
	Flood plain deposits predominant, a few scattered ponds. Divides low, streams small.	Local relief low. Source area of streams low.	Moderate, evenly distributed throughout year.

FIG. 2. Chart of environmental conditions during the Clear Fork. The thicknesses of formations are only approximate.

that the water courses became somewhat saline as a result of tidal effects. Local relief appears to have been low both in the delta and on adjacent land during the Arroyo. Rainfall appears to have been moderate and rather evenly distributed throughout the year.

The Vale was marked by a rapid increase in development of major stream channels, as indicated by deposits of conglomerate and coarse, dipping marginal flood plain deposits. Clay-pebble conglomerates became dominant in the upper half. Relief appears to have increased markedly, first in the non-deltaic areas and then on the delta. Sedimentation suggests that seasonality of rainfall increased, with eventual establishment of monsoonal conditions.

The Choza opened with rapidly decreasing local relief and by mid-Choza the surface of the delta lay near to sea level. Rainfall decreased, with seasonality persisting, to produce rather arid conditions accompanied by development of evaporite basins. Such basins are formed where deposition of salts takes place from waters that have become highly saline through evaporation.

The Chronofauna and Its Subzonal Assemblages

The chronofauna considered in the analyses that follow includes genera and species for which information on identity, frequency, and distribution is sufficient to be meaningful. Vertebrates not specifically considered certainly played a part in some of the changes witnessed, but it is impossible to evaluate their importance. Twenty-one genera and thirty-two species of fish, amphibians and reptiles are included. The genera are listed in figure 3 along with stratigraphic data for each. Environmental distributions and the habits of the animals are more important in the present considerations than classification of the genera into appropriate higher categories, and the latter will not figure prominently in the discussions.

Fish are represented by a fresh water, predaceous shark, and by lung fish. It is certain that many of the waters were inhabited by small fish, probably palaeoniscoids for the most part, and that these fish formed an important part of the diet of other water living animals. Both labyrinthodont and lepospondylous amphibians figure prominently in the fauna. Reptiles are all relatively primitive types, being distributed among the Seymouriamorpha, Diadectomorpha, Captorhinomorpha, and Pelycosauria. Only passing reference will be made to the plants, which have a fair fossil record, and invertebrates, which have almost no record. It is not to be supposed, however, that they are not considered important in the evolution of the vertebrates.

The problem of analysis of the interrelationships of the species in the various environmental subzones and between subzones is intricate. As a first step in analysis, the assemblage of each subzone will be examined without reference to environmental interrelationships.

Upland Subzone: This subzone, alone among the four, is thought to have continued with little change into the land areas adjacent to the delta. It is believed, furthermore, that the inhabitants occupied the subzone, insofar as their particular niches were available, throughout its geographic extent. Genera known from equivalent deposits in Oklahoma tend to support this conclusion. The most important points concerning this subzone are as follows:

1. *Dimetrodon gigashomogenes* is present in the Arroyo, Vale, and lower and middle Choza. *Captorhinus aguti* occurs from the base of the Arroyo through the upper Vale. There are traces of the genus in the lower and middle Choza, but remains are so fragmentary that species determination is uncertain. No differences from *Captorhinus aguti* have been noted and the specimens have been tentatively referred to this species. The only break in stratigraphic continuity of *Dimetrodon gigashomogenes* and *Captorhinus aguti* is at the red shale zone at the top of the Arroyo.

2. Two species, *Dimetrodon major* and *Diadectes tenuitectus,* disappear by the end of the Arroyo, both diminishing in numbers prior to their disappearance.

3. A new, larger species of *Captorhinus* (possibly a new genus) unnamed as yet, appears in the Vale in the same subzone as *Captorhinus aguti.* It is probable, on the basis of size and dentition, that the dietary habits of the two were somewhat different. This species has not been found in the Choza.

4. *Casea broilii* appears in the record at the beginning of the Vale. Two spe-

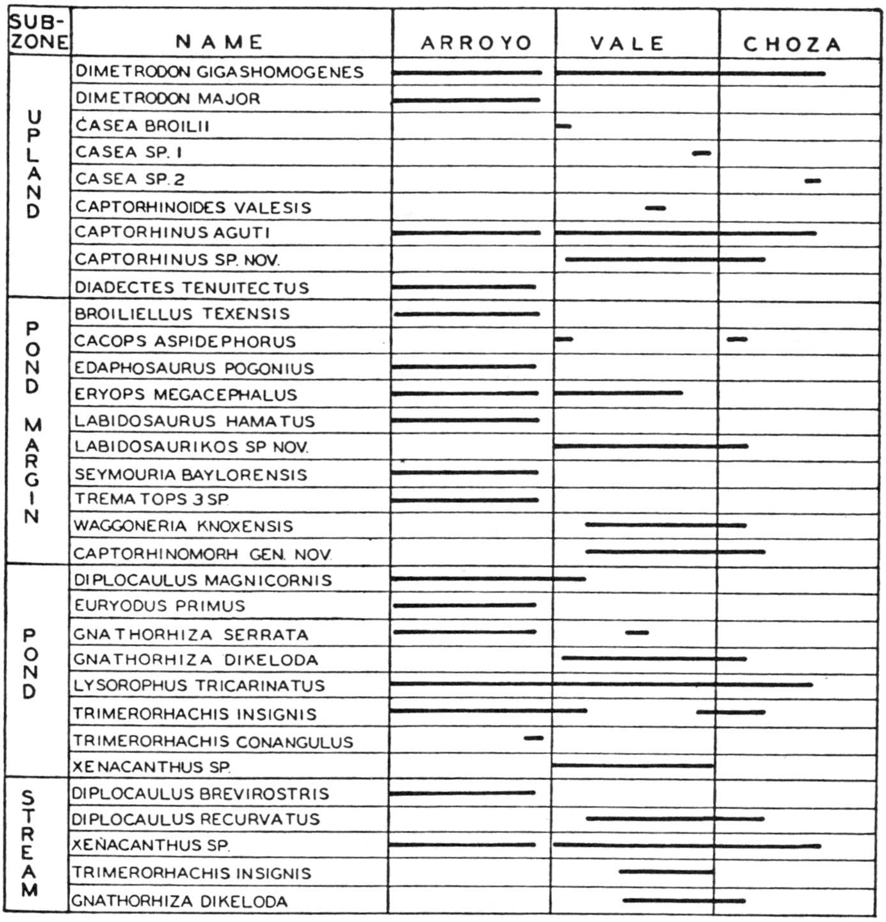

FIG. 3. Stratigraphic and subzonal distribution of genera and species of vertebrates during the Clear Fork.

ies, successively larger, are found, one in the upper Vale and the other in mid-Choza.

Pond Margins. This environmental subzone is believed to have been largely, if not entirely, restricted to the delta. The following points are of particular importance:

1. Only one Arroyo species, *Eryops megacephalus,* persisted into the Vale. It appears to have diminished greatly in numbers from the mid-Arroyo on, and is unknown above the base of the upper Vale. Two species of captorhinomorph reptiles, one seymouriamorph and one dissorophid amphibian appear in the subzone during the lower Vale.

Ponds. Ponds and lakes of the types encountered are not known to have existed beyond the limits of the delta. The following items are of especial significance.

1. The pond species that persisted into the upper part of the Arroyo consistently passed the Arroyo-Vale boundary without change.

2. These species failed to persist in the pond environment, except the burrower, *Lysorophus tricarinatus,* beyond the lower part of the middle Vale.

3. *Trimerorhachis insignis* reappeared in the upper Vale in the ponds. It is not known to have been associated with *Xenacanthus* at this level.

4. *Xenacanthus,* a predaceous shark, is found in ponds in the lowest Vale. It is associated with *Diplocaulus magnicornis* and *Trimerorhachis insignis,* aquatic amphibians. It persisted in this environment until the upper Vale, but it is not present in all of fossiliferous pond deposits of that time. It has not been found in ponds of Choza age, although it is common in the stream deposits.

5. A large species of lung fish, *Gnathorhiza dikeloda,* appeared in pond deposits shortly after the beginning of the Vale. It seems to have been derived from

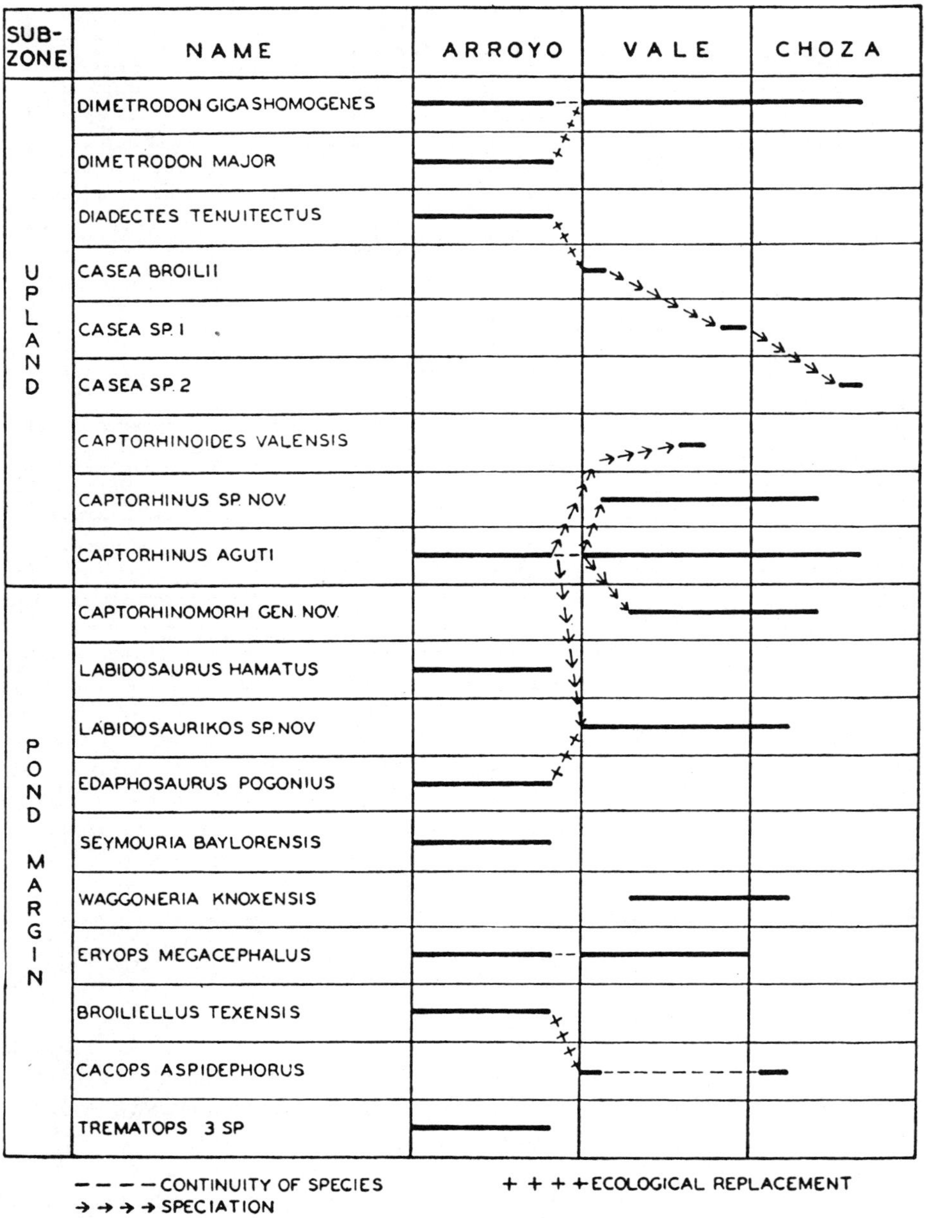

FIG. 4. The genera and species of the upland and pond margin subzones showing the histories and interrelationships during the Clear Fork.

Gnathorhiza serrata, and to have coexisted with this species until the beginning of the upper Vale.

Streams. Although there is no doubt that the streams of the delta were fed by those of adjacent, higher lands, the fate of the species inhabiting the deltaic streams strongly suggests that there were no effective migrations between the nondeltaic and deltaic phases of the streams. Pertinent items are as follows:

1. *Diplocaulus brevirostris* and *Xenacanthus* sp. are the only well known inhabitants of the streams during the Arroyo. Coprolites of the latter contain scales of paleoniscoids, indicating that these small fish were a source of food.

2. There is a break in the record of stream deposition during the time of deposition of the upper Arroyo red shale.

3. *Diplocaulus brevirostris* did not persist beyond the red shale zone.

4. *Diplocaulus* sp., *Trimerorhachis insignis* and *Gnathorhiza dikeloda* came to inhabit the streams during the Vale. The first was derived from *Diplocaulus magnicornis,* a typical pond animal of the Arroyo. *Trimerorhachis insignis* was a common pond form in the Arroyo, not known from stream deposits. It is abundant in stream deposits in the Vale and very rare in ponds. *Gnathorhiza dikeloda,* derived from a pond species of the Arroyo, was successful in the streams of the Vale and *Xenacanthus* continued to be present in the streams in large numbers.

Interpretation of the Evidence

The most vital part of the basic data upon which analyses of the evolution of the vertebrate assemblages of the delta may be based has now been briefly summarized in the figures and text. To reduce the complexities that are involved in treating the different subzones and their changing occupants together, analyses will be made by citing, and illustrating by examples, a series of points under two general headings: (I) the persistence of species, and II patterns of modification.

I. The persistence of species:

1. In the same subzone during a period of time that witnessed changes both in physical and biological factors of environment.

Dimetrodon gigashomogenes and *Captorhinus aguti,* both of the upland subzone, illustrate this pattern (fig. 4). They both persisted from the beginning of the Arroyo into the Choza without detectable morphological change. During this time, topography, rainfall, both in amount and seasonal distribution, and the associated animals of the subzone changed. The two species appear to have lived outside the delta during the Arroyo-Vale transition. In spite of these events, the two species persisted and their success is attested by the relative abundance, particularly in the case of *Dimetrodon gigashomogenes,* in rocks deposited throughout much of the period of time involved.

2. In transition from one subzone to another.

Persistence of species under this circumstance is best illustrated by *Trimerorhachis insignis,* an aquatic amphibian, in its shift from ponds to streams early in the Vale and its reappearance in large numbers in ponds in the upper part of the formation (fig. 5). *Gnathorhiza dikeloda* and probably *Xenacanthus* sp. represent other cases. Both stratigraphic and biological evidence concerning *Trimerorhachis insignis* has been derived from an extensive series of observations. Statistical analyses of reasonable samples of skulls, vertebrae and limb bones have revealed no significant differences in the animals involved in the transitions cited. Whatever changes may have occurred in physiology or soft anatomy without affecting the osteology cannot, of course, be determined.

3. In a special, limited environment.

The highly specialized, elongated, burrowing, aquatic amphibian, *Lysorophus tricarinatus* persisted without apparent change from near the base of the Arroyo through the lower one-third of the Choza (fig. 4). It is abundant throughout, suggesting consistently large populations. It lived, during this time, in ponds, most of which appear to have been dry during part of the year. Specimens are found coiled and in concentrations in most instances, suggesting that many died while in burrows. Although association with other species is uncommon, uncoiled specimens have been found in pond clays along with other aquatic vertebrates. Presumably these met death during non-aestivating phases of their existence. Apparently the associated forms, different at various times during the Clear Fork, did not exercise selection pressure sufficient to initiate any evolutionary changes in *Lysorophus tricarinatus.*

II. Patterns of modification:

1. Phyletic evolution of a species in a single environmental subzone.

Evidence for this type of change is found in the rather meager remains of the genus *Casea* (fig. 4). The three species, lower Vale, upper Vale, and mid-Choza, respectively, show rapid increase in size, and accompanying proportional modifications of various skeletal parts. There appears to be a drift toward larger size, probably through selection of larger members of successive populations. *Casea* was herbivorous and size increase was, perhaps, related to the advantage that size would give in warding off the attacks of carnivores. It is interesting to

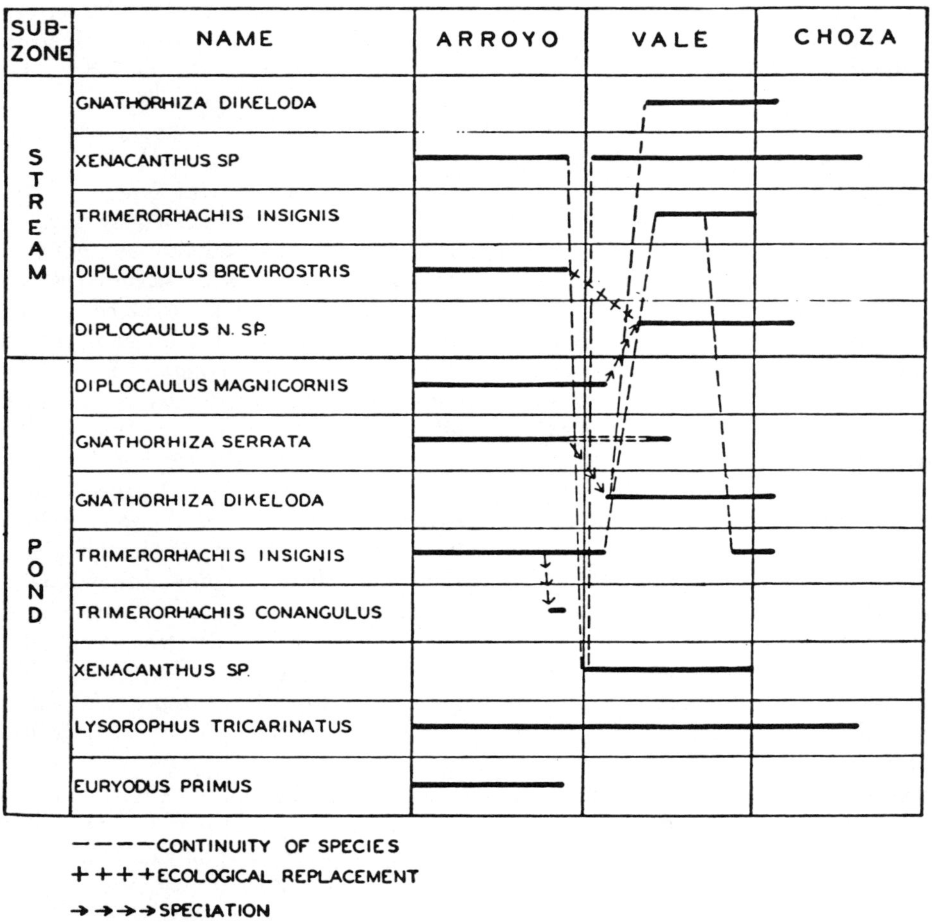

FIG. 5. The genera and species of the stream and pond subzones showing the histories and interrelationships during the Clear Fork.

note that there is no commensurate increase of size in the principal large predator of the upland zone, *Dimetrodon gigashomogenes*.

2. Numerical decrease and final disappearance of species during a period of time marked by no recognizable pronounced change in biological or physical environment.

This pattern is not infrequent in paleontological studies. It is best shown in the fauna under consideration by *Diadectes tenuitectus* (fig. 4). There is no evidence that the diminution in numbers and extinction were dependent upon outside physical or biological factors. If this is accepted as showing that there were, in fact, no external influences, the decrease in numbers and extinction must have resulted from internal factors. This is, of course, not out of the question and such factors have been appealed to in many instances to explain extinctions. There is, however, considerable danger in drawing conclusions from such negative evidence, especially in cases such as this in which it is inevitable that various factors that might have been operative are not likely to have left a clear record. At present it seems possible only to note that *Diadectes tenuitectus* disappeared and, so far as is known did not reappear, with reasons for the presumed extinction unknown.

3. Environmental change resulting in unoccupied environmental niches.

The evidence that this phenomenon occurred is drawn primarily from events affecting the pond margin subzone. With onset of very low lands at the end of the Arroyo, this subzone appears to have disappeared. Obviously, since the ponds themselves continued to exist, pond margins were physically present. Their existence as a life subzone comparable to that existing during the Arroyo, however, probably was dependent upon a balance with the pond subzone on the one hand and the upland subzone on the other, for both physical and biological features were somewhat transitional. With the disappearance of the upland subzone, this balance was destroyed, and, it would appear, the pond margins were unable to support the type of organisms that had previously found them suitable. Temporarily, thus, this subzone and its niches became impoverished. The failure of the species of the subzone to reappear when the habitat was reestablished in the lower Vale indicates that no refuge areas existed beyond the limits of the delta; that is, the subzone was confined to the delta. Thus, when the subzone did reappear, it offered open ecological niches. The same applies, in part, to the other subzones but the pattern of occupancy tends to be less clear. The series of events just described is related to points 4 and 5 considered below.

4. Speciation related to occupancy of niches and subzones not occupied by the parent species.

This and the next point concern modes of occupancy of vacant niches in the various subzones, created as indicated in the previous point. The most striking case is the development of new species and genera of captorhinomorph reptiles both in the upland and pond margin subzones (fig. 4). *Captorhinus aguti* appears to have been the source of this radiation, but, of course, this cannot be conclusively demonstrated. On the basis of its morphology it could have occupied this position, something not true of any other known species. Temporally it is appropriately placed. The assumption is therefore made that it represents the actual ancestor. Changes were principally in size, proportions of skeletal parts, and dentition. Within the subzone occupied by *Captorhinus aguti* arose a new species of *Captorhinus,* or possibly a new genus, as yet unnamed. Presumably isolation related to occupancy of an ecological niche, different from that of the parent species, occurred, and the principal changes seem to have been related to diet. Adults were about one-third larger than the largest adults assigned to *Captorhinus aguti.* A new genus in the same subzone, *Captorhinoides valensis,* differs from *Captorhinus,* most importantly, in auditory structure (Olson, 1951a). The adaptive significance of this change is obscure. Two genera, *Labidosaurikos* and an unnamed genus, developed in this radiation, invaded the pond margins in the lower Vale, occupying the niches left open by the depletion of the Arroyo occupants, and became fairly abundant in this subzone by mid-Vale. Both are characterized by highly modified dentitions, derivable from the *Captorhinus* pattern, and moderately large size. Both appear to have been herbivores.

This radiation occurred during a short period of time. Intermediate stages between *Captorhinus aguti* and the genera and species believed to have been derived from it are not certainly known.

5. Occupancy of ecological niches of a subzone, impoverished by temporary change and then re-established, by species unable to penetrate the delta when the subzone was inhabited.

This appears to have occurred in a number of instances in which animals that had lived in non-deltaic areas found suitable habitats in niches of the delta that developed after the time of the deposition of the red shale near the end of the Arroyo (figs. 4 and 5). Such animals were not, so far as known, derived from species existing on the delta during the Arroyo, and it is thus inferred that they, or their ancestors, lived elsewhere, presumably in the higher lands away from the delta. *Waggoneria,* a seymouriamorph, *Cacops,* a dissorphid amphibian, and *Casea,* a pelycosaur, are examples. This phenomenon has not been recognized in the stream or pond habitats, probably because of the almost immediate occupancy of the environment by predaceous sharks in the first instance and the persistence of the habitat across the Vale-Arroyo boundary in the second.

6. Modifications of faunules, migrations, and speciation related to occupancy of a subzone by a species not previously represented.

This point concerns a rather complex series of events and involves an intimate knowledge of a complex and its physical environment for interpretation. For illustration, one such series pertaining to the ponds and streams will be summarized (fig. 5).. *Xenacanthus,* the predaceous shark, inhabited streams but is unknown from the ponds during the Arroyo. It appeared in the ponds in the lowest part of the Vale. This transition is thought to have resulted from the pressure exerted by the virtual disappearance of the stream habitat, perhaps as a result of tidal penetration, during the red shale interval. During the Arroyo, *Xenacanthus,* was associated with *Diplocaulus brevirostris,* a stream-dwelling species of aquatic amphibian, but not with *Diplocaulus magnicornis* or *Trimerorhachis insignis,* Arroyo pond dwellers. *Diplocaulus brevirostris* disappeared during the deposition of the red shale and did not reappear. *Xenacanthus,* on the other hand, is found in lower Vale pond deposits in association with *Diplocaulus magnicornis, Trimerorhachis insignis,* and the lung fish, *Gnathorhiza dikeloda.* Soon after this association had been realized, *Diplocaulus* and *Trimerorhachis* disappeared from ponds only to appear in the streams. It is believed that these amphibians were unable to persist in the quiet pond waters in association with the predaceous shark, but were able to meet this competition in the streams. *Trimerorhachis* shows no change in the course of transition but *Diplocaulus* is represented by a new species in the streams.

The "horns" of the somewhat triangular skull of *Diplocaulus* are strongly recurved in the new species, a character that may be called "crook" for ease of reference. In this character, the skulls of the new species somewhat resemble those of the stream species of the Arroyo, *Diplocaulus brevirostris.* All other characters of the skull, however, resemble those of the pond species of the Arroyo, *Diplocaulus magnicornis,* very closely, so that there can be no doubt that this was the source of the new species (Olson, in press). The character of crook does occur in *Diplocaulus magnicornis* with a frequency of 16 in 100 (based on a sample of 50). Field evidence indicates that this character was randomly distributed throughout the population both in space, at any particular time, and during the time through which the species is known to have existed, i.e., the time represented by the deposition of the Arroyo formation. The distribution of crook and straight "horn" in the sample of *Diplocaulus magnicornis* is suggestive of a morphological expression of equilibrium attained under conditions expressed by the Hardy-Weinberg law.[8] It seems probable that population size and changing environmental conditions, offering some selective advantage, would have vitiated complete conformity to this principle. In spite of these factors, the equilibrium concept is valuable in an interpretation of the population structure as it has been observed.

The character of crook is present in all known specimens of the new species of *Diplocaulus* from the Vale in which the postero-lateral part of the skull is preserved (about 20 individuals). The presence of this character in *Diplocaulus magnicornis* gives, in one sense, a partial preadaptation to stream life. It appears that the members of the population of *Diplocaulus magnicornis* that possessed this character were the source of the new species. They had the ability to penetrate streams, and thus to become isolated during a time when existence of the genus in the ponds became precarious and eventually impossible.

Whereas the two amphibians left the pond environment, the lung fish, *Gnathorhiza,* seems to have met the onslaught of *Xenacanthus* successfully. It is possible that *Gnathorhiza,* presumably a bottom feeder, did not come into direct competition with *Xenacanthus.* The Arroyo species of *Gnathorhiza, G. serrata,* which is rare throughout the Arroyo, is known to have crossed the Arroyo-Vale boundary, for it occurs in the middle Vale. It has been found in pond deposits. In the lower Vale, another species, *G. dikeloda,* larger in size and with more highly developed sectorial teeth, appeared in the ponds (Olson, 1951b). It is thought to have arisen from *Gnathorhiza serrata,* the only possible known source, but intermediates have not as yet been identified. This appears to be a case of speciation induced by the introduction of a new predator in which marginal survival of the old species continued while the new species not only thrived in the initial subzone, presumably in a different ecological niche, but was able to penetrate a new subzone, the streams.

7. Interdependence of faunules of two or more subzones.

No doubt such relationships existed in a complex fashion between all adjacent subzones and their various subdivisions. Only one of several known instances in the Clear Fork will be cited to illustrate the part that this interdependence may play in the evolution of the chronofauna (fig. 5). It has been noted that the assemblage of the pond margins of the Arroyo was composed, in large part, of carnivores, *Edaphosaurus,* being the exception. The food source must have been in part the animals of the ponds. It has also been shown that, shortly after the beginning of the Vale, the pond assemblages changed markedly with the result that potential prey for marginal predators was greatly reduced. The captorhinomorph radiation and immigrations that took place early during deposition of the Vale, repopulating the pond margins, produced a great preponderance of herbivores. The only known carnivores are *Eryops megacephalus,* a species that persisted from the Arroyo, and *Cacops,* an amphibian adaptively very similar to *Broiliellus.* Both are low in frequency of occurrence. It seems justifiable to infer a causal relationship between the reduction of potential prey in the ponds and the preponderance of herbivores in the newly developed faunule of the pond

[8] See Stern (1943) for a brief exposition of this concept.

margins. This chain of events and reasoning may be carried further to include the effects of the change of the marginal animals on those of the upland, for again there is a predator-prey relationship, and to the part that this played in modifications of the assemblage of the streams. What has been said, however, will be sufficient to indicate the type of evidence that is available and one instance in which an interpretation has been made.

8. Disappearance of a chronofauna with gradual but intense changes in physical environment.

Heretofore, attention has been directed to patterns of change that did not alter the basic characteristics of the chronofauna, cases in which physical changes were not sufficient to have drastic effects. The present case is the story of the Choza, the last page in the history of the Clear Fork chronofauna on the delta. Physical changes were primarily a continued lowering of the land and decrease of rainfall to the point of incipient aridity. Fossil remains tend to occur increasingly in local concentrations, either in deposits laid down in streams or ponds. The number of ecological niches available to vertebrate life appears to have been successively reduced, with elimination of some species and intense competition between those whose niches persisted. Changes appear to have been such that for the most part they could not be met by speciation; in only one line, that of *Casea,* does a new species appear. This represents continuation of a tendency toward larger size initiated earlier. The only other evidence of change is the appearance of a new eryopid amphibian of somewhat uncertain affinities. Thus, under changes that exceeded the tolerance of many species, there seems to have been extinction rather than modification to meet the changes. In the mid-Choza, all known specimens of animals and plants have come from ponds, their margins, and deposits of small channels leading into these ponds. Only a few such deposits are known in the upper one-third of the Choza and they are unfossiliferous. To the best of our knowledge the vertebrates had gone from the delta. It seems improbable that those elements that were strictly deltaic in habitat left direct descendants, unless it were possible for them to invade the non-deltaic areas either with or without change. The events noted at the Arroyo-Vale boundary suggests that this would have been improbable in most cases.

The only later vertebrate fauna known from the Permian of North America comes from the San Angelo formation that overlies the Clear Fork disconformably. As yet this recently obtained fauna has not been studied in detail. No trace of any Clear Fork genus, except *Xenacanthus,* has been encountered.

Conclusions

What follows is an attempt to generalize from the specific problems considered previously. An element of speculation, of course, enters in since the data are incomplete. The generalizations concern patterns of interaction within a chronofauna. To establish the position of this level, I will allude briefly to a discussion by Dobzhansky (1941, pp. 152–154) in "Genetics and the Origin of Species" in which he differentiates the factors active at two levels of organization: the physicochemical principles involved in the origin of variation and the principles of physiology of populations at the population level of organization. The basis of the present study, is an extension to a third, or ecological level, that of the chronofauna involving the interaction of populations. The behavior of individual populations is injected into this field of action, but basic factors are different from those concerned exclusively with populations. Certain of these factors are outlined and discussed in "Principles of Animal Ecology" by Allee, Emerson, Park, Park, and Schmidt (1949).

It has been shown that various species had different histories during the time represented by the Clear Fork. The course followed by each is the function of a complex series of variables, some of which can be recognized and evaluated, as noted in the preceding sections, others that can only be surmized and some, undoubtedly, that remain undetected. Most important is the point of view that all species are interrelated and that the constitution of the complex of animals at any given moment is the resultant of the patterns of interrelationship. This concept introduces additional variables that relate to interdependence and interactions of populations, particularly as they modify the significance of such items as population size, variability of species, ecological niches, environmental subzones, etc., in contrast to the way that they are usually viewed when single populations are considered.

In the preceding paragraphs an effort was made to interpret interactions of species and their environments whenever they could be related to positive events. From these interpretations we may now attempt a synthesis. Points that seem to derive from observations and interpretations that have been made are as follows:

1. A chronofauna tends to remain virtually unchanged as long as environmental conditions, or possibly factors of population structure unrelated to environment, are not modified sufficiently to result in the elimination of one or more of its component species. Such identity may be preserved in the face of changes that enlarge or restrict ecological niches and thus affect the population size of the occupying species. Minor changes in species may occur through phyletic change of a population within a particular niche without isolation or significant change in variability (see point 5 below).

2. Environmental changes that occur over a relatively short span of time may tend to result in modifications of structure through:

(1) Alteration of faunal balance by elimination of one or more species, rather than by initiation of adaptive modification within those populations whose niches are affected.

(2) Initiation of change in habitat of the part of a population that may be preadapted to another available habitat:

(a) The affected part may be virtually the whole population, in which event the shift may be made without marked morphological or physiological change or modification of variability (see for example, *Trimerorhachis insignis,* p. 10).

(b) It may involve only a fraction of the total population, the part that is morphologically preadapted (for example, the presence of Crook in 16 per cent of the sample of *Diplocaulus magnicornis,* see p. 14). In this event, the shift involves selection and reduction of variability. This can be recognized in fossil vertebrates only if selection involves osteological features.

3. Open ecological niches may be established through physical changes as follows: the niches, and their inhabitants, that exist when physical change begins, are destroyed as a result of the change. Re-establishment of physical conditions not greatly different from those that existed at the onset of change tends to result in re-establishment of potentially similar, but unoccupied niches. Such re-established niches tend to be filled rapidly by speciation of members of the chronofauna and immigration from adjacent environmental subzones. Re-establishment of balance can produce moderate changes in a chronofauna over a very short span of time.

4. Entrance of a new species into an environment, such as a subzone, occurs only:

(1) When an ecological niche is open, or:

(2) When a niche is very inefficiently occupied with respect to the usurper's potential.

Such an occurrence may disrupt the balance within a subzone. Through this avenue, the total balance may be affected sufficiently that considerable reorganization takes place within a chronofauna.

5. Phyletic change of the occupant of a particular ecological niche may occur when all niches of a subzone are occupied, as long as adaptation to the specific niche alone is perfected, without materially altering other parts of the chronofauna. Such change need involve neither isolation nor reduction of variability. Successive stages in a phyletic series, viewed discretely, correspond in all respects to species, as, for example, in the genus *Casea,* p. 11. In this sense phyletic change involves speciation. The mode of initiation of change, however, may be different from that in which virtually nontemporal isolation of parts of a single population occurs and initiation of change is directly related to this isolation.

Within these five points lie the principal factors of modification of a chronofauna that have been witnessed in the current study. Stability under the influence of slow and moderate physical change stands out. Evolution may occur under such circumstances but seems to have little effect upon internal balance. It can become important if it contributes to a significant modification of the environment of other populations, either by direct invasion of ecological niches or by indirect impingement upon the niches. This has not appeared as an important factor in the problem studied.

The elimination of parts of a chronofauna through environmental change and subsequent speciation or immigration by new or remaining elements into new and/or re-established subzones and niches are the most important factors in the modification of a chronofauna. Evolution of populations through successive small changes (that may tend toward adaptations to new circumstances as they arise) appear to have only a minor effect upon equilibrium. An extinction or migration of one or more species, which may result from a change in the environment that is inconsequential in its direct effect on most species can be effective in establishing a lack of balance and in producing subsequent drastic changes as equilibrium is re-established. When such an event occurs at a time when changing conditions are tending to produce an environment that approaches the limits of tolerance of an appreciable number of species, extinction of much or all of the chronofauna is a probable result. This probability follows logically from the concept of interdependency of species and the apparent inability of many species to adapt to extensive change over a short period of time. The events of the Choza discussed earlier in this paper provide illustration of this series of events.

It is tempting to speculate beyond the limits of this case and to suggest that widespread physical change, as appears to have occurred at various times in the geological history of the earth, might initiate loss of equilibrium over widespread areas. Extinction of many species would be expected as chains of interrelationships were destroyed by the loss of a few. Most likely to succeed under these circumstances and to provide a basis for re-establishment of equilibrium would be those populations that in one way or another were preadapted to circumstances not dependent upon the particular linkages with which these organisms previously had been associated. There would result new chronofaunas, very different from those that existed prior to the initiation of the changes.

Summary

An analysis of the interrelated changes of the species of a chronofauna with time and altering environmental conditions has been based upon studies of an Early

Permian fauna of fish, amphibians and reptiles from beds of Clear Fork age of Texas. This chronofauna evolved with only minor influences of migration from adjacent areas. It existed on a delta that underwent modifications in topography and rainfall during the period of time studied.

Four ecological subzones inhabited by the populations of the chronofauna have been recognized from physical criteria: streams, ponds, pond margins, and flood plains and divides, called uplands. The changes of the subzones and their inhabitants through the Clear Fork show the importance of considering evolution from an ecological viewpoint. Stability of certain species and changes of others with respect to habitat and/or morphology are related to biological and physical events correlated with interrelated courses followed by populations under environmental change.

The general tendency for stability of a chronofauna under moderate environmental change stands out. As long as the various ecological niches are effectively occupied little evolution occurs. Conversely, changes that have a marked effect upon but one or a few species may initiate a chain of events that can alter the internal balance radically and result in rapid evolution. Moderately drastic changes of physical environment do not appear to have been met by adaptive modification but rather to have set in motion a series of events that led to the extinction of the chronofauna through elimination of several species essential to the maintenance of its internal balance.

References

Allee, W. C., Emerson, A. E., Park, O., Park, T., and Schmidt, K. P. 1949. Principles of animal ecology. W. B. Saunders Co., Philadelphia. 803 pp., 263 figs.

Dobzhansky, T. 1941. Genetics and the origin of species. Columbia Univ. Press. New York, N. Y. 428 pp., 24 figs.

Olson, E. C. 1951a. Vertebrates from the Choza formation, Permian of Texas. Jour. Geol., **59**: 178–181, fig. 1.

——. 1951b. Fauna of the upper Vale and Choza: 1–5. Fieldiana, Geology, Chicago Nat. Hist. Mus., **10**: 89–128.

——. (In press.) Fauna of the upper Vale and Choza: 6. *Diplocaulus.* Fieldiana, Geology, Chicago Nat. Hist. Mus.

Stern, C. 1943. The Hardy-Weinberg Law. Science, n.s., **97**: 137–138.

21

Reprinted from *Systematic Zoology* **15:**261-278 (1966)

The Extinction of the Multituberculates

LEIGH VAN VALEN AND ROBERT E. SLOAN

Abstract

The multituberculate mammals declined in numerical abundance from the late Cretaceous to their extinction in the late Eocene. Their maximal diversity, however, occurred in the late part of the middle Paleocene, suggesting an increase in specialization in the face of competition from placental mammals, which diversified very rapidly in the Paleocene. Various lines of evidence lead to the conclusion that first condylarths, then primates, and finally rodents contributed to the gradual extinction of multituberculates, at least in part by evolution in the same region of competing resource requirements. A phylogeny of the Cenozoic Multituberculata and their ancestors is given, the mammal faunas of the Paleocene are listed by more or less their absolute ages, and the fauna of Bear Creek is analyzed in detail. A theory of replacement of supraspecific taxa is presented and applied.

Introduction

The multituberculates were a group of mammals phyletically far from the marsupials, placentals, and monotremes; they are placed in a subclass of their own (unless, expanding a suggestion of Van Valen, 1960, the Therapsida are enlarged to include all non-therian mammals) and have no close living relatives. In the Paleocene as well as earlier they were the most abundant and varied order of mammals. Their extinction in the Eocene is one of the more dramatic events in the history of mammals. Several authors (Jepsen, 1949; Wilson, 1951; Simpson, 1953; McKenna, 1961; Wood, 1962) have proposed that the extinction of the multituberculates in the Eocene was due to competition with rodents, which first diversified at that time and are similar to the multituberculates in size and probable broad adaptations. Landry (1965) has recently argued against this view. We believe that Landry's presentation is deficient ecologically and paleontologically, and will give the evidence now available on the problem in some detail. Landry ignored several pertinent paleontological references cited in the present paper; our differences on ecology will be apparent from the following discussion.

Theory of Replacement

In order for two species to be in competition, it is a minimal requirement that they share a common resource that is in short supply to one or both species. For the operation of competitive exclusion, it is further necessary that this resource limit the population density of the species that is inferior in the competition. This limitation need not be density-dependent if extinction of local populations is prevalent. "Resource" can be regarded as food in the present discussion, and the sufficient conditions for competitive exclusion are rather more complex than the two necessary ones just given. It is not necessary that the species be similar or interact in any other respects than the two given. For example, fungi and grasshoppers can compete with mice and men for grain, although there are also other ecological interactions among these species and most have alternate foods. Contemporaneous evolving mammals could have caused the extinction of the much larger dinosaurs by competition, although other explanations are also possible. Stone et al. (1957) report competition for fruit between *Drosophila* and *Rattus* on Eniwetok.

The notion of niche or adaptive zone can be made relatively precise. To do this, consider the various kinds of resources of a species: nourishment, space, dispersal or pollination agencies, or whatever a particular species requires, and also the variables of the physical environment: temperature, moisture, radiation, soil slope, substrate composition, wind or current velocities, temporal variations in all of these, etc. Competitors, predators, and parasites are explicitly excluded. Then, mainly following Hutchinson (1957), a hyperspace is constructed in which each relevant variable determines a dimension. Along each axis thus determined, the maximum and minimum (if both exist) values in which the species can exist are then plotted. The hyper-rectangle formed by the intersection of these values contains the segment of the environment in which the species can survive. The existence of correlations and interactions among the variables makes the actual viable segment smaller than this rectangular one, and its actual shape will be irregular, with outpocketings and indentations at various places. Nourishment and any other subdivisible variables can be split into separate dimensions to any degree that may be ecologically relevant. When this is done, the final irregular boundary determines the segment of the resource space in which the species can survive, i.e., its niche. An additional dimension can be added to indicate the fitness of the species in each such environment, as was first done by Wright (1932) for a genotypic hyperspace and by Simpson (1944) for a phenotypic one. The adaptive zone of a supraspecific taxon is equal to the union of the niches of its component species; again an additional dimension is possible for the mean fitness at each point in the adaptive zone. The concept of a resource hyperspace or topography is also applicable to the different ecological requirements of different genotypes and phenotypes within a species.

When two taxa enter into competition with each other for some density-limiting segment of a common resource space, there are several possible outcomes. (1) One taxon may retire from the competition by ceding the resource to the other, either by specialization in a previously broad adaptive zone or by expansion or shift of a previously narrow adaptive zone. (2) One taxon may be eliminated by the other. (3) The competition may continue indefinitely because at least the weaker competitor has an alternative form of the limiting resource that the other competitor cannot reach (or is inferior in). (4) The competition may continue indefinitely because of alternating dominance of the two taxa, caused by more or less cyclic frequency-dependent selection for competitive ability (Pimentel et al., 1965, and references therein). It has not been shown that an equilibrium (as opposed to a stable or semistable pattern of fluctuation) would be established in this case, as Pimentel believes. (5) The segment of resource space may be partitioned between the two taxa, ending the competition. These alternatives are not all sharply distinct or mutually exclusive. For example, (5) may be followed or accompanied by (3), which may be followed by (2) because of genetic or environmental change. Geographical extension of the taxa permits more varied responses but not different total results from the ones given.

These alternatives apply to two supraspecific taxa as well as to two species. For supraspecific taxa, the structure of the response is likely to be more complex than in the case of species, because more reproductively isolated units are involved, and for any segment of the resource space the several competitively superior species may not all belong to the same higher taxon. Probably the most likely alternatives are (2) and (5). In the latter, each higher taxon becomes (or remains) superior to the other in some part of the resource space, with any species that happens to be in the zone of the other taxon having a greater probability of extinction without replacement from its own taxon than it would if it were in the zone of its own relatives. The evolutionary divergence of herbivores and carnivores in many groups of animals exemplifies mainly outcome (5).

In outcome (2), extinction of one higher taxon by competition cannot occur until all the species of this taxon are inferior to species of the other taxon in all accessible and reasonably permanent regions of the resource space. It is probable that attainment of such a situation would often be a matter of several million generations for taxa of moderate size, for each species would have to be eliminated individually, and these species would themselves usually experience selective pressure to compete better with their rivals as these impinge. It is well known (see Simpson, 1953, for reasons) that it is difficult to dislodge a well-adapted species by evolution sympatric with it. And niches evolve as well as species. Species well adapted to relatively isolated niches could perhaps withstand the presence of a generally better adapted higher taxon for considerable periods or even indefinitely. If extinction for reasons entirely unrelated to competition ever occurs for widespread species, this would accelerate replacement of one higher taxon by another because species relatively isolated from their relatives in the resource space would occasionally be eliminated randomly.

At least relatively localized species can of course become extinct through other environmental changes than the evolution or immigration of superior competitors. It is even possible for them to "commit suicide" by prezygotic selection of deleterious genes. The evidence for extermination of a supraspecific taxon by competition is its extinction upon the appearance of an at least locally new taxon with apparently similar resource needs, or a correlative reduction in diversity of one such taxon with an increase in diversity of the other. Such evidence is subject to the usual restraints on inferring a pattern of causation from a correlation, but it is improved if the taxa are relatively large (because then the segment of resource space is also relatively large and is less subject to local perturbations) and if no other cause is apparent.

Control by predators or parasites leads to a somewhat more complex situation, although to the same general conclusions in most cases. Parasites are equivalent to predators, here. Three distinctions must be made. A regulating factor (one of the density-dependent controls which keeps a population from expanding indefinitely) may be different from a key factor (the major cause of mortality, on the average, over several generations). Only the former is relevant to competitive exclusion, although the latter is more important for intrapopulation selection. Either or both of these kinds of factors can be seasonal (even in a rain forest) and therefore not always operative. Furthermore, the immediate regulating factor need not be the same as the factor which determines the density that can be supported indefinitely by the environment. For example, the latter can be food and the former can be an internal population regulation, varieties of which have been reviewed by Wynne-Edwards (1962) and Christian (1963). Only food is relevant to competitive exclusion in this case, and that the resource limit in general is relevant should be clear from the sequel.

If two species are regulated by predators, but the predators are different and specific, then the density of each predator–prey system is itself directly or indirectly regulated by the available resources. The two predator–prey systems therefore behave as two species with respect to competitive exclusion, although if the regulation is indirect then replacement is correspondingly slowed. Patterns in the abundance of predator–prey systems are obvious from the most cursory inspection of communities: ladybird beetles and aphids are more abundant that wolves and deer, and insectivorous birds are more abundant with respect to individuals per species, as well as to species, in forests than in deserts. Such patterns clearly depend on energy availability and energy requirements (and on other resource interactions) and thus demonstrate direct or indirect regulation by resources.

If two species are regulated by the same predator, an important difference arises only if the predator searches preferentially for individuals of whichever prey species is more common, i.e., if the probability of being eaten is greater for an individual of the more common species as well as for the species itself. This situation is probably relatively common for vertebrate predators. In the other cases (prey taken in proportion to their abundance, or one prey species always preferred) the total density of the two prey species is what is regulated by the predator. Because any initial equilibrium of relative fitness is unstable if present, one prey species is eliminated whether or not its niche importantly overlaps that of the other prey species, even if the latter species cannot increase to the original abundance of both. In more complex (and more common) cases (e.g., more than one predator, predators dependent on other prey also, evolution within the species) the conditions for coexistence are less stringent. However, it is clearly possible in some cases for a prey species to be eliminated by a predator that would not have eliminated it without the presence of another prey species (for a near example of this, see Pearson, 1966). We do not choose to call this interaction between the prey species "competition," although there is some similarity. Because it is irrelevant to resources, this interaction is presumably not disproportionately frequent between species that are similarly adapted as opposed to those that are dissimilarly adapted but of similar size and general habitat. Its importance in general is questionable as a cause of extinction rather than of cycles, and because the extinction of the multituberculates appears from evidence given below to have been related to the evolution of similarly adapted animals that tended to occupy somewhat different habitats from the multituberculates, prey interaction will be ignored in the remainder of the paper.

If there is a substantial amount of unused food (not needed for the perpetuation of the food species) even in poor years and seasons, there is room for the immigration, perhaps after a period of evolution, of species from elsewhere to utilize it. At least on a geological time scale, and probably more rapidly, such an expansion of a fauna to more or less fill the accessible food vacuum is reasonable, and there is evidence (Elton, 1958, and the more recent papers on species abundance and island faunas, mainly in the American Naturalist, partially reviewed by MacArthur, 1965) that at least tetrapod niches are in fact more or less contiguous and mutually restrictive.

Mammals of the Paleocene

Upon considering the diversity of food habits of recent species with teeth generally similar among the species, we are impressed with the impossibility in the foreseeable future of reconstructing in any detail the diets of Paleocene mammals. When there is a phyletic trend towards a more herbivorously adapted dentition, it is probably valid to say that most of the later members of this clade were herbivorous, or more herbivorous than their ancestors. There are very few Paleocene mammals of any kind for which a more detailed statement on diet is possible.

In the Paleocene there were a number of more or less common small mammals that were, to the extent that tooth structure and the direction of its evolution permit a judgment, at least largely herbivorous. These are some or all members of the condylarth families Arctocyonidae, Hyopsodontidae, Phenacodontidae, Didolodontidae, and Periptychidae, the primate families Paromomyidae, Carpolestidae, and Plesiadapidae, and possibly the insectivore families Mixodectidae and Adapisoricidae (in particular *Adapisorex*) among the placentals, and probably all the multituberculates. The latter group in particular was considerably more diversified than the published evidence indicates.

If it is correct that detailed inferences on the food habits of Paleocene mammals is usually impossible, it follows that we do not know from morphological evidence which Paleocene placental mammals were closest adaptively to some or all multituberculates, and which multituberculates were closest to any placentals. Enlarged and

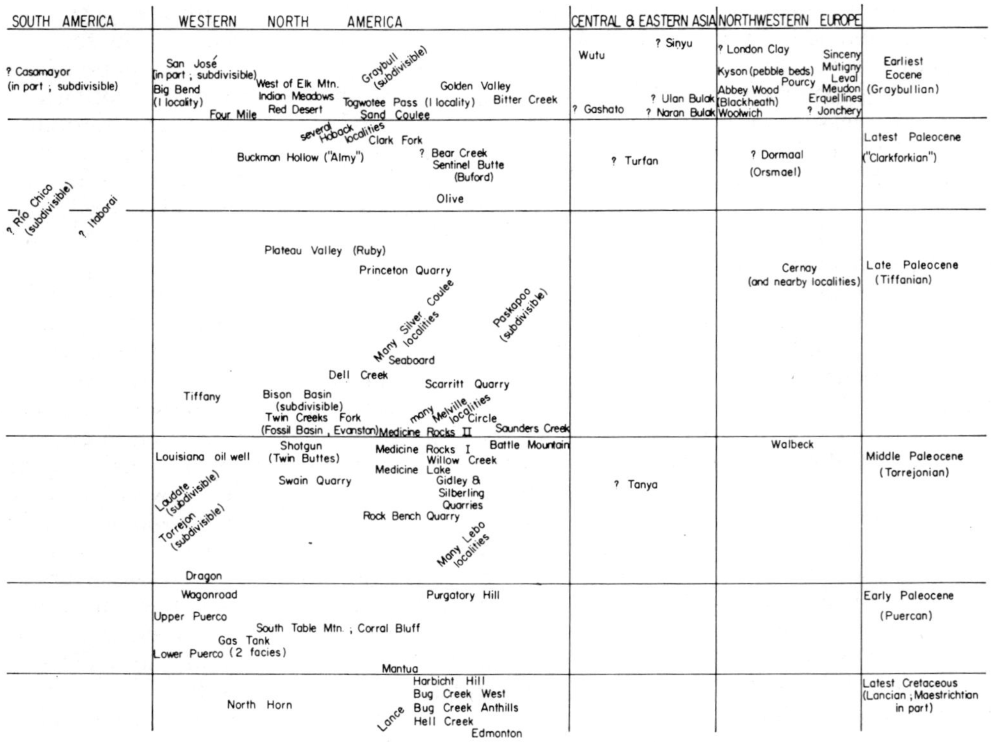

FIG. 1. Approximately chronological sequence of all published Paleocene (and bordering) mammal localities and faunas. Circle and Olive are unpublished faunas from Montana now being studied by us. Gas Tank is Locality 2 of Spieker (1960).

procumbent lower incisors were present in the Paramyidae, Plesiadapidae, Carpolestidae, Esthonychidae, Microsyopidae, Mixodectidae, and some but not all members of the Paromomyidae (Van Valen, 1965) in addition to the Multituberculata, but this evidence is insufficient to warrant the conclusion that the foods of these taxa were more similar to each other than to those of the small condylarths. In taeniolabidoid multituberculates, as opposed to ptilodontoids (the other Cenozoic group: Sloan and Van Valen, 1965), the enamel on the lower incisors is, after initial wear, limited to the anterior and in part the labial surfaces, and the lower incisors move forward in the jaw in compensation for attrition, although they are rooted. In contrast, taeniolabidoid upper incisors are short, rooted, and bicuspid, and do not move forward in the alveolus as the tooth wears. Although the upper incisors are completely covered with enamel, a self-sharpening, medially tapering wear facet develops on them because the labial enamel is slightly thicker than the lingual enamel. In ptilodontoids, attrition of the incisors is negligible compared with the rate of wear of molars and the premolar blades.

It will be useful to consider the distribution in time and space of apparently herbivorous mammals in the Paleocene. For present purposes the relative numbers of individuals are more important than the numbers of species. Persistently rare species have not dominated common ones in competition. Species numbers are however useful in a secondary way, in giving rough

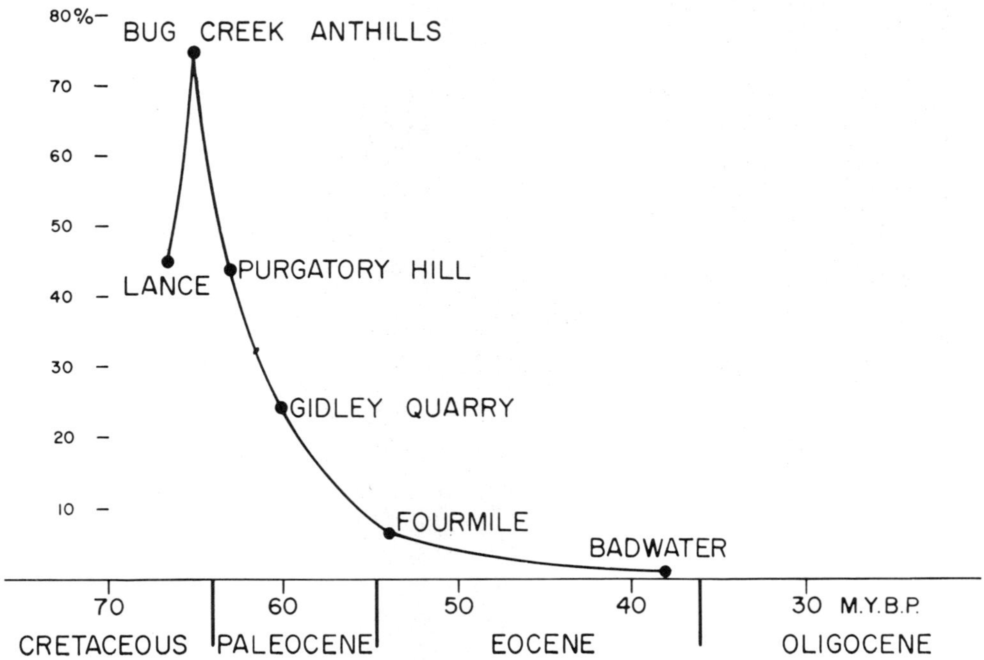

FIG. 2. Multituberculates as a proportion of the total number of mammal teeth in several late Cretaceous, Paleocene, and Eocene faunas of the north central United States. The figures are based on minimum numbers of individuals except for Bug Creek Anthills, where the true percentage of multituberculates would be at least as high as that shown (cf. Van Valen and Sloan, 1965, note 10). Time is in million years before present.

indications of the width of the adaptive zone (more strictly, the hypervolume of the resource space occupied) and of the species present in communities other than the one mainly sampled in a deposit.

The available data on both species abundance and species numbers for Paleocene mammals are now inadequate and are unlikely to be adequate in the near future. Numerical tables comparing several faunas are therefore likely to be misleading. Collecting methods have differed widely and have different biases. There are, in collections already made, many undescribed taxa, which are unequally distributed among the orders. The multituberculates are probably the most deficient in described species both absolutely and relatively, with insectivores and condylarths next, in inverse relation to physical size. Specimen counts have been made for only a few of the 40 or so Paleocene mammal faunas now known (not all published), and some of these are inaccurate. However, one or both of us have examined most of the specimens of Paleocene mammals that have been collected in the world, and some patterns are sufficiently pronounced that they may be regarded as real. The published faunas are listed in Figure 1. The time scale for this figure, and for Figures 3 and 4, has been estimated from published and unpublished potassium-argon dates, thickness of sediments in areas of more or less continuous

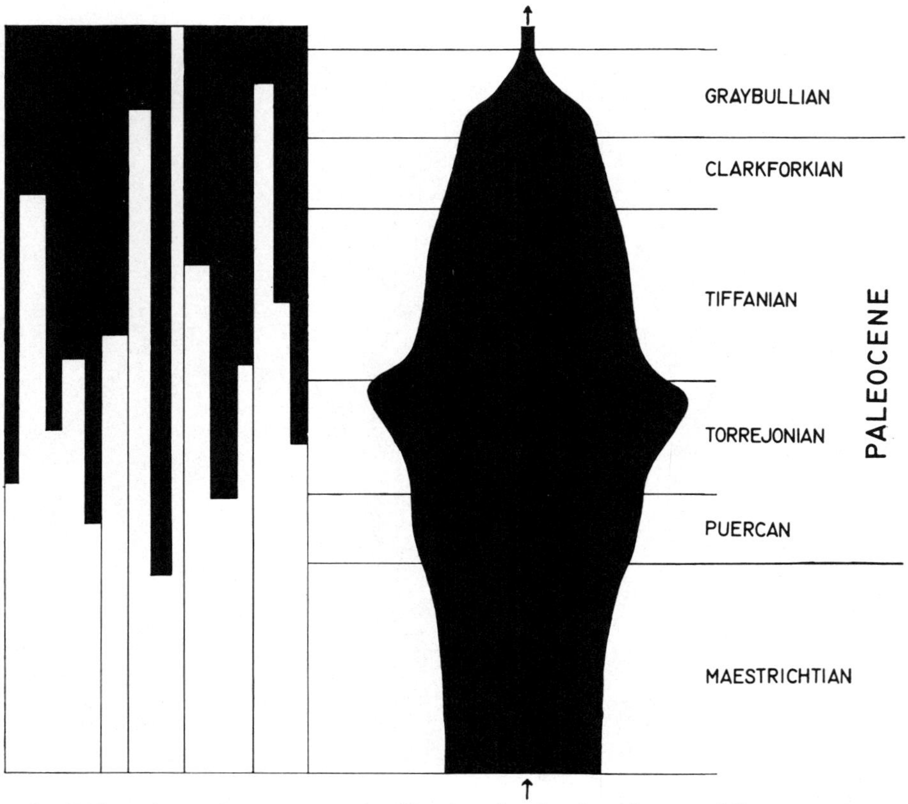

FIG. 3. Right: Approximate taxonomic diversity of multituberculates at different times, based on number of species known and degree of knowledge of faunas. Left: Schematic and hypothetical representation of the history of partitioning of the adaptive zone of multituberculates. Vertical lines separate niches of individual species; blackened niches are filled by placentals. The niches are represented as unchanging, which is undoubtedly false.

deposition of a column of similar rocks, and biostratigraphic placement of all known faunas. In Figure 1, the late Cretaceous and early Eocene are not bounded away from the Paleocene.

As at other times, including the Recent, the distribution of mammalian species in the Paleocene can conveniently be regarded as resulting from three levels of factors: continental, provincial (e.g., in the Paleocene and late Cretaceous, Wyoming, Montana, the Dakotas, and southern Alberta on the one hand and New Mexico, Colorado, Utah, and probably Louisiana on the other, with California probably a third), and facies-dependent (e.g., forest *vs.* riparian *vs.* savannah, or possible influences of soil type). Evidence for a fourth level (above that of facies), that of local topography (e.g., river distribution), is lacking in the Paleocene, in part because of the relatively small number of collections but also because its effects are difficult to distinguish from some of those of facies. Different methods of collection and deposition produce a bias that can be qualitatively evaluated.

The principal bias is the mode of collection, as can be seen from a comparison of the structure of many fossil "faunas" and recent communities. In a normal Recent terrestrial vertebrate community, the population size of a species is inversely related to the physical size of an individual of that species, as is the number of species of a given physical size range, and for any physical size group of species (squirrel-sized, etc.) herbivores, omnivores, and carnivores are present in decreasing abundance. We assume that the degree of bias in the representation of a community by a fossil collection can be very roughly indicated by the degree of agreement with the above relationships. For example, on this basis, of all the known Clarkforkian faunas, only the Bear Creek Local Fauna seems to be little biased by collecting, and this fauna is clearly of peculiar facies compared with both earlier and later well-sampled faunas. The local faunas collected by surface prospecting over a restricted area and stratigraphic interval are likely to be biased in favor of the medium-sized and large mammals. On the other hand, the local faunas collected by washing or extensive quarrying of a single rich bed are likely to be biased by the relative scarcity of the largest mammals (of tapir-size and larger), since the bone concentrations worked are as a rule the result of a hydraulic size-sorting process, and identifiable teeth and bones of the largest mammals are not likely to be deposited in their correct abundance at the same sites as the elements of smaller mammals. In terms of biomass and especially of community structure, this is a less serious bias than that present in the surface prospecting samples. On this basis only, a few North American Paleocene local faunas are adequately known for an analysis of changes in ecology with time. These are, in approximate stratigraphic order, Mantua Quarry, the lower Puerco of Tsosie Arroyo, Purgatory Hill, Wilson's Quarry 9 and Quarry 13 faunas of the Torrejon, Rock Bench Quarry, Swain Quarry, Gidley Quarry, Shotgun, Medicine Rocks, the Saddle Locality of Bison Basin, Scarritt Quarry, Princeton Quarry, and Bear Creek. Only Gidley and Scarritt quarries are as yet more or less adequately published; adequate accounts of at least Tsosie, Purgatory Hill, and Bison Basin should be published in the near future. The faunas of these Paleocene localities, when compared with those of the Lance and Bug Creek Cretaceous localities and the Eocene Four Mile Fauna, give a picture of a progressive elaboration of a varied series of mammalian communities, with a progressive and relatively rapid increase in species number and diversity and an elaboration of niches. The rate of morphological and taxonomic change and of change in community structure during this interval is greater by at least an order of magnitude than in the comparable time interval in the late Cretaceous from the Judith River Fauna to the Lance Fauna, and is also much greater than in the Eocene.

Despite the fact that the taxonomy connected with these faunas is largely unpublished, certain trends are obvious from examination of collections and published data. Concomitant with the reduction in reptilian herbivores, principally dinosaurs, from the Lance to the Bug Creek facies during the late Cretaceous, the proportion of multituberculate individuals in the total mammal fauna increased greatly. Subsequently, even though the known taxonomic variety of multituberculates increased very rapidly, the proportion of multituberculate individuals in the total mammal fauna decreased more or less steadily (Fig. 2).

Figure 3 gives a summary of the taxonomic diversity of multituberculates at different times, subjectively corrected for the different degrees of ignorance of the multituberculate faunas at different times. Taken in conjunction, Figures 2 and 3 show that the multituberculates decreased in relative abundance in the early and middle Paleocene, while they apparently became more diverse taxonomically. This pattern suggests that the adaptive zone available to the multituberculates was partitioned more finely in the middle Paleocene than it was in the late Cretaceous, presumably at least in part because portions of the zone were invaded by the rapidly diversifying condylarths. The abundant multituberculates of the late Cretaceous were therefore probably for the most part occupants of broader niches than their Paleo-

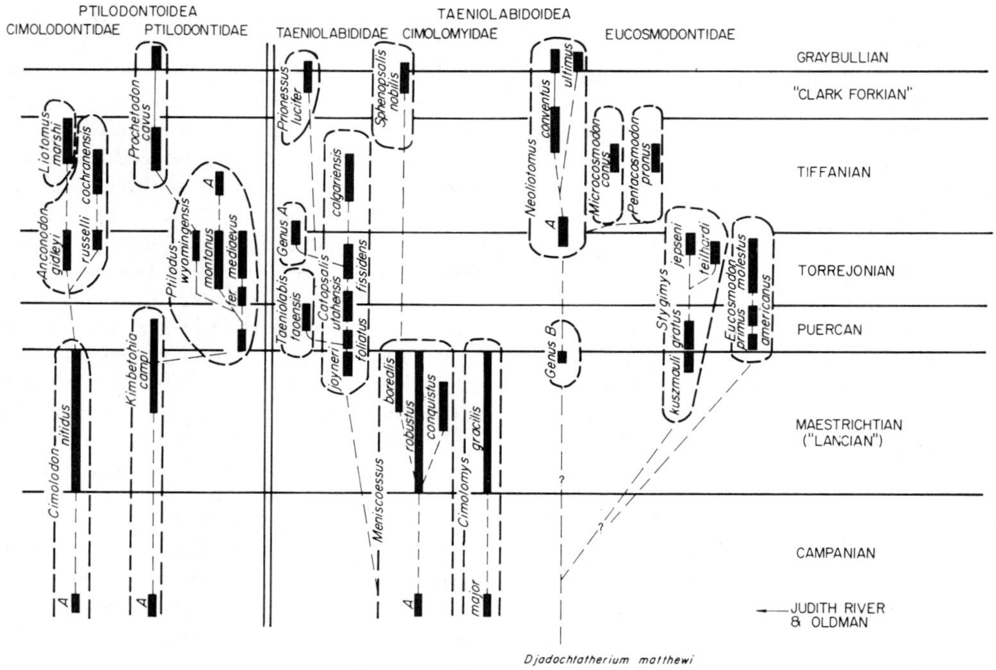

FIG. 4. Current estimate of the phylogeny of the Taeniolabidoidea and of the non-ectypodont families of the Ptilodontoidea (original, R. E. Sloan).

cene descendants; i.e., the latter were ecologically more specialized. These more specialized forms themselves then gradually died out with increasing diversification of the placentals, as will be discussed below.

It is possible that the multituberculates themselves broke up their adaptive zone into smaller units without pressure by infiltration from outside. There are, however, rather compelling arguments against this possibility. First, the niches were relatively broad in the late Cretaceous, after scores of millions of years of multituberculate evolution, during which specialization could have been evolved as easily as in the three million years or so in the first part of the Paleocene. This is particularly true because the late Cretaceous climate was warmer than that of the early Paleocene, and niches tend to be narrower in more tropical regions. Secondly, as measured by the number of distinguishable kinds of pollen, the food of multituberculates became not more but substantially less diverse between the late Cretaceous and the middle Paleocene. Thirdly, the rapidly diversifying and apparently herbivorous condylarths must have eaten something. The multituberculates were the only abundant small mammalian herbivores in the late Cretaceous before the diversification of condylarths, and, unless all the condylarths ate foods that no multituberculate had eaten, there must have been a reduction in the portion of the resource space available to multituberculates. In fact, because both the diversity of plants decreased and the floral composition changed moderately, the multituberculates would have had to eat new foods to maintain their previous share of the resource space.

The large collections from the Paleocene localities listed above have permitted one of us (R.E.S.) to construct an improved picture of mutituberculate phylogeny. As long as collections made by classic methods were all that were available, multituberculates, particularly the small and most

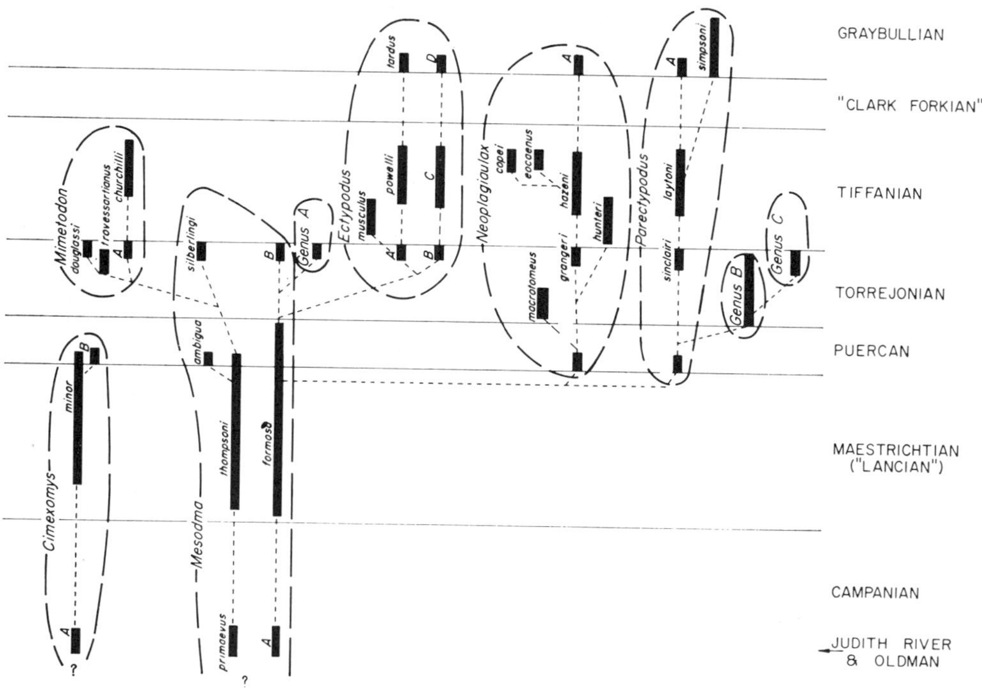

FIG. 5. Current estimate of the phylogeny of the longest-lived family of the Ptilodontoidea, the Ectypodontidae (original, R. E. Sloan).

abundant species, the members of the family Ectypodontidae (emended from Ectypodidae Sloan and Van Valen, 1965), were rare fossils, and no obvious ancestor-descendant relationships were apparent; indeed, most species were known only from the sites of their first discovery. The early quarrying localities and the other localities later worked by the washing technique have changed this situation. Multituberculates turn out to be common animals, and species are likely to recur in any well collected locality of appropriate age, province, and facies. Figures 4 and 5 show a tentative phylogeny of most of the known late Cretaceous and early Cenozoic multituberculates. Many of these species and some genera are undescribed and are simply listed here by arbitrary letter designations or none. In addition, there are some undescribed species not shown at all, due to lack of knowledge of their placement in these lineages.

Some conclusions can be drawn from the charts. Ectypodonts in particular, but members of other families as well, underwent a small radiation during the time dinosaur communities were becoming extinct. This radiation continued until the end of the middle Paleocene, at which time a number of lineages of multituberculates became extinct. This is the time at which two or three groups of primates developed diastemata and procumbent, more or less gliriform incisors: *Plesiadapis* as opposed to its ancestor *Pronothodectes, Phenacolemur* as opposed to its ancestor *Paromomys*, and possibly *Carpodaptes* as opposed to its ancestor *Elphidotarsius*. *Plesiadapis* was especially gliriform (the rodents probably originated from the Plesiadapidae) and was abundant and rather diversified in the late Paleocene in both Europe and North America. The process of extinction was not abrupt but continued through the late Paleocene and into the Eocene. The different multituberculate families were morphologically about as diverse as rodent suborders; the rodent family Paramyidae was probably not in direct and close competition with all of them. The sole known late Eocene species (Robinson, Black, and Dawson, 1964), from the Badwater Fauna, was derived from *Parectypodus laytoni* via an undescribed species from the Four Mile Local Fauna. The extinction process thus lasted for about 20 million years, and probably at least this many multituberculate generations.

Sequence of Faunas

In the early to middle Paleocene of western temperate to subtropical North America, the abundant groups of herbivores in both species and individuals (and dominant also in biomass) were the multituberculates, the condylarth families Periptychidae, Arctocyonidae, and Hyopsodontidae, and to some degree the Phenacodontidae. Their abundance relative to each other is moderately dependent on province and facies. The only early or middle Paleocene fauna known outside this region is from the fissure-filling of Walbeck, Germany, of about late middle Paleocene age. The Walbeck fauna (Russell, 1964) is remarkable in being one of the few Paleocene faunas in which multituberculates are absent, others including Bear Creek (see below) and the late Paleocene Buckman Hollow Local Fauna (which consists of only 73 relatively large specimens collected on the surface). The peculiar Shabarakh Usu ("Djadochta") mammals of Mongolia, probably Cretaceous but possibly from redeposited sediments of as late as Paleocene age (for discussion of these views see Van Valen, 1966), consist of a multituberculate and four genera of placentals, at least three of which have enlarged lower incisors. Probably not all these (and perhaps none of the placentals) were mainly herbivorous.

There was a moderate decline in the numbers of multituberculates in the late Paleocene. This decline affected both ptilodontoids and taeniolabidoids. In western temperate North America in the late Paleocene, the periptychids declined rapidly to extinction, the hyopsodonts declined moderately, but the arctocyonids and their descendants the phenacodonts remained abundant. The Plesiadapidae, a rodent-like family of primates that probably gave rise to the rodents, rapidly differentiated and was usually common. The Mixodectidae and to a lesser extent the Paromomyidae and Carpolestidae were moderately common in the middle and late Paleocene. Two groups of large herbivores, the Dinocerata (uintatheres) and the Pantodonta, evolved from arctocyonids in about the early Paleocene.

Outside western temperate North America, several possibly or surely late Paleocene faunas are known. That of Cernay, in France, had a moderate proportion of multituberculates, but these were outnumbered by plesiadapids and by two apparently herbivorous genera probably belonging to the aberrant condylarth family Meniscotheriidae. Arctocyonids, the blunt-toothed *Adapisorex* (probably an erinaceoid), and to a lesser extent hyopsodonts were also common. Two faunas in South America of about late Paleocene age lack multituberculates, which are unknown from that continent. The marsupial family Polydolopidae was dentally very similar to the multituberculates, however.

It is possible that the fauna of Dormaal (formerly Orsmael), in Belgium, was contemporaneous with that of the Clark Fork rather than with that of the Gray Bull. Several primitive species of rodents are present at Dormaal (Wood, 1962; Quinet, 1964), but perissodactyls and artiodactyls are conspicuously absent. Multituberculates are also absent in the collections thus far gathered at Dormaal and probably would have been found despite their small size, if they were abundant. This argument is as yet based on facies and may well be incorrect.

The Gashato (Hashiato) and Naran-Bulak faunas in Central Asia are of either late Paleocene or early Eocene age. Taeniolabidoid multituberculates were moderately common in the Gashato fauna (that of Naran-Bulak is incompletely described), in which small notoungulates were the most common element. A lagomorph and a lagomorph-like genus were also not rare, and a pantodont, a uintathere, and the large condylarth, *Phenacolophus*, were important in energy flow. Rodents were absent but are not yet known in any part of Asia until the late Eocene. Isolated finds in the Paleocene of China are not yet relevant to ecological considerations. Paleocene mammals are unknown from regions not mentioned above. Cenozoic multituberculates are known outside North America only from Cernay, Gashato, and Naran-Bulak (personal communication from K. Kowalski for the latter), and late Cretaceous multituberculates are known outside of North America only from Shabarakh Usu.

The two Northern Hemisphere Paleocene faunas that lack multituberculates will be considered in somewhat more detail. In the Walbeck fissure-filling, between 50 and 60 per cent of the specimens are of two species of arctocyonids. *Adapisorex* constitutes another 20 or 25 per cent. *Plesiadapis*, the possibly adapisoricid insectivore *Adunator*, and hyopsodonts are also moderately common. One additional gliriform primate, *Saxonella*, is present although rare. As this is the only European mammal fauna yet discovered in the middle Paleocene, and one of two or three in the entire Paleocene of Europe, it is uncertain whether the absence of multituberculates is due to continental, provincial, or facies-dependent reasons. The only American supposed fissure-filling in the Paleocene, Mason Pocket in the Tiffany, has a moderately abundant species of multituberculate, but this deposit is in insoluble shale (the cavity was thus presumably not produced by solution) and perhaps resulted from the filling in a backwater of a burrow or a similar hole in semi-consolidated sediments. The moderate abundance of multituberculates in the later Cernay fauna suggests that their absence at Walbeck may not be directly related to their extinction.

The Bear Creek Fauna

Bear Creek, however, is a different situation. Here is a moderately well collected fauna, with many small elements, in a region and at a time where multituberculates were abundant. The fauna is badly in need of restudy, although no important collections have been made since Simpson's descriptions (1929 and earlier). The following remarks are based on observations by Van Valen, mainly on American Museum specimens, and do not substitute for an adequate revision.

There are 16 species of mammals now known from Bear Creek. About half the specimens belong to the oldest probable dermopteran, *Planetetherium mirabile*, the food of which is uncertain. It is more primitive than its relative *Plagiomene* in the early Eocene Gray Bull. The insectivore, *Protentomodon ursirivalis*, is less common and is of uncertain affinities, but may belong to the Pentacodontidae. The rare insectivore, *Apatemys kayi*, is more or less comparable to early Eocene species and is more advanced than its known Paleocene relatives (cf. McKenna, 1963). The erinaceoid insectivore, *Entomolestes siegfriedti*, is moderately common and is more primitive than the congeneric species in the basal Eocene Four Mile Fauna (McKenna, 1960). Two upper molars (A.M.N.H.[1] Nos. 22175 and 22221) are referable to the pantolestid insectivore *Palaeosinopa* and may be conspecific with the Gray Bull species *P. didelphoides*. They are more advanced than the upper teeth from the Scarritt Quarry referred to "*Palaeosinopa*" *senior*. The type of *P. senior* is however a palaeoryctid. Two isolated teeth (A.M.N.H. Nos. 22244 and 22245), a P_4 and a P^4 respectively, represent an undescribed genus of apheliscine hyopsodonts. This genus also occurs in the Gray Bull (A.M.N.H. No. 16935, P^3–M^1) as a different and smaller species. The Bear Creek species is similar in grade to the Clark Fork-Sand Coulee species, *Apheliscus nitidus*. The

[1] American Museum of Natural History.

multituberculate-like primates, *Carpolestes aquilae* and *C. nigridens*, are moderately common, are comparable to the congeneric species from the latest Paleocene Clark Fork, and are more advanced than the early late Paleocene genus *Carpodaptes* (probably ancestral) from Tiffany and Scarritt Quarry. One primate M_3 (A.M.N.H. No. 22154), referred by Simpson to *Plesiadapis,* belongs to *Paromomys* or *Phenacolemur* and is useless stratigraphically. The rodent, *Paramys atavus*, is known from several specimens (see especially McKenna, 1961a, and Wood, 1962) and is probably more primitive than the primitive Gray Bull species of *Paramys* and *Franimys*. Van Houten (1944:180) recorded a rodent incisor from the Clark Fork, but the validity of this record may be questioned until details are given. *Lampadophorus lobdelli*, a relatively rare taeniodont and a moderately large herbivore, seems slightly more advanced than the late Paleocene species *L. expectatus* from Plateau Valley (for which see Patterson, 1949), but the two populations may not be distinct at the specific level. The relatively rare herbivorous arctocyonid, *Thryptacodon "pseudarctos,"* is indistinguishable from *T. antiquus* of the Gray Bull and Clark Fork and may be synonymized with this species. One tooth of the herbivorous phenacodont, *Phenacodus*, is comparable to Clark Fork and Gray Bull species. A small hyopsodontid condylarth, represented by A.M.N.H. No. 22181, may belong to *Haplaletes*, elsewhere known from the middle and late Paleocene. The lack of closely comparable species elsewhere precludes its use in detailed correlation. The perhaps predaceous mesonychid, *Dissacus*, is known from one tooth and is useless stratigraphically. One tooth (A.M.N.H. No. 22220) of a small predator seems referable to either the Hyaenodontidae or the Palaeoryctidae, more probably the former (Van Valen, 1966). It represents a form more advanced than other intermediate species from Scarritt Quarry and the Puerco but is smaller than any Gray Bull species.

The above observations suggest that Bear Creek is of latest Paleocene (Clark Fork) age, but it could be somewhat earlier or even later. Crocodilians and turtles are abundant as usual, and a champsosaur and a salamander (Auffenberg, 1961) are present. The sedimentary environment (carbonaceous shale just above a coal seam; the matrix with at least American Museum specimens is dark and often bituminous) and the faunal composition (especially the preponderance of probably or possibly aboreal forms) suggest mainly or entirely a forest community at the edge of a variably boggy pond. Such an inferred community is elsewhere unknown in the Paleocene (as is, e.g., *Planetetherium*, much the most abundant taxon at Bear Creek). In addition to multituberculates, plesiadapids and miacids are conspicuously absent at Bear Creek, and condylarths are outnumbered by both insectivores and non-plesiadapid primates as well as by *Planetetherium.* These are features strongly different from other Paleocene faunas.

The Early Eocene

At about the beginning of Sand Coulee or Four Mile time (conventionally regarded as the beginning of the Eocene in North America), but with some changes occurring earlier and later, the communities sampled by fossil deposits in the temperate Rocky Mountain region and in northeastern France changed radically. These are the only two regions in the Northern Hemisphere in which both late Paleocene and early Eocene mammals are known. Rodents became ubiquitous, numerous in individuals, and with a moderate number of species. Primitive artiodactyls appeared, as did in somewhat greater abundance several related kinds of primitive perissodactyls. New groups of carnivores appeared for the first time, at least in any abundance: the Miacinae, Hyaenodontidae, and Oxyaenidae. Lemuroid primates and the Microsopidae, primate-like insectivores or insectivore-like primates, first appeared and were relatively abundant. Other changes

TABLE 1. FREQUENCY OF MAMMAL GROUPS (SEE TEXT) IN QUARRIES OF THE FOUR MILE FAUNA, EARLIEST EOCENE OF COLORADO. FOR EACH QUARRY, NUMBER OF SPECIMENS IS TO THE LEFT AND FREQUENCY PER THOUSAND TO THE RIGHT.

	Quarry													
	Anthill		Kent		Despair		Alheits		Sand		V-5423		Timberlake	
Multituberculates (1)	0	0	1	7	6	9	31	24	16	27	0	0	31	24
Insectivores and marsupials (2)	13	35	41	270	36	53	76	57	18	30	4	38	83	63
Primates (3)	34	92	14	92	39	58	116	88	48	81	2	19	81	62
Cynodontomys (4)	47	127	2	13	22	33	45	34	19	32	8	77	33	25
Rodents (5)	127	343	65	428	226	335	209	158	165	277	59	567	584	445
Hyopsodonts (6)	109	295	11	72	237	351	543	410	198	332	15	144	386	294
Large condylarths (7)	12	32	2	13	11	16	16	12	7	12	1	10	12	9
Carnivores (8)	10	27	4	26	23	34	38	29	22	37	1	10	14	11
Perissodactyls (9)	12	32	0	0	46	68	230	174	79	133	14	135	62	47
Artiodactyls (10)	6	16	12	79	29	43	20	15	24	40	0	0	26	20

also occurred. The vegetation may well have been undergoing a change of similar magnitude in these regions, but this has not yet been studied. The relative importance of facies (community) changes and migration is unknown.

The Four Mile Fauna, described by McKenna (1960), actually consists of samples from several sites that represent in part rather different communities and perhaps slightly different ages. All identifiable specimens (and many others) were collected from various volumes of matrix. One of us (L.V.) has analyzed the specimen counts published by McKenna to see what evidence, if any, may be obtainable on mutual interference among the several higher taxa. The difficulties discussed elsewhere (Van Valen, 1964) do not seem applicable to simple comparisons among frequencies in the different samples. The data are those of McKenna (1960), as modified slightly by Van Valen (1964). Later collections are partly unsorted and have not been used. The total number of specimens rather than minimum number of individuals has been used, as justified elsewhere (Van Valen and Sloan, 1965). The multituberculates are underrepresented relative to the other taxa because they are known mainly from P_4s, but this is true of all samples and has been assumed to be equally true of all.

The mammalian fauna was divided into the following ten groups, based only on taxonomy and apparent adaptive similarity: multituberculates; marsupials and insectivores (including *Didelphodus*); primates; *Cynodontomys;* rodents; carnivores (including *Pachyaena*); hyopsodonts and *Apheliscus*; larger condylarths (including *Esthonyx* and arctocyonids) and *Coryphodon*; perissodactyls; and artiodactyls. The number of specimens of each group from each locality (East and West Alheit are combined) and the relative frequency per thousand are given in Table 1.

Simple comparisons of relative frequencies are improper because each relative frequency is dependent on all others. To obtain independent estimates for each pair of taxa, we divided the number of specimens of one taxon by the total minus the number of specimens of the other taxon being compared at that time, i.e., $N_1/(N_T - N_2)$, and similarly for the second taxon. The frequencies so obtained are of course different for each pair of taxa. For the 13 comparisons that seemed potentially most interesting on adaptive grounds, we calculated the correlations between the corrected abundances in the seven localities used. In several cases two or three of the above groups were combined.

Both positive and negative correlations may be obtained. A positive correlation would suggest that the taxa did not particularly interfere with each other (at the time represented by the sample) because they tend to be most abundant in the same communities. For example, different groups of grazers would be expected to be positively correlated in a mosaic of grass and forest communities. A negative correlation between similarly adapted taxa would suggest interference in the occupation of similar habitats. (It is important that the taxa be similarly adapted. Whales and artiodactyls would probably have a nearly perfect negative correlation, but this does not mean that one group excludes the other, even though this is possible in the marginal case of potential evolution from hippopotami and river dolphins.) The estimates possible are very rough and, particularly because it is impossible to determine the relative frequency of the different communities represented in each sample, are mainly of qualitative significance.

The results in Table 2 suggest that most pairs of taxa adapted for generally similar food did not in fact strongly interfere with each other at this (early Eocene) time. The standard errors of the correlations are large since all the correlations are based on a sample size of seven, and only the largest correlation is significantly different from zero, taken individually. Nevertheless the general pattern is probably real, and the negative correlations are probably more significant than the usual test indicates because of interference by the community effect mentioned above. The only indications of interference are between large condylarths and the more advanced ungulates, between multituberculates and rodents, and between multituberculates and primates. Interference on a less coarse scale than the community level used is presumably operative in some other cases, but the nature of the evidence does not permit a test of this possibility.

It is interesting that one group in each of the three pairs with moderately negative correlations, the multituberculates (1) and large condylarths (7), became extinct or greatly reduced in abundance relatively soon after Four Mile time. This may well have been due to the competition implied by the negative correlations.

TABLE 2. CORRELATIONS (r) BETWEEN ABUNDANCES OF GROUPS OF MAMMALS, NUMBERED AS IN TABLE 1, FROM THE FOUR MILE FAUNA. FOR FURTHER EXPLANATION, SEE TEXT.

Groups	Correlation
(9) – (10)	+0.61
(7) – (9 + 10)	–0.23
(1) – (7 + 9 + 10)	+0.58
(6) – (7 + 9 + 10)	+0.81
(5) – (7 + 9 + 10)	+0.08
(1) – (3 + 4)	–0.38
(1) – (5)	–0.38
(1) – (6)	+0.64
(3 + 4) – (5)	+0.45
(3 + 4) – (6)	+0.52
(5) – (6)	–0.09
(2) – (3 + 4)	+0.24
(3) – (4)	+0.14

There is therefore suggestive evidence that both primates (together with or including the dentally similar *Cynodontomys*) and their collateral descendants, the rodents, contributed to the great reduction of the multituberculates in the early Eocene. It is probably also not a coincidence that both of these placental groups expanded in the late Paleocene, when the multituberculates began their decline in taxonomic diversity. The primates appear to have expanded first and were probably the most important cause of the crisis for the multituberculates at the end of the middle Paleocene. As the more abundant and more rapidly diversifying group in the Eocene, the rodents were probably more effective at least at this later time. Although the gross Four Mile data give no evidence of this, it is also possible that the later reduction of primates in Holarctica (and the extinction of condylarths, unless, as is moderately justifiable, *Orycteropus* or even the hyracoids are regarded as surviving condylarths; it is not yet adequately shown that *Orycteropus* is a condylarth derivative) was at least in part due to the expanding adaptive zone of rodents.

As collections from the Clark Fork and demonstrable equivalents are still inadequate, it is unknown whether the low frequency of multituberculates in these collections in part reflects a decline at that time or is entirely a result of the different methods of collection employed and the facies of Bear Creek.

The only multituberculate reported from sediments later than the early Eocene (middle Gray Bull) is a species of the primitive family Ectypodontidae from the late Eocene Badwater Fauna of Wyoming (Robinson, Black and Dawson, 1964), found together with remains of several rodents. As frequently happens, its more obviously specialized relatives had become extinct earlier. In view of the numbers of small insectivores and rodents recovered from the middle Eocene of the Bridger Basin, Wyoming, it is unlikely that multituberculates were numerous there in the middle Eocene, in the communities mainly sampled.

Discussion

It is easier to ask than to answer what the reasons might be for the inferred competitive inferiority of multituberculates as compared to rodents and primates. The dentition is apparently not critical in a comparison with primates, at least, for the dentition of the carpolestid primates is remarkably like that of multituberculates. The polydolopid marsupials of the Paleocene and Eocene of South America also developed a similar dentition (they became extinct by the Oligocene, when rodents are first known in South America, but this is perhaps only coincidental), as did the surviving macropodid marsupials and their Pleistocene collateral relative *Burramys*, of Australasia. The rare South American abderitine marsupials developed, probably but not certainly independently, a less specialized replica of the polydolopid dentition and survived with rodents into the Miocene. A detailed functional analysis of these dentitions and those of their competitors will be required to determine which, if any, is appreciably superior (cf. Simpson, 1933).

It is conceivable that if multituberculates lacked tooth replacement (as postulated by Jepsen, 1940) they would be at some disadvantage for this reason (although this does not seem to be true for the virtually monophyodont marsupials and shrews), but Szalay (1965; suggested by Clemens, 1963) has clearly shown that replacement occurs.

Several morphological features seem to be appreciably inferior in multituberculates as compared to therian mammals. The cochlea is straight and short, roughly one sixth the length of that of Cretaceous to Eocene placentals of comparable size; this suggests, but does not prove, weaker powers of pitch discrimination for multituberculates. The acetabulum of the multituberculate pelvis is completely open dorsally; the femur is held in articulation only by muscles in this direction. This suggests a near or complete lack of ricochetal or gliding habits among multituberculates and may have permitted at most only a low possibility of arboreal activity. All torsional rotation of the forearm about its longitudinal axis had to be accomplished by the radius, as rotation of the ulna was completely restricted to the ulnar-humeral plane by their articulating surfaces. This unique character of the humeral-ulnar articulation probably did not allow sufficient freedom of movement for evolution of effective mole-type burrowing habits or desman-type swimming.

In the absence of the whole supraspinous group of muscles originating on the scapula (McKenna, 1961b), multituberculates are significantly more cynodont-like than are therian mammals.

In the Cretaceous braincases now known from Bug Creek, the cerebrum of multituberculates (known in *Mesodma thompsoni*, *Stygimys kuszmauli*, and *Catopsalis joyneri*) is smooth, implying the absence of a neopallium. It is still smooth in *Ptilodus montanus* from the middle Paleocene, although the cerebrum of the Cretaceous leptictid *Procerberus formicarum* (structurally almost ancestral to all later placental mammals) is fissured.

However, competitive inferiority is not really demonstrated by apparent primitiveness in individual characters. Only more or less complete knowledge of the total biology of all relevant elements of a fauna would permit such a conclusion, and such knowledge is not foreseeable for Paleocene mammals. We do not even know why echidnas have survived a wide radiation of advanced marsupials, nor why oppossums (primitive in many ways even if not in their reproduction) have recently expanded their range in the placental-infested continent of North America.

From the above evidence it may be concluded that, as far as now can be determined, the decline of the multituberculates was probably initiated by competition with condylarths, increased by primates, and completed by rodents. This decline was not catastrophic (extinctions of higher taxa seldom and perhaps never are), but it occurred more or less simultaneously in western Europe and in western central North America. It probably involved both immigration of superior competitors and a slower, more or less sympatric evolution in condylarths, primates, and rodents of an expansion of their zones in the resource space, thus compressing the region of the resource space available to multituberculates. Such a niche expansion is in general aided by environmental changes (as in the vegetation) to which both competing groups must adapt and in which the previous occupancy of the changing niches is less important than in a secularly more stable environment. Niche expansion for a supraspecific taxon can involve either increase in width of the niches of the component species or increase in the number of species. The latter aspect seems to have been dominant in the decline of multituberculates. The discovery of a late Eocene multituberculate proves that faunal immigration was not the only cause of multituberculate extinction, and niche evolution may have been at least as important as immigration in the Paleocene and early Eocene.

Acknowledgments

This study was supported in part by grant GB-2213 from the National Science Foundation to Sloan. We are grateful to Dr. J. S. Mellett and Mr. S. J. Gould for comments.

REFERENCES

Auffenberg, W. 1961. A new genus of fossil salamander from North America. Amer. Midland Nat. 66:456–465.

Christian, J. J. 1963. Endocrine adaptive mechanisms and the physiologic regulation of population growth. *In* W. V. Mayer and R. G. Van Gelder [eds.], Physiological mammalogy, vol. 1, p. 189–353. Academic Press, New York.

Clemens, W. A. 1963. Fossil mammals of the type Lance Formation, Wyoming. Part I. Introduction and Multituberculata. Univ. California Publ. Geol. Sci. 48:1–105.

Elton, C. S. 1958. The ecology of invasions by animals and plants. Methuen, London.

Hutchinson, G. E. 1957. Concluding remarks. Cold Spring Harbor Sympos. Quant. Biol. 22: 415–427.

Jepsen, G. L. 1940. Paleocene faunas of the Polecat Bench Formation, Park County, Wyoming. Part I. Proc. Amer. Phil. Soc. 83:217–341.

———. 1949. Selection, "orthogenesis," and the fossil record. Proc. Amer. Philos. Soc. 93:473–500.

Landry, S. 1965. The status of the theory of the replacement of the Multituberculata by the Rodentia. Jour. Mammalogy 46:280–286.

MacArthur, R. H. 1965. Patterns of species diversity. Biol. Rev. 40:510–533.

McKenna, M. C. 1960. Fossil Mammalia from the early Wasatchian Four Mile fauna, Eocene of northwest Colorado. Univ. California Publ. Geol. Sci. 37:1–130.

1961a. A note on the origin of rodents. Amer. Mus. Novitates No. 2037:1–5.

1961b. On the shoulder girdle of the mammalian subclass Allotheria. Amer. Mus. Novitates No. 2066:1–27.

1963. Primitive Paleocene and Eocene Apatemyidae (Mammalia, Insectivora) and the primate-insectivore boundary. Amer. Mus. Novitates No. 2160:1–39.

Patterson, B. 1949. Rates of evolution in taeniodonts. *In* G. L. Jepsen, E. Mayr, and G. G. Simpson [eds.], Genetics, paleontology, and evolution, p. 243–278. Princeton Univ. Press.

Pearson, O. P. 1966. The prey of carnivores during one cycle of mouse abundance. Jour. An. Ecol. 35:217–233.

Pimentel, D., E. H. Feinberg, P. W. Wood, and J. T. Hayes. 1965. Selection, spatial distribution, and the coexistence of competing fly species. Amer. Nat. 99:97–109.

Quinet, G. E. 1964. Morphologie dentaire des Mammifères éocènes de Dormaal. Bull. Groupement Internat. Rech. Sci. Stomatologie 7:272–294.

Robinson, P., C. Black, and M. Dawson. 1964. Late Eocene multituberculates and other mammals from Wyoming. Science 145:809–811.

Russell, D. E. 1964. Les Mammifères du Paléocène d'Europe. Mém. Mus. Natl. Hist. Nat. (N. S.) (C) 13:1–324.

Simpson, G. G. 1928. A new mammalian fauna from the Fort Union of southern Montana. Amer. Mus. Novitates No. 297:1–15.

1929. Third contribution to the Fort Union fauna at Bear Creek, Montana. Amer. Mus. Novitates No. 345:1–12.

1933. The "plagiaulacoid" type of mammalian dentition, a study of convergence. Jour. Mammalogy 14:97–107.

1944. Tempo and mode in evolution. Columbia Univ. Press.

1953. The major features of evolution. Columbia Univ. Press.

Sloan, R. E., and L. Van Valen. 1965. Cretaceous mammals from Montana. Science 148: 220–227.

Spieker, E. D. 1960. The Cretaceous-Tertiary boundary in Utah. Internat. Geol. Congr., Rept. 21 Session, 5:14–24.

Stone, W. S., M. R. Wheeler, W. P. Spencer, F. D. Wilson, J. T. Neuenschwander, T. G. Gregg, R. L. Seecof, and C. L. Ward. 1957. Genetic studies of irradiated natural population of *Drosophila*. Studies in the Genetics of Drosophila, 9 (Univ. Texas Publ. 5721):260–316.

Szalay, F. S. 1965. First evidence of tooth replacement in the subclass Allotheria (Multituberculata). Amer. Mus. Novitates No. 2226: 1–12.

Van Houten, F. B. 1944. Stratigraphy of the Willwood and Tatman formations in northwestern Wyoming. Bull. Geol. Soc. Amer. 55: 165–210.

Van Valen, L. 1960. Therapsids as mammals. Evolution 14:304–313.

1964. Relative abundance of species in some fossil mammal faunas. Amer. Nat. 98:109–116.

1965. A middle Palaeocene primate. Nature 207:435–436.

1966. Deltatheridia, a new order of mammals. Bull. Amer. Mus. Nat. Hist., 132:1–126.

Van Valen, L., and R. E. Sloan. 1965. The earliest primates. Science 150:743–745.

Wilson, R. W. 1951. Evolution of the early Tertiary rodents. Evolution 5:207–215.

Wood, A. E. 1962. Early Tertiary rodents of the family Paramyidae. Trans. Amer. Philos. Soc. 52:1–261.

Wright, S. 1932. The roles of mutation, inbreeding, crossbreeding, and selection in evolution. Proc. VI Int. Congr. Genetics 1:356–366.

Wynne-Edwards, V. C. 1962. Animal dispersion in relation to social behaviour. Oliver and Boyd, Edinburgh.

LEIGH VAN VALEN is Research Fellow in the Department of Vertebrate Paleontology, American Museum of Natural History, New York. From January 1, 1967, he will be Assistant Professor of Anatomy in the University of Chicago. **ROBERT E. SLOAN** is Associate Professor of Geology and Geophysics in the University of Minnesota, Minneapolis.

22

Reprinted from pages 349-364 of *Pleistocene Extinctions, the Search for a Cause*, P. S. Martin and H. E. Wright, eds., Yale University Press, New Haven, Conn., 1967

THE PLEISTOCENE EXTINCTION OF MAMMALS IN EUROPE

K. Kowalski

Abstract

In Europe—which, because of its geographic situation had only limited possibilities of faunal migrations between north and south—the Ice Age caused the impoverishment of fauna. This depletion affected all groups of mammals (except, perhaps, bats), but it especially influenced the larger mammals, which were not able to survive in the small refuges of southern Europe. Their limited populations approached more easily the point at which the lack of genetic plasticity, accidental change of sex ratio, epidemics, and other factors could bring about their extinction.

The hunting activity of primitive man, even over a very long period, does not necessarily bring the extinction of his prey. Extensive literature concerning the hunting habits of European Paleolithic man indicates that all large phytophagous mammals were hunted, but it is very difficult to judge the size of the human population and the degree to which hunting influenced the population of particular species. Reindeer, red deer, aurochs, wild horse—all extensively hunted—were able to survive, whereas some mammals, which were probably more difficult for primitive man to kill (mammoth, woolly rhinoceros, cave bear) or were of no interest to him (lion, hyena), became extinct. When the density of the prey dropped, man with his primitive weapons could not get enough game and began to hunt other animals, migrated, or finally died of hunger. Not only in the forests of Europe but also in the steppes of Asia, numerous herds of large herbivorous mammals were able to persist in spite of centuries of hunting, until in modern times they were decimated with firearms. The extinction (local, at least) of many mammals was caused only by the development of agriculture and/or the raising of domestic animals, which make the human population independent of game animals.

In the forest zone of Europe, only extensive deforestation in the late Middle Ages brought the danger of extinction to such species as the aurochs and the bison. No doubt, the preservation of the European bison, as well as of other mammals (such as beaver, elk, red deer, and brown bear) is now possible only through active protection.

The extinction of large mammals at the end of the last glaciation is less dramatic in Europe than in North America. Its main cause, in Europe, was the disappearance of the specific habitat represented by the "steppe–tundra" of the Pleistocene. The species connected with this biotope became extinct, while others were able to survive in the zone of warm steppes (horse and saiga) or of tundra (reindeer).

The extinction of large mammals in Europe needs further research, especially in radiocarbon dating and the stratigraphy of Paleolithic sites, chiefly in caves.

A comparison of Europe and North America can be of interest in the study of the Pleistocene extinction of animals. Although the climatic and geologic conditions were similar, man coexisted much longer with the fauna in Europe than in North America, where the primitive human population appeared dramatically at the decline of the Ice Age. Knowledge of the composition of the European Quaternary mammalian fauna is extensive but still far from complete.

Using Guilday's conclusions (this volume) I shall try to distinguish between extinction through evolution—the disappearance of a species through its transformation into another—and true extinction—the disappearance of an evolutionary line without descendants. The first type of extinction proves the success of a phylogenetic branch, and the second, its failure; the types are therefore antithetical. Sometimes, however, this distinction is not simple. In the fauna of European small mammals, the rodent genus *Mimomys* disappeared in the early-Pleistocene and was replaced in northwestern Europe by a similar genus, *Arvicola*, representing a higher stage in the development of molar teeth. It was recently discovered, however, that in southeastern Europe *Arvicola* appeared earlier and coexisted during a long period with the latest species of *Mimomys*. *Arvicola* apparently developed from one of the kinds of *Mimomys*, probably outside Europe. In some regions, then, *Mimomys* experienced "extinction through evolution." In regions characterized by the "true extinction" of the genus, *Arvicola* appeared when some species of *Mimomys* still existed and perhaps caused their extinction through active competition. The same systematic unit can therefore disappear partly through evolutionary and partly through true extinction.

When discussing the Quaternary fauna of European mammals and its changes, we must keep in mind the peculiar geographic situation of that continent. Europe, which could be considered a peninsula of Asia, is wide open to migrations from the east. At the same time, the Mediterranean Sea and the European mountain ranges that run mostly from east to west limit to a great degree the possibility of migration from north to south. In the cold periods typical of the Pleistocene, elements of the thermophilous fauna could find shelter in the forest regions of Southern Europe, but the Pyrenean and Apennine regions and the Balkan peninsulas were limited in space and were isolated.

In spite of many studies, we still do not know enough about the change in the European Quaternary faunas to be able to correlate them with the pattern of four glaciations described by geologists. Many paleozoologists, primarily in unglaciated regions of Europe, prefer therefore to use special names for the periods of the faunal development, independent of the names of glaciations (Kretzoi, 1965). Farther north the connection of the faunal changes with the course of glaciations, especially later ones, is evident.

The Ice Age, as a whole, caused the extinction of the rich Pliocene fauna of Europe, with its numerous subtropical elements. In the oldest Pleistocene (Villafranchian) a successive extinction of many evolutionary lines occurred (e.g. of mastodonts, tapirs, gazelles, *Parailurus*, and many groups of deer). The same is true for small mammals and also for many plants (*Azolla*, *Tsuga*, *Pterocarya*, *Magnolia*). Even in the Mediterranean regions, where the modern climate is similar to that of Central Europe in the late Pliocene, the fauna is much poorer as a result of Pleistocene extinction. The family Soricidae, for example, is represented in the Pliocene of central Europe by the genera *Sorex*, *Neomys*, *Crocidura*, *Petenyia*, *Beremendia*, *Blarinoides*, *Petenyiella*, *Soriculus*, and *Allosorex*, with numerous species. In similar climatic conditions in southern Europe, the same family recently contained only four genera: *Sorex*, *Neomys*, *Crocidura*, and *Suncus*, with few species. This depletion of the fauna of Europe has not yet been compensated, because migrations in the Holocene were somewhat restricted by geographical barriers. In eastern Asia, where the migrations from north to the south were not impeded, the fauna of Insectivora is much more diversified—comparable to that of preglacial Europe. Many ecological niches are therefore unoccupied in the recent fauna of Europe, particularly in glaciated islands (e.g. Ireland, where moles, voles, and many other mammalian species are lacking, and their niches apparently empty).

The spatial limitations of the refuges of European thermophilous fauna during the Ice Age had different effects on different species. In many cases large mammals became extinct, whereas small mammals were able to survive in limited territories but were sometimes no longer able to expand when conditions improved. For example, water-moles (Desmaninae), widely distributed over Europe in the Pliocene and early Pleistocene (Fig. 1), now live only in two refuges, the Pyrenees (*Desmana*

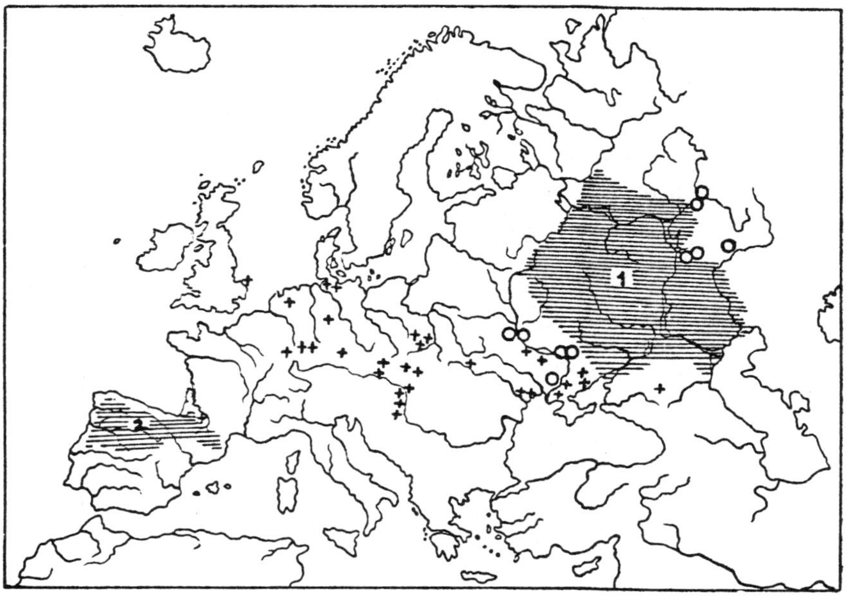

FIG. 1. Distribution of *Desmana* in Europe. 1, *D. moschata;* 2, *D. pyrenaica;* +, fossil localities; O, localities of recent artificial introduction. Once continuous areas are now limited to refuges because of the Ice Age.

pyrenaica) and the Ukraine (*D. moschata*). Other species, when the ice retreated, managed to spread again into central and northern Europe. The moles (*Talpa europaea*) of Europe are clearly differentiated into three forms—western, central, and eastern—each of which expanded in postglacial time from other refuges (Fig. 2) (Stein, 1963).

Thus the Pleistocene brought the gradual extinction of many mammalian elements of the "warm" Tertiary fauna. Particular species died out during different glaciations. The extinction affected large as well as small mammals, although the latter had a better chance of surviving in refuges; sometimes their isolation during cold periods brought the evolution of new subspecies or even species.

The influence of the last glaciation (Würm) on the fauna of Europe is better known than that of earlier glaciations. The small-mammal fauna of the last interglacial (Eemian) in Europe is very similar to the recent one. Not a single species of micromammalia of the European Eemian forest fauna is completely extinct today, but the fauna of the large mammals contains extinct as well as extant species, the former including the forest elephant *Palaeoloxodon antiquus*, the rhinoceros *Dicerorhinus kirchbergensis*, the hippopotamus *Hippopotamus amphibius*, and fallow deer *Dama dama*. The hippopotamus and fallow deer

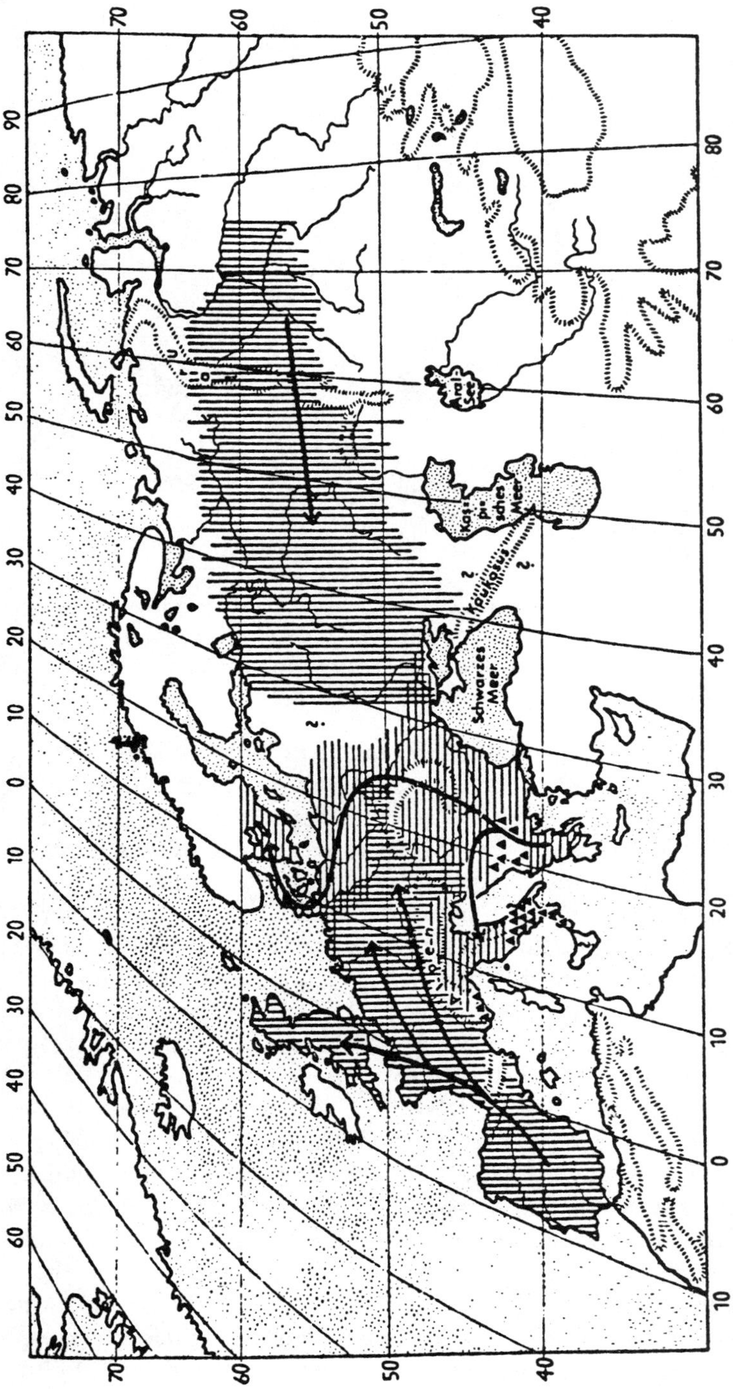

FIG. 2. Distribution of *Talpa europea* in Europe. Vertical lines, populations with narrow skulls; horizontal lines, populations with broad skulls; solid triangles, *T. e. romana*. Repopulation of central Europe after glaciations starting from different refuges (after Stein, 1963).

disappeared in Europe but still exist outside its limits (fallow deer was later reintroduced in Europe for hunting). The extinction of these species can be explained by changes in climate. The last glaciation limited the suitable biotope of enumerated species to a few areas in southern Europe. The population of elephant and rhinoceros in each of the refuges was so small that even a local factor could bring extinction. Man was present at that time in Europe, but there is almost no evidence for his role in extermination of animals; at such a primitive stage of culture he probably was not a real danger for animals as large as elephant or rhinoceros. The disappearance of these large mammals at the end of the last interglacial is the last episode in the great extinction of European fauna caused by Quaternary climatic changes in conjunction with the geographical configuration of the continent. According to the quantity of their populations and the extent of individual areas, large and small mammals were affected differently by the climate.

The Würm glaciation brought a great extension of the Scandinavian ice sheet and the total destruction of the fauna of northern Europe. The European climate became increasingly continental as a permanent center of high atmospheric pressure developed above the ice sheet and the extent of the sea diminished (because of the glaciation of the Baltic and North seas and by the isostatic lowering of the world oceanic level). The great area of the "steppe–tundra" was formed, connecting central and western Europe with the steppe regions of western and central Asia, with the rich fauna of open country. The composition of the rodent fauna, as well as pollen analysis, shows that the character of Pleistocene vegetation at the time was different from that of the recent Arctic tundra in northern Europe and from that of the recent dry steppes in southeastern Europe and southwestern Asia. In the Pleistocene of central Europe the lemmings *Lemmus* and *Dicrostonyx*, which are now limited to the tundra, lived side by side with the small hamsters (*Cricetulus*), now confined to the steppes far south of the lemming area. *Microtus gregalis*, a characteristic rodent of the Würm glaciation in Europe, lives today in the tundra and steppe regions of central and eastern Asia. In general, Pleistocene vegetation from the time of the Würm glaciation is most analogous to the vegetation of eastern Asia, which developed under the influence of a cold continental climate in middle latitudes.

At the decline of the Würm glaciation in Europe, small mammals of the steppe and tundra retired toward the north or southeast, unable to live in rapidly developing forests. During the same period six species of the larger mammals became extinct: mammoth (*Mammuthus primigenius*), woolly rhinoceros (*Coelodonta antiquitatis*), Irish deer (*Megaceros giganteus*), cave bear (*Ursus spelaeus*), cave lion (*Panthera spelaea*), and cave hyena (*Crocuta spelaea*). The extant European bison (*Bison bonasus*) may be the direct offspring of the extinct primitive bison (*Bison priscus*). The musk-ox (*Ovibos maschatus*) became extinct in the Old World but still lives in America. The wild horse (*Equus przewalskii*) lives in Central Asia; the saiga antelope (*Saiga tatarica*) persisted in Asia as well as on the outskirts of Europe; and the reindeer (*Rangifer tarandus*) was preserved partly as a wild, partly as a domestic animal, in extreme northern Eurasia.

Mammoth (*Mammuthus primigenius*) developed from the steppe elephant (*Elephas trogontherii*) in the penultimate (Riss) glaciation. Its makeup is well known, because, besides numerous skeletal remains, many specimens with soft parts are preserved in the permafrost of Siberia, and one in a layer of silt, impregnated with petroleum and salt in Starunia on the northern slope of the Carpathians (Garutt, 1964). The mammoth's adaptation to the cold climate is provided, for example, by its long hair, small ears, and tusks shaped to drive away snow in the search for food in winter. The geological situation of mammoth fossil remains, and the accompanying fauna and flora indicate that it lived in an open, woodless landscape.

The mammoth, whose Old World fossil remains are distributed from the Atlantic to the Pacific and from the Arctic to the Mediterranean and to Central Asia, probably had a more limited range at any one time. Carbon-14 dates of mammoth corpses discovered in Siberia range between 32,500 and 39,000 years B.P. (Heinz and Garutt, 1964) and thus suggest correlation with the relatively warm oscillation (Göttweig) within the last glaciation. The mammoth from Taimyr was dated by the above authors as 11,450 ± 250 years B.P. (an indication of the Allerød warmer oscillation at the end of the last glaciation). The stratigraphy of the European sites with mammoth also suggests its disappearance at the end of the Würm.

There is no doubt that Upper Paleolithic man hunted the mammoth. Probably it was done mostly by stampede, but pitfalls were also in use. The important role of mammoth in the engravings and sculptures of primitive man in Europe also suggests that it was hunted (Fig. 3).

The problem of the extinction of mammoth has been widely discussed. Earlier hypothetical explanations—e.g. that its extinction was a consequence of its poor adaptation to the conditions of life (Neuville, 1921)—now have only historical significance. The suppositions about a larger number of specimens with pathological characteristics immediately before extinction find no confirmation in an analysis of broader materials (Kubiak, 1965). The role of Paleolithic hunters in the extinction of mammoth is difficult to estimate, but mammoth was probably never

FIG. 3. Paleolithic drawings of mammoth (*Mammuthus primigenius*) (from different sources, after Garutt, 1964).

the principal game of human groups, and the traces of human colonization of Siberia at that time are very scarce.

The most logical explanation of the extinction of mammoth seems to be that it was caused by the disappearance of its habitat at the end of the Würm glaciation. The Pleistocene vegetation during this glaciation was different from that of the northern tundra as well as from that of the present steppes in Europe and western Asia. The adaptation of mammoth to the cold climate precluded life in the southern steppes, with their warm and dry summers. The present tundra, north of the Polar Circle, is very different from the steppe–tundra of the late Pleistocene; long polar nights and abundant snowfall in winter create quite peculiar conditions of life. Central and western Europe, independent of temperature changes in the Pleistocene, had the same geographical latitude as today and, of course, no polar night; the snowfall was probably slight. The development of forests in the postglacial of Europe and Siberia (Vangengeim, 1961) restricted the habitat of mammoth and finally caused its extinction.

The history of the woolly rhinoceros (*Coelodonta antiquitatis*) is similar. It is also known from the corpses with soft parts preserved in the permafrost of Siberia and particularly in the salt- and petroleum impregnated silt in Starunia on the northern slope of the Carpathians (Fig. 4). Hypsodont molars, reduced front teeth, and the anatomy of the upper lips prove that it was a grass-eating animal; the long hair and the paleontologic and geologic data suggest its adaptation to cold climate. The woolly rhinoceros did not develop until the late Pleistocene. The remains found on the Indigirka River in Siberia are radiocarbon-dated at 38,000 years B.P. (Heinz and Garutt, 1964); the species became extinct in Europe at the end of the Würm glaciation. Although *Coelodonta antiquitatis* was sometimes represented in Paleolithic art, few data suggest that it was hunted in the older Stone Age. In all probability its extinction may be explained by the disappearance of its proper habitat, as in the case of mammoth.

The Irish deer (*Megaceros giganteus*) also developed in the late-Pleistocene. Its older forms (*M. g. antecedens*), provided with smaller antlers, could have been forest animals, but later populations (*M. g. giganteus*) lived in an open landscape, as is suggested at least by the enormous antlers, which probably would have made movements in forests impossible. The Irish deer, probably confined to milder climate than was the contemporaneous reindeer, was still living in the Allerød, when the warming climate caused the expansion of forests, and died out in the beginning of the Dryas period about 10,000 years B.P. (Guenther, 1960).

The European primitive bison (*Bison priscus*) undoubtedly inhabited open, treeless country of the late Pleistocene. Its relation to the extant forest bison (*B. bonasus*) needs further study (Degerbøl and Iversen, 1945). *B. priscus* might be the ancestor, or both forms might have lived side by side in different habitats, and *B. priscus* might have disappeared with the decline of grassland in Europe at the end of the last glaciation, while the other persisted. The steppe bison was important game for primitive man and was often represented in Paleolithic art.

Musk-ox (*Ovibos moschatus*) developed from a species adapted to mild climate and was not confined to an arctic environment until the late-Pleistocene. Its extinction in the Old World is difficult to explain,

FIG. 4. Woolly rhinoceros (*Coelodonta antiquitatis*), specimen preserved with soft parts in Starunia, now in the museum of the Institute of Systematic Zoology, Kraków (photo by L. Sych).

because it seems to have been well adapted to life in the Eurasian tundra. Few data indicate that it was hunted by Paleolithic man; its distribution was probably always limited to the sparsely inhabited periglacial regions.

In spite of voluminous literature devoted to the wild horse, a clear picture of its extinction in Europe is difficult to obtain. In the open areas of the late Pleistocene, the wild horse was very common and was a principal prey of Paleolithic hunters. In the postglacial, the range of the wild horse contracted, beginning with western Europe, and *Equus przewalskii* now lives only in the semideserts of Central Asia (Mohr, 1959), on the verge of extinction. Historical data prove the existence of wild horses in the Ukrainian steppes as late as the middle of the nineteenth century. These horses were described as a separate species, *Equus gmelini*, but they were more probably feral. Vetulani (1933) supposed that the peculiar forest form of *E. gmelini* was extant in western Europe in the Middle Ages and in Poland as late as the eighteenth century. Primitive domestic horses in Poland were, according to him, the direct descendants of these wild horses, and he started to reconstruct this form by breeding the Polish primitive strain in wild conditions (Vetulani, 1948). In these continuing experiments, the animals have developed some apparently primitive characters (not necessarily related to ancestral genotypes). The existence of a sylvan form of the horse seems ecologically improbable, and the supposed "forest tarpan" was probably a feral domestic horse. The postglacial development of forests made the existence of the wild horse in western and central Europe impossible, and the final limitation of its area to the semideserts of central Asia was the result of predation by man.

The extinction of the cave bear (*Ursus spelaeus*), the remains of which are very abundant in European caves, has been widely discussed (Kurtén, 1958). The cave bear appeared during the Mindel Glaciation, survived two later glaciations and two interglacials, and became extinct at the end of the Würm glaciation; its latest remains are from Würm III. It coexisted in Europe with the brown bear (*U. arctos*), but these two species were seldom found together. The cave bear can be distinguished from the brown bear by its larger dimensions, by many details of the skeleton and dentition, and by its probably strictly phytophagous habits.

Abel's suggestion (1929) that the extinction of the cave bear was anticipated by the appearance of degenerated, pathological specimens and by the excess of male births was rejected by Kurtén (1958) as being based on fragmentary or erroneously interpreted material. Another hypothesis holds Paleolithic man responsible for the extermination of the cave bear; according to some archaeologists, tribes of Mousterian man living mostly in mountains specialized in hunting bears. These suppositions also lack an adequate factual basis, and they concern the time of the Eemian Interglacial, long before the process of extinction of cave bear began. Kurtén (1958) stated that the extinction of cave bear was gradual, and that in many regions the species disappeared long before its final extinction elsewhere. Before extinction, isolated populations with a limited number of specimens revealed the effects of inbreeding in the diminished variation of morphological characters.

There is no defense for Kurtén's explanation that cave bears disappeared because of the occupation by Paleolithic man of caves used as bear dens. Most of the remains of cave bear were found in caves only because of the much better conditions for the preservation of bones in cave sediments than outside, and not because of a narrow specialization of cave bear to wintering in caves. In the Tatra Mountains of Poland, where brown bears still exist, their subfossil remains are found exclusively in caves. The animals hibernate sometimes in caves, but even in the calcareous Tatras, where caves are abundant, the hibernation places are generally located below uprooted trees in the forests. The preservation of bones outside caves, especially in mountains where erosion prevails over accumulation, is very unlikely. Because of the highly irregular distribution of caves in Europe, it is hard to imagine a species of large mammals that depended on the caves for its existence. In the same period, Paleolithic settlement of caves, except for some regions in France and Spain, was irregular and sporadic. In Poland the caves with bear remains have only scarce traces of human habitation, and the high mountains (e.g. Tatra), where cave bear remains are numerous, had no Paleolithic habitation at all. The competition for shelter between Paleolithic man and cave bear, therefore, cannot be the principal cause of the reduction of the once numerous population of this animal to scattered small groups as a prelude to extinction. Here again, the disappearance of the proper biotope seems to be the most probable cause of extinction.

Cave lion (*Panthera spelaea*) and cave hyena (*Crocuta spelaea*) became extinct in the final stages of the Würm glaciation. They were probably never regularly hunted by Paleolithic man.

Most large Pleistocene forest mammals able to survive the last glaciation still live in Europe: the elk (*Alces alces*), red deer (*Cervus elaphus*), roe deer (*Capreolus capreolus*), wild boar (*Sus scrofa*), and others. Two of them have faced extinction in historical times: the aurochs (*Bos primigenius*), which disappeared in the wild and survived only as domestic cattle, and the European bison (*Bison bonasus*), which was saved through active protection when it was on the verge of extinction.

Aurochs appeared in Europe during the Holsteinian or Great Interglacial (Mindel–Riss) and survived two glaciations. It was principally a member of the forest community but was also able to live in open areas. Aurochs was distributed through Europe except for the northernmost regions, northern and temperate Asia and northern Africa. Important game of Paleolithic man, it was hunted also in later prehistoric times, in antiquity, and in the Middle Ages. The rapid process of its extinction began in the late Middle Ages, probably as a result of extensive deforestation. In the thirteenth century aurochs were still present in France, Germany, Sweden, Poland, and Russia. In the fourteenth and fifteenth centuries they virtually disappeared as wild animals (Fig. 5). In Poland the hunting of aurochs was early reserved for monarchs and sovereigns, who prohibited general hunting of the animal and commanded their protection (e.g. Prince Boleslaus of Masovia in 1298, Prince Ziemovit of

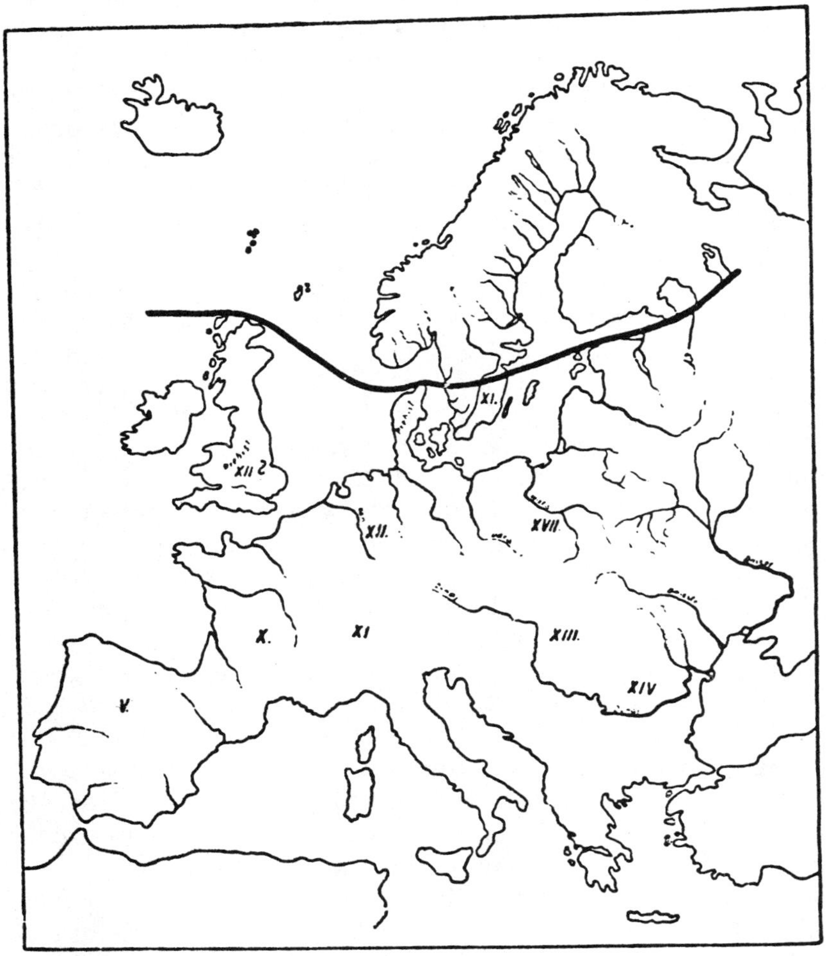

Fig. 5. The dates (by century) of extinction of aurochs (*Bos primigenius*) in different countries in Europe. Solid line, northern limit of distribution (after Lukaszewicz, 1952).

Masovia in 1359, and others). In the sixteenth century, aurochs existed only in the Jaktorov primeval forest near Warsaw, where, under official protection, they were guarded by the inhabitants of surrounding villages, who gave them hay in winter. Reports about the aurochs reservation in the Jaktorov Forest—prepared by royal officers between 1510 and 1630 and published recently by Lukaszewicz (1952)—offer interesting data not only on the final extinction of *Bos primigenius* in spite of its protection but also on their biology and on the ecological conditions in which the last of them lived. In 1557 there were still about fifty in Jaktorov Forest, and in 1562 there were thirty-eight (eleven of them bulls). At that time all hunting was forbidden, but the deforestation and the grazing of domestic cattle in the forest were disastrous for the population. In 1599, in spite of numerous decrees commanding their protection, there were only twenty-four aurochs living in Jaktorov. In 1601, a plague killed all but four, only one of them a female. In 1630 the royal officer noted that he found no aurochs in Jaktorov, but was told that the last animal, a female, died in 1627 (Fig. 6).

It is interesting to note that in Germany experiments were started to "reconstruct" aurochs through the hybridization and selection of some primitive forms of cattle resembling their extinct ancestors (Lengerken, 1953).

The European bison (*Bison bonasus*) was also on the verge of extinction. Whether it is the descendant of *B. priscus* or the latest stage of the independent line of forest bison, it appears in the postglacial and is

FIG. 6. A drawing of aurochs (*Bos primigenius*) from A.D. 1557 (after Lengerken, 1953).

limited in its distribution to Europe. In antiquity it was common on the whole continent. The process of its extinction began in the Middle Ages, and in the eighteenth century it disappeared everywhere but in the Bialowieza primeval forest and in the Caucasus. The lowland population from Bialowieza and the mountain population in the Caucasus were in contact via the forested valley of the Don as late as the Middle Ages. The Caucasian bison is smaller and of darker color.

In 1803 Bialowieza, then in the Russian empire, was declared the private hunting ground of the czars. The herd of bison numbered 300 to 500. In 1860 there were as many as 1,500 bison there, but at the beginning of World War I the population once more declined to 750. The main reason for the weak birth rate and the later decline of the bison population was the destruction of forest vegetation in Bialowieza by an excessive number of red deer, maintained there for hunting purposes. During the nineteenth century some of the bison from Bialowieza were sent to different places as gifts from the czars. Four specimens presented to Pszczyna in Silesia propagated in 1920 to a herd of seventy head.

During World War I, when the Russian army retreated, German soldiers killed about 500 bison; the remainder were exterminated by poachers when Bialowieza passed successively into the hands of the Lithuanian, Russian, and Polish armies. At the same time, the Pszczyna herd, the largest outside Bialowieza, was decimated, and only three specimens survived; other bison lived in zoological gardens. There were at that time no more than fifty purebred European bison in the world. Their breeding in Poland, Germany, Sweden, and the Netherlands increased the number in 1939 to ninety-four. During World War II some centers of bison breeding were destroyed, but the herd in Bialowieza survived. Now there are several hundred bison in Europe, 213 of them in 1966 in Poland, where some live in reservations, and three herds live freely in large forest regions such as Bialowieza.

The Caucasian herd of bison numbered about a thousand specimens in the second half of the nineteenth century. It diminished gradually, and the last known specimen was killed in 1926. Only one Caucasian bison, a bull, was transported from Russia before World War I; it mated with the lowland bison from Bialowieza and left numerous hybrid offspring. The issue of this line is kept separately and now forms quite a large population in Russian and Polish reservations.

We can hope that the European bison has been saved as a species, although it is almost completely dependent on continuous protection from man.

References

Abel, O., 1929, *Paläobiologie und Stammesgeschichte:* Jena, Fischer, 423 p.

Degerbøl, M., and Iversen, J., 1945, The bison in Denmark: København, *Danmarks geol. unders.* R. *II*, no. *73*, p. 1–62

Garutt, W. E., 1964, Das Mammut: Wittenberg, *Neue Brehm-Bücherei,* no. *331*, p. 1–140

Guenther, E. W., 1960, Funde des Riesenhirsches in Schleswig-Holstein und ihre zeitliche Einordnung, p. 201–06, in *Steinzeitfragen d. Alten u. Neuen Welt;* Zotz-Festschrift: Bonn, L. Röhrscheid Verlag

Guilday, J. E., Differential extinction during late-Pleistocene and Recent times (this volume)

Heinz, A. E., and Garutt, W. E., 1964, Determination with radioactive carbon of the absolute age of fossil remains of mammoth and woolly rhinoceros found in permanently frozen ground of Siberia (in Russian): Moscow, *Doklady Akad. Nauk SSSR*, v. *154*, p. 1367–70

Kretzoi, M., 1965, Die Nager und Lagomorphen von Voigtstedt in Thüringen und ihre chronologische Aussage: Berlin, *Paläont. Abh., A*, v. *II*, no. *2/3*, p. 585–660

Kubiak, H., 1965, Examples of abnormalities in the dentition of fossil elephants (in Polish): Krakow, *Folia. Quatern.*, v. *19*, p. 45–61

Kurtén, B., 1958, Life and death of the Pleistocene cave bear: *Acta. zool. fennica*, v. *95*, p. 1–59

Lengerken, H., 1953, Der Ur und seine Beziehungen zum Menschen: Wittenberg, *Neue Brehm-Bücherei*, no. *105*, p. 1–80

Lukaszewicz, K., 1952, The Ure-ox (in Polish): Krakow, *Ochrona Przyrody*, v. *20*, p. 1–33

Mohr, E., 1959, Das Urwildpferd: Wittenberg, *Neue Brehm-Bücherei*, no. *249*, p. 1–144

Neuville, H., 1921, On the extinction of the Mammoth: *Smithson. Inst. Ann. Report for 1919*, p. 327–38

Stein, G., 1963, Unterartengliederung und Nacheiszeitliche Ausbreitung des Maulwurfs, Talpa europaea L.: Berlin, *Mitteil. zool. Mus.*, v. *39*, p. 379–402

Vangengeim, E. A., 1961, Paleontologic materials for the stratigraphy of the Quaternary sediments of northeastern Siberia (in Russian): Moscow, *Trudy geol. inst. Akad. Nauk SSSR*, v. *48*, p. 1–182

Vetulani, T., 1933, Zwei weitere Quellen zur Frage des europäischen Waldtarpans: Berlin, *Zeit. Säugetierkunde*, v. *8*, p. 281–82

Vetulani, T., 1948, Premières observations sur la régénération du Tarpan sylvestre européen dans la forêt vierge de Bialowieza: *Bull. int. Acad. Pol. Sci. Lettres, Cl. Sci. Math. Nat.*, B, 1947, p. 1–22

CONCLUDING REMARKS

The preceding selections have addressed some of the major topics that have been, and continue to be, of concern to vertebrate paleontologists: the general history of vertebrate life; biostratigraphy and biogeography; comparative anatomy, classification, and phylogeny; and functional anatomy, paleobiology, and paleoecology of vertebrates. But what are the newest developments in the field, and what lies in store for vertebrate paleontology? I will briefly discuss some of the latest developments and venture a few speculations as to future research trends.

Perhaps the most important recent development in vertebrate paleontology is a new emphasis on rigor in thought, logic, and scientific procedure. In the past, vertebrate paleontology was often thought of as a soft, or inexact, science by workers in other fields. Many vertebrate paleontologists no longer view it as sufficient to collect facts and then to formulate a plausible scenario to explain the facts. Instead, Popper's (1968) hypotheticodeductive method of scientific inquiry has become increasingly popular. Testable hypotheses are formulated, and then the worker tries to falsify these hypotheses with either the data at hand or by collecting additional data. In the subdiscipline of phylogenetic analysis and reconstruction, this methodology has been exemplified by the application of cladistic methodology (see Wiley, 1981). Adherents of the cladistic school of phylogeny reconstruction assert that hypotheses of phylogenetic relationships can be rigorously tested on the basis of the distribution of shared-derived (synapomorphous; i.e., advanced, specialized, derived, evolved, or homologous) character states (usually morphological traits in the case of fossils) among the organisms concerned (for a brief introduction to cladistics, see Patterson, 1980, and Cracraft, 1983). This new rigor has extended to subdisciplines other than phylogeny reconstruction in vertebrate paleontology. However, it is particularly important that studies concerning the basic relationships of the vertebrates be based on well-founded premises. Such phylogenetic studies often lay the groundwork for other, more derived studies. Rigorous phylogenetic studies will be of increasing importance in the future.

Recently there has been a renewed interest in, and awareness of, geological effects on fossil deposition and preservation (Badgley, 1982; Behrensmeyer, 1982; Clark, Beerbower, and Kietzke, 1967; McKinney and Schoch, 1983) and the interpretation of fossil soils in terrestrial sediments relative to vertebrate fossils (Bown and Kraus, 1981*a*, 1981*b*; Retallack, 1983*a*, 1983*b*). Paleontologists now realize that they must be more cautious than they have sometimes been in the past. The stratigraphic record is generally very incomplete and one cannot directly read evolutionary history from the rocks. Rather, one must carefully distinguish "fossil patterns" as they occur in the rocks from true "evolutionary patterns" (Schoch, 1982; McKinney and Schoch, 1983; Schopf and Hoffman, 1983). In the future we can look forward to a healthy mutual exchange of information, techniques, and ideas among vertebrate paleontologists, geologists, and biologists.

With neontologists, paleobotanists, and invertebrate paleontologists, vertebrate paleontologists have been actively concerned with the ongoing study of the pattern and process, tempo and mode of macroevolutionary (i.e., evolution above the species level) change in organisms. This has been one of the most controversial subjects in paleontology in the 1970s and early 1980s. Much of the recent work has centered around rates of evolution (Gingerich, 1983), concentrating on the question of whether "punctuated equilibria" or "phyletic gradualism" is the dominant pattern of macroevolutionary change (Eldredge and Gould, 1972; Gingerich, 1976; Gould and Eldredge, 1977). To characterize phyletic gradualism: Classic, Neo-Darwinian evolutionists have assumed a priori that evolution occurs through the accumulation of slow, steady, and gradual change. Given a complete succession of fossils through time, they will show a continuous gradation, and all morphological gaps will be filled between any two end forms, be these end forms closely related species or the first Precambrian heterotroph and man (Mayr, 1982; Vrba, 1980). In contrast, those who espouse punctuated equilibria, such as Eldredge and Gould (1972; Gould and Eldredge, 1977) in particular, have asserted that evolution tends to be episodic rather than gradual. They believe that extremely high rates of evolution occur in spurts, preceded and followed by evolutionary stasis during which a species does not appreciably evolve or change, but rather fluctuates mildly around a basic morphology. Vertebrate paleontologists have added much fuel to this ongoing debate (see, for example, Gingerich, 1976; Gould and Eldredge, 1977; Stanley, 1979; Vrba, 1980 and references cited therein).

Although not definitively resolved, argument over the punctuated equilibria versus phyletic gradualism controversy seems to be abating. It appears that both tempos may have occurred in this history of life (Hoffman, 1981). But, concurrent with this work has come a new questioning of, and interest in, basic evolutionary theory (Gould, 1982). Topics of current interest include how long-term evolutionary patterns (such as trends, radiations, and extinctions of groups of species) may be brought about (Vrba, 1983), and the role of adaptation (Gould and Lewontin, 1979), development (Rachootin and Thomson, 1981), and ecology and environment (Van Valen, 1973; Wei and Kennett, 1983) in evolutionary processes. It has been suggested that many macroevolutionary patterns may be explained by differential speciation or extinction rates within groups of closely related organisms (Gould, 1982; Stanley, 1975). In particular, based substantially on vertebrate paleontologic data, Vrba (1983) has suggested that specialist organisms (stenotopes) may be prone to higher rates of speciation and extinction than broadly adapted generalists (eurytopes). Complementing Vrba's (1983) hypothesis, West-Eberhard (1983, p. 155) has suggested that "rapid divergence and speciation can occur between populations with or without ecological differences under selection for success in intraspecific social competition." This suggestion may be particularly important to vertebrate paleontologists as many vertebrates are social organisms. However, it is unclear exactly what effect social competition has on the long-term evolution of organisms (Schwartz and Armitage, 1980). Finally, the recently formulated concept of "success-extinction"—that a species may bring about its own extinction, and perhaps that of subsidiary species as well, through the unchecked exhaustion of the resources of its niche (Meredith, 1982; cf. Thomson, 1976)—may prove useful in the interpretation of the history of vertebrate life.

On another front, the late 1960s and 1970s was a time of intensive work on the paleophysiology, paleobiology, and paleoecology of dinosaurs (Dinosauria, *sensu stricto,* and Theropoda). Many workers concluded that dinosaurs could be characterized as "warm-blooded" (see reviews and references in Thomas and Olson, 1980; Hopson and Radinsky, 1980). Also of recent interest have been studies of the evolutionary dynamics of entire paleofaunas (e.g., Bonaparte, 1982; Rose, 1981; Van Valkenburgh, 1982), building on the work of Olson (Paper 20), and studies of large-scale patterns of vertebrate diversification through time (e.g., Thomson, 1976, 1977). Finally, as in the past, contemporary vertebrate paleontologists are interested in elucidating the origins of major adaptations and groups (Padian, 1982), and also in explaining paleobiogeographical patterns (Cracraft, 1983).

Concluding Remarks

Vertebrate paleontology today is still a discipline in which much basic data gathering remains to be done. Active field work and collecting needs to be continued because no one can predict what might be found in the future. Only recently, for example, a number of important dinosaurs and Cretaceous mammals were found in the Gobi Desert (Kielan-Jaworowska, 1969); the first fossil land mammal was found in the Antarctic (Woodburne and Zinsmeister, 1982); a new suborder of Thecodontia, the Doswelliina, was discovered in the Triassic of Virginia (Weems, 1980); and nests of juvenile dinosaurs have been found in Montana (Horner and Makela, 1979). (I had the privilege of helping Horner and Makela collect more of these juvenile dinosaurs during the summer of 1980.) Also, many vertebrate paleontological specimens that have never been studied, or that need thorough restudy, are languishing in museums. There is an ongoing need for detailed, monographic studies of particular specimens, groups, and problems (specifically, systematic and functional morphologic studies are needed). Such studies lay the foundation for large-scale, synthetic work. Over the last two centuries, vertebrate paleontologists have discovered the broad outlines of vertebrate history and evolution. The challenge to vertebrate paleontology is now to both explain the known facts and to fill in the remaining major gaps (for example, the origin of vertebrates, the fish-amphibian transition, and the amphibian-reptile transition), along with numerous details, in the story of vertebrate life.

REFERENCES

Badgley, C., 1982, How Much Time is Represented in the Present?: The Development of Time-averaged Modern Assemblages as Models for the Fossil Record, *North American Paleontol. Conv., Third, Proc.* **1:**23–28.

Behrensmeyer, A. K., 1982, Time Sampling in the Vertebrate Fossil Record, *North American Paleontol. Conv., Third, Proc.* **1:**41–45.

Bonaparte, J. F., 1982, Faunal Replacement in the Triassic of South America, *Jour. Vert. Paleontology* **2:**362–371.

Bown, T. M., and M. J. Kraus, 1981*a*, Lower Eocene Alluvial Paleosols (Willwood Formation, Northwest Wyoming, U.S.A.) and Their Significance for Paleoecology, Paleoclimatology, and Basin Analysis, *Palaeogeography, Palaeoclimatology, Palaeoecology* **34:**1–30.

Bown, T. M., and M. J. Kraus, 1981*b*, Vertebrate Fossil-bearing Paleosol Units (Willwood Formation, Northwest Wyoming, U.S.A.): Implications for Taphonomy, Biostratigraphy, and Assemblage Analysis, *Palaeogeography, Palaeoclimatology, Palaeoecology* **34:**31–56.

Clark, J., J. R. Beerbower, and K. K. Kietzke, 1967, Oligocene Sedimentation, Stratigraphy, Paleoecology, and Paleoclimatology in the Big Badlands of South Dakota, *Fieldiana: Geology Memoirs* **5:**1–158.

Cracraft, J., 1983, Cladistic Analysis and Vicariance Biogeography, *Am. Scientist* **71:**273–281.

Eldredge, N., and S. J. Gould, 1972, Punctuated Equilibria: An Alternative to Phyletic Gradualism, in *Models in Paleobiology,* T. J. M. Schopf, ed., Freeman, Cooper & Co., San Francisco, pp. 82-115.
Gingerich, P. D., 1976, Paleontology and Phylogeny: Patterns of Evolution at the Species Level in Early Tertiary Mammals, *Am. Jour. Sci.* **276:**1-28.
Gingerich, P. D., 1983, Rates of Evolution: Effects of Time and Temporal Scaling, *Science* **222:**159-161.
Gould, S. J., 1982, Darwinism and the Expansion of Evolutionary Theory, *Science* **216:**380-387.
Gould, S. J., and N. Eldredge, 1977, Punctuated Equilibria: The Tempo and Mode of Evolution Reconsidered, *Paleobiology* **3:**115-151.
Gould, S. J., and R. C. Lewontin, 1979, The Spandrels of San Marco and the Panglossian Paradigm: A Critique of the Adaptationist Programme, *Royal Soc. (London) Proc., ser. B,* **205:**581-598.
Hoffman, A., 1981, Biological Controls on the Punctuated Versus Gradual Mode of Species Evolution, in *International Symposium on Concept and Method in Paleontology,* vol. 1, J. Martinell, ed., University of Barcelona, Spain, pp. 57-63.
Hopson, J. A., and L. B. Radinsky, 1980, Vertebrate Paleontology: New Approaches and New Insights, *Paleobiology* **6:**250-270.
Horner, J. R., and R. Makela, 1979, Nest of Juveniles Provides Evidence of Family Structure Among Dinosaurs, *Nature* **282:**296-298.
Kielan-Jaworowska, Z., 1969, *Hunting Dinosaurs,* MIT Press, Cambridge, Mass., and London, 192p.
McKinney, M. L., and R. M. Schoch, 1983, A Composite Terrestrial Paleocene Section with Completeness Estimates, Based upon Magnetostratigraphy, *Am. Jour. Sci.* **283:**801-814.
Mayr, E., 1982, *The Growth of Biological Thought,* Harvard University Press, Cambridge, Mass., and London, 986p.
Meredith, A., 1982, Devolution, *Jour. Theor. Biol.* **96:**49-65.
Padian, K., 1982, Macroevolution and the Origin of Major Adaptations: Vertebrate Flight as a Paradigm for the Analysis of Patterns, *North American Paleontol. Conv., Third, Proc.* **2:**387-392.
Patterson, C., 1980, Cladistics, *Biologist* **27:**234-240.
Popper, K., 1968, *The Logic of Scientific Discovery,* Harper & Row, New York, 480p.
Rachootin, S. P., and K. S. Thomson, 1981, Epigenetics, Paleontology and Evolution, in *Evolution Today,* G. G. E. Scudder and J. L. Reveal, eds., Proceedings of the Second International Congress on Systematic and Evolutionary Biology, Carnegie-Mellon University, Pittsburgh, Pa., pp. 181-193.
Retallack, G. J., 1983*a,* A Paleopedological Approach to the Interpretation of Terrestrial Sedimentary Rocks: The Mid-Tertiary Fossil Soils of Badlands National Park, South Dakota, *Geol. Soc. America Bull.* **94:**823-840.
Retallack, G. J., 1983*b, Late Eocene and Oligocene paleosols from Badlands National Park, South Dakota,* Geol. Soc. America Spec. Paper 193, Geological Society of America, Boulder, Colo., 90p.
Rose, K. D., 1981, The Clarkforkian Land-Mammal age and Mammalian Faunal Composition across the Paleocene-Eocene Boundary, *Michigan Univ. Papers Paleontol.,* No. 26, 208p.
Schoch, R. M., 1982, Gaps in the Fossil Record: Fossils and Stratigraphy, *Nature* **299:**490.

Schopf, T. J. M., and A. Hoffman, 1983, Punctuated Equilibrium and the Fossil Record, *Science* **219:**438–439.
Schwartz, O. A., and K. B. Armitage, 1980, Genetic Variation in Social Mammals: The Marmot Model, *Science* **207:**665–667.
Stanley, S. M., 1975, A Theory of Evolution above the Species Level, *Natl. Acad. Sci. (USA) Proc.* **72:**646–650.
Stanley, S. M., 1979, *Macroevolution: Pattern and Process,* W. H. Freeman & Co., San Francisco, 344p.
Thomas, R. D. K., and E. C. Olson, eds., 1980, *A Cold Look at the Warm-blooded Dinosaurs,* AAAS Selected Symposia Series No. 28, American Association for the Advancement of Science, Washington, D.C., 514p.
Thomson, K. S., 1976, Explanation of Large Scale Extinctions of Lower Vertebrates, *Nature* **261:**578–580.
Thomson, K. S., 1977, The Pattern of Diversification among Fishes, in *Patterns of Evolution,* A. Hallam, ed., Elsevier Scientific Publishing Co., Amsterdam, pp. 377–404.
Van Valen, L. M., 1973, A New Evolutionary Law, *Evol. Theory* **1:**1–30.
Van Valkenburgh, B., 1982, Evolutionary Dynamics of Terrestrial, Large, Predator Guilds, *North America Paleontol. Conv., Third, Proc.* **2:**557–562.
Vrba, E. S., 1980, Evolution, Species, and Fossils: How Does Life Evolve?, *South African Jour. Sci.* **76:**61–84.
Vrba, E. S., 1983, Macroevolutionary Trends: New Perspectives on the Roles of Adaptation and Incidental Effect, *Science* **221:**387–389.
Weems, R. E., 1980, An Unusual Newly Discovered Archosaur from the Upper Triassic of Virginia, U.S.A., *Am. Philos. Soc. Trans.* **70:**1–53.
Wei, K. -Y., and J. P. Kennett, 1983, Nonconstant Extinction Rates of Neogene Planktonic Foraminifera, *Nature* **305:**218–220.
West-Eberhard, M. J., 1983, Sexual Selection, Social Competition, and Speciation, *Quart. Rev. Biol.* **58:**155–183.
Wiley, E. O., 1981, *Phylogenetics: The Theory and Practice of Phylogenetic Systematics,* John Wiley & Sons, New York, 454p.
Woodburne, M. O., and W. J. Zinsmeister, 1982, Fossil Land Mammal from Antarctica, *Science* **218:**284–286.

AUTHOR CITATION INDEX

SUBJECT INDEX

About The Editor

ROBERT MILTON SCHOCH is a curatorial assistant in the Division of Vertebrate Paleontology, Peabody Museum of Natural History, Yale University, New Haven, Connecticut. He is also affiliated with the Department of Geology and Geophysics at Yale. Born and raised in the Washington, D.C. area, he received the B.A. (1979) in anthropology and the B.S. (1979) in geology from George Washington University. He received the M.S. (1981), the M.Phil. (1981), and the Ph.D. (1983) in geology and geophysics from Yale University, specializing in vertebrate paleontology. Dr. Schoch's specialty is Paleocene and Eocene mammals and he is an authority on the extinct mammalian order *Taeniodonta*. He has contributed many papers to the scientific literature, published in such journals as *Journal of Paleontology, Journal of Mammalogy, Journal of Vertebrate Paleontology, Discovery, Bulletin of the Geological Institutions of the University of Uppsala, Postilla (Yale University), American Journal of Science, Copeia,* and *Geological Society of America Bulletin.* Dr. Schoch has participated in geological and paleontological expeditions to Idaho (1979), New Mexico (1980), Montana (1980), and Pakistan (1981–1982).

Most recently (July 1984) Dr. Schoch has pursued fieldwork on Devon and Ellesmere Islands (Canadian High Arctic) and has been appointed assistant professor in the Division of Science, College of Basic Studies, Boston University. He resides with his wife, Cynthia, in Watertown, Massachusetts.